D0990040

HARCOURT MATHEMATICS 11
Functions/Relations

Text Editor
Ronald Dunkley

General Editor
Enzo Carli

Authors
Ronald Green
Gordon Nicholls
Loraine Wilson

Copyright © Harcourt Canada Ltd.

All rights reserved. No part of this publication may be reproduced or transmitted in any form or by any means, electronic or mechanical, including photocopy, recording, or any information storage and retrieval system, without permission in writing from the publisher. Reproducing passages from this book without such written permission is an infringement of copyright law.

Requests for permission to photocopy any part of this work should be sent in writing to: Schools Licensing Officer, CANCOPY, 1 Yonge Street, Suite 1900, Toronto, ON, M5E 1E5. Fax: (416)868-1621. All other inquiries should be directed to the publisher.

Every reasonable effort has been made to acquire permission for copyright material used in this text, and to acknowledge such indebtedness accurately. Any errors or omissions called to the publisher's attention will be corrected in future printings.

Canadian Cataloguing in Publication Data

Green, Ronald (Ronald E.)
    Harcourt math 11 : functions/relations

Includes index.
ISBN 0-7747-1452-2

1. Functions. 2. Set theory. I. Nicholls, Gordon T. II. Wilson, Loraine.
III. Dunkley, Ronald G. IV. Carli, E. G. V. Title.

QA331.3.G73 2001   511.3'3   C00-932956-0

Text Editor
Ronald Dunkley
Centre for Education in Mathematics and Computing
University of Waterloo

General Editor
Enzo Carli
formerly Mathematics Department Head
Forest Heights Collegiate Institute
Waterloo Region District School Board

Authors
Ronald Green
formerly St. Benedict Catholic High School
Waterloo Catholic District School Board

Gordon Nicholls
formerly Mathematics Department Head
Preston Collegiate Institute
Waterloo Region District School Board

Loraine Wilson
formerly Francis Libermann Catholic High School
Toronto Catholic District School Board

Reviewers
Jacquie Brodsky
Medway High School
Thames Valley District
School Board

Susan Brooks
Elmira District High School
Waterloo Region District
School Board

Ed D'Andrea
Father John Redmond
Catholic Secondary School
Toronto Catholic District
School Board

Elizabeth Fraser
Glebe Secondary School
Ottawa-Carleton District
School Board

Alexis Galvao
John Cabot Catholic
Secondary School
Dufferin-Peel Catholic
District School Board

Pat Grew
Frontenac Secondary School
Limestone District School
Board

David McKay
Westdale Secondary School
Hamilton-Wentworth District
School Board

Linda Obermeyer
Notredame Secondary School
Halton Catholic District
School Board

David Rushby
Martingrove Collegiate
Institute
Toronto District School Board

Rod Yeager
Orangeville District High
School
Upper Grand District School
Board

Project Manager: Deborah Davidson
Editors: Lynn Cowan, Ray MacDonald, Brett Savory
Senior Production Editor: Sharon Dzubinsky
Production Editor: Maraya Raduha
Production Co-ordinator: Jonathan Pressick
Photo Researchers: Karen Becker, Mary Rose MacLachlan
Interior Design, Page Composition, Technical Art, and Illustrations: Brian Lehen • Graphic Design Ltd.
Page 7 illustration: Sami Suomalainen
Cover Design: Brian Lehen • Graphic Design Ltd.
Cover Image: Peter Griffith/Masterfile
Chapter-Opener Images: **1** Holly Harris/Stone; **2** Albert Normandin/Masterfile; **3** Bob Torrez/Stone; **4** Peter Griffith/Masterfile; **5** Jim Craigmyle/Masterfile; **6** Mark Harwood/Stone; **7** Brad Wrobleski/Masterfile; **8** Guy Grenier/Masterfile; **9** Bryan Reinhart/Masterfile; **10** Didier Dorval/Masterfile; **11** Lloyd Sutton/Masterfile
Printing and Binding: Friesens

⊗ This book is printed in Canada on acid-free paper.
1 2 3 4 5   05 04 03 02 01

Contents

Foreword

We prepared this text with two primary goals in mind. First, we conscientiously addressed the presentation of material to ensure the curriculum published by the Ministry of Education was covered. Second, we wanted to provide the best possible foundation in preparing students for continuation of mathematics studies through grade 12 onward into post-secondary mathematics courses. This second goal is synonymous with the first goal to an extent; however, we included the optional topic on the trigonometric compound angle functions, which are within the ability of students to master and are essential in preparing them for the calculus of trigonometric functions.

Every aspect of this book reflects our belief that a mathematics text should include the following:

◆ an exploration of the uses and the history of mathematics. We have provided numerous examples of applications of the content and summaries of the historical development of mathematics.

◆ an indication of technological applications and problems requiring use of technology.

◆ sets of problems that will lead students to mathematics concepts and to an appreciation of the intellectual component of mathematics education. Many of the problems have been included in contests of the Canadian Mathematics Competition. Others were chosen to extend the understanding of concepts by students.

◆ comprehensive sets of cumulative reviews at the conclusion of each topic.

With respect to the use of technology, we have assumed that teachers and students are familiar with the technical capabilities of graphing calculators and do not need to be instructed in their use. Where there is need for detail, it is given in the Teacher's Guide. We have further assumed that we need not give specific instructions for the employment of technology. It is important that students learn that there are appropriate places in which to employ technology and that these differ from student to student. It is left for teachers and students to determine where technology is to be employed, except for situations in which such use is required by the curriculum.

Discussion of student work should be an integral part of classroom activity. We have included many questions in the exercises for this express purpose. In addition, while some of the Part C exercise problems are designed to test the abilities of dedicated students, many are extensions of the curriculum material and provide rich material for discussion. Questions in the problems sets are quite accessible to most students and will provide a basis for a lively interchange of ideas.

It takes many people to produce a text. This one, because it represents a new approach to text production, has involved persons from two organizations. The Centre for Education in Mathematics and Computing was established with the purpose of developing activities to enrich and enhance the study of the two disciplines. It created the Canadian Mathematics Competition and has published numerous books on problem solving. With this text, it broadens its mandate into the preparation of student classroom texts, a logical extension. Harcourt Canada has a long-established reputation for publishing popular mathematics texts, primarily at the senior, secondary-school level. Together, the two organizations offer a text that we believe will provide every student with a sound understanding of the curriculum.

Thanks are due to the reviewers for their helpful comments and suggestions.

Special mention should be made of a few people involved. Mr. Frank Rachich reviewed the entire manuscript and made numerous suggestions for improvement. He also provided us with solutions to all questions in the text. Manuscript production was done by Bonnie Findlay and Linda Schmidt, who not only entered the script but also made numerous suggestions for improvements. The staff at Harcourt Canada made further improvements in design and presentation.

This text will not by itself make anyone a good student. It provides material that if used wisely will help a dedicated student move one more step along the path of life-long education. We hope that many students will gain a sense of the intellectual joy that can be found in this magnificent discipline.

Ronald Dunkley
Enzo Carli
Ronald Green
Gordon Nicholls
Loraine Wilson

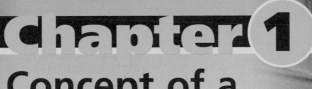

Chapter 1

Concept of a Function

Functions are used to describe relationships between two quantities. The amount of money you have for college may depend somewhat on the number of summer jobs you have before graduating from high school and the rate of pay at each job. The amount of milk left in the refrigerator may depend on, or vary with, the number of people who get home earlier than you do. Functions can provide a model for relations as diverse as production costs, prices, distance and time for vehicles, age and heart rates, units of measurement, altitudes and temperatures, and interest on investments or loans. As you learn how to represent functions algebraically, verbally, and graphically, you will develop strategies for making and testing predictions and drawing conclusions. You will also become more aware of the role that functions play in everyday communications.

IN THIS CHAPTER, YOU CAN . . .

- define the term *function*;
- describe functions and determine their properties;
- substitute into and evaluate functions;
- represent transformations of functions algebraically and graphically;
- identify the domain and range of relations and functions;
- represent inverse functions;
- study the relationship between a function and its inverse.

1.1 Relations

We frequently make statements in which some condition or rule relates one thing to another. *Robert is a brother of Maria* relates Maria to Robert by the rule "is a brother of." *The number n is three times the number q* relates the number *n* to the number *q* by the rule "is three times."

Each of these examples describes a rule whereby, given a number or thing, another number or thing is assigned.

Jason has a summer job cutting lawns. He is paid $10 per lawn, so the amount he earns is related to the number of lawns he cuts.

These data can also be displayed as a set of ordered pairs, (1, 10), (2, 20), (3, 30), (4, 40), (5, 50), (6, 60), in which the first coordinate represents the number of lawns that Jason cuts and the second coordinate represents the amount of money that he earns. This is an example of a **relation.**

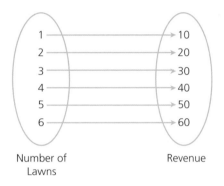

Number of Lawns   Revenue

A **relation** is a rule that associates each element *x* in a set *A* with one or more element(s) *y* in a set *B*. This rule creates a set of ordered pairs (*x*, *y*).

The set *A*, which is the set of first coordinates in the ordered pairs, is called the **domain** of the relation. The set *B*, which is the set of second coordinates in the ordered pairs, is called the **range** of the relation.

In the example above, the domain is {1, 2, 3, 4, 5, 6} and the range is {10, 20, 30, 40, 50, 60}.

EXAMPLE 1

Temperature readings for a number of cities on January 15 are given in the table.

Ottawa	−14°C
Kingston	−12°C
Toronto	−8°C
London	−3°C
Windsor	+2°C

a. List the ordered pairs of cities (*x*, *y*) where *x* is the city with the lower temperature.

b. List the domain and range of this relation.

SOLUTION

a. Let A represent the relation.

$A = \{$(Ottawa, Kingston), (Ottawa, Toronto), (Ottawa, London), (Ottawa, Windsor), (Kingston, Toronto), (Kingston, London), (Kingston, Windsor), (Toronto, London), (Toronto, Windsor), (London, Windsor)$\}$

b. The domain of the relation is {Ottawa, Kingston, Toronto, London}.
The range of the relation is {Kingston, Toronto, London, Windsor}.

From the graph of a relation, it is usually possible to state the domain and range.

EXAMPLE 2

A pool at a fitness centre is being drained. The number of kilolitres of water N in the pool after an elapsed time t, in minutes, is given by the formula $N = 100 - 0.25t$. Graph this relation and state its domain and range.

Photo: COMSTOCK/R. Digiacomo

SOLUTION

Using a graphing calculator or a computer program, you can obtain the graph shown.

From the graph, the formula is valid for values of t from 0 to 400 and for values of N from 0 to 100.

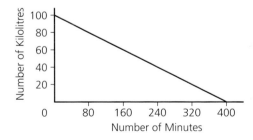

The domain of this relation is the set of times t between 0 and 400 inclusive.

Algebraically this is written $0 \le t \le 400, t \in R$.

The range of this relation is the set of the number of kilolitres N between 0 and 100 inclusive.

Algebraically this is written $0 \le N \le 100, N \in R$.

EXAMPLE 3

State the domain and range of each of the following relations.

a.

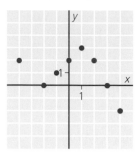

b.

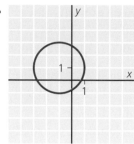

SOLUTION

Reading from the graph,

a. the domain is $\{-4, -2, -1, 0, 1, 2, 3, 4\}$ and the range is $\{-2, 0, 1, 2, 3\}$;

b. the domain is $-3 \leq x \leq 1, x \in R$, and the range is $-1 \leq y \leq 3, y \in R$.

EXAMPLE 4

What is the domain and range of the relation given by the equation $y = x^2$?

SOLUTION

The result of squaring any real number is either a positive number or zero. The domain of this relation is any real number, written $x \in R$. The range of this relation is all real numbers greater than or equal to zero, written $y \geq 0, y \in R$.

EXAMPLE 5

State the domain and range of the relation $y = \sqrt{x - 1}$.

SOLUTION

In order for the square root to be a real number, the radicand must always be greater than or equal to zero. In this case $x - 1 \geq 0$ or $x \geq 1$.

The domain of the relation is $x \geq 1, x \in R$.

Since y is the positive square root of a number equal to or greater than 0, for every $x \geq 1$ the range is $y \geq 0, y \in R$.

Summary

A relation can be given

◆ by a table of values,

◆ by a set of ordered pairs,

◆ in words,

◆ by an equation,

◆ by a graph.

Exercise 1.1

PART A

1. The height of a tree is related to the diameter of its trunk. The table shows the height and the diameter of five maple trees.

Diameter (cm)	112	120	122	132	140
Height (m)	27	28	29	31	33

    a. State the ordered pairs (D, H) of this relation.

    b. State the domain and range of this relation.

2. State the domain and range of each relation.

    a. $\{(0, 2), (1, 5) (3, 11), (4, 14), (5, 15)\}$

    b. $\{(-3, 4), (2, 1), (4, 4), (9, 5)\}$

3. a. In Example 2, why is the formula $N = 100 - 0.25t$ not valid for t less than 0 ($t < 0$) or for N less than 0 ($N < 0$)?

    b. How many kilolitres of water are in the pool before it is drained?

    c. How long does it take to drain the pool?

4. The graph shows the distance d travelled by every member of the Hiking Club in time t.

    a. What is the domain and range of this relation?

    b. What does the horizontal line segment of the graph represent?

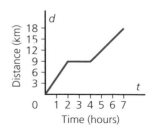

5. State the domain and range of each of the following relations.

a.

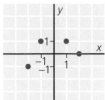

b.

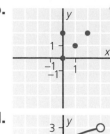

c.

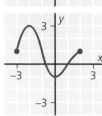

d.

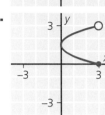

6. State the domain and range of the relation defined by the equation $y = 2x + 3$, $x \in R$.

PART B

7. The mass, in kilograms, of a number of students in a mathematics class is given in the table.

Hinta	73 kg
Karen	62 kg
Shenif	65 kg
Henry	74 kg

a. List the ordered pairs in the relation defined by the statement "is lighter than." For example, the ordered pair (Hinta, Karen) means that Karen is lighter than Hinta.

b. List the domain and range of this relation.

8. a. Using only the single-digit positive integers, list the ordered pairs of the relation "is a divisor of." For example, the ordered pair (6, 2) means that 2 is a divisor of 6.

b. List the domain and range of this relation.

9. The four children of the Collin family in order of birth are Vicki, Elizabeth, Chris, and David.

a. List the ordered pair of the relation "is younger than." For example, the ordered pair (Vicki, Elizabeth) means that Elizabeth is younger than Vicki.

b. List the domain and range of this relation.

10. Use your graphing calculator or a computer program to help you graph each of the following relations. Write the domain and range of each relation.

   a. $2x + 3y = 12$ **b.** $y = 3x^2$ **c.** $y = \sqrt{x + 4}$

   d. $y = \dfrac{1}{x}$ **e.** $y = x^2 + 3$ **f.** $y = \dfrac{3}{x - 2}$

11. A baseball player hits a ball toward right field. The path of this ball can be represented by the formula $h = -0.005x^2 + 0.6x + 1.25$, where h represents the height of the ball and x the distance of the ball from home plate in metres.

   a. Use a graphing calculator to graph this relation. Set the domain $-5 \le x \le 140$ and the range $-5 \le y \le 25$.

   b. What is the highest point reached by the ball?

   c. If the fence is 110 m from home plate and 3 m high, will the ball land outside the ballpark?

12. Draw a graph to describe the following situation.

The height of a flag as it is being hoisted to the top of the flagpole is related to the time it takes to raise the flag.
What is the domain and range of this relation?

13. a. Summarize the different ways in which we can describe a relation.

   b. Describe how you can find the domain and range of a relation.

PART C

14. Determine the domain and range of each relation.

   a. $x^2 + y^2 = 25$ **b.** $x^2 + 4y^2 = 25$

   c. $x^2 - 4y^2 = 25$ **d.** $(x - 2)^2 + y^2 = 25$

1.2 Definition of a Function

In some relations, one value of the domain gives one value of the range. In others, one value of the domain gives more than one value of the range. For example, if $y = x^2$, substituting $x = 4$ gives $y = 16$; if $y^2 = x$, substituting $x = 4$ gives $y^2 = 4$ and $y = \pm 2$.

Relations in which each value of the domain gives exactly one value of the range are of sufficient importance that we identify them under the term **function.** We now say that the second number (or thing) is a function of the first number (or thing).

For example, if a store is having a sale and taking 35% off all prices marked, then we say that the sale price is a function of the original price. The phrase "a function of " in this example could be replaced by " related to."

> A **function** f is a relation that assigns to each element in the domain
> exactly one element in the range.

This means that if we think in terms of ordered pairs, each of the ordered pairs has a different first coordinate.

The height of a football is related to the time since it was kicked. The football is never at two different heights at the same time and so this is a function.

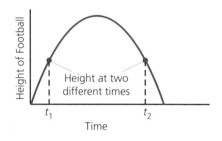

Consider the relation that associates a person to a telephone number.

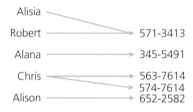

This relation is *not* a function because there is an element in the first set, namely Chris, matched with two elements in the second set, his telephone numbers.

EXAMPLE 1

State, with reasons, whether or not each of the following relations is a function.

a. $\{(1, -1), (2, 0), (3, 1), (-1, 3), (-2, -3)\}$

b. $\{(3, 3), (2, -1), (4, 1), (3, -2), (1, 2)\}$

a. Since every ordered pair has a different first coordinate, the relation is a function.

b. Since two of the ordered pairs, (3, 3) and (3, −2), have the same first coordinate, the relation is not a function.

Each ordered pair in the relations above is associated with a point on a graph. Graphing these relations on separate grids will illustrate another way of determining if a relation is a function.

a.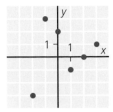

No two points on the graph lie on the same vertical line.

This relation is a function.

b.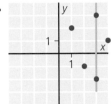

The points (3, 3) and (3, −2) have the same first coordinates and lie on the same vertical line.

This relation is not a function.

This illustrates the **vertical line test** for functions.

> If no two points on the graph of a relation lie on the same vertical line, the relation represented by the graph is a function.

Consider the function f that squares every real number. The function's action is indicated by letting the function operate on a number. For example, the square of −2 is 4. The notation $f(-2)$ tells us to follow the action described by the function f.

We can think of a function as a machine. In our example, the function f represents the action of squaring.

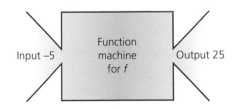

$f(-5) = 25$ (illustrated at the right by the function machine)

In general, $f(x) = x^2$.

The input value is the **independent variable** and the output is the **dependent variable.** If we put in a specific value for the independent variable, the function will produce a unique value for the dependent variable.

The function that describes the squaring of a real number, $f(x) = x^2$, can also be written as an equation, $y = x^2$. The notations $f(x)$ and y are interchangeable, and either can be used as is convenient. One advantage of the $f(x)$ notation is the simplicity of indicating a replacement value for x. In writing $f(7)$, we indicate that x is replaced by 7 in the function, leading to a specific function value.

Another function g adds the number 3 to any real number.

$$g(2) = 2 + 3$$
$$= 5$$
$$g(-4) = -4 + 3$$
$$= -1$$

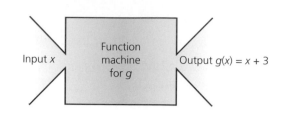

Input x — Function machine for g — Output $g(x) = x + 3$

In general, $g(x) = x + 3$.

At the beginning of this section, we considered the relation that a store was having a sale and taking 35% off the price on all items.

If x represents the original price of an item and y the sale price, then written as an equation

$$y = x - 0.35x$$
$$y = 0.65x.$$

Written in function notation

$$f(x) = 0.65x.$$

EXAMPLE 2

If $f(x) = x^2 - 3x + 1$, determine $f(3)$ and $f(-2)$.

SOLUTION

$$f(3) = (3)^2 - 3(3) + 1$$
$$= 9 - 9 + 1$$
$$= 1$$

$$f(-2) = (-2)^2 - 3(-2) + 1$$
$$= 4 + 6 + 1$$
$$= 11$$

EXAMPLE 3

Is the relation $y^2 = x^2 - 9$ a function?

SOLUTION

To determine values for y, we can only substitute values for x that will make $x^2 - 9$ equal to or greater than zero, that is $x \geq 3$ or $x \leq -3$.

Since $y^2 = x^2 - 9$, $y = \pm\sqrt{x^2 - 9}$.

For any $x > 3$ or $x < -3$, $x \in R$, we get two values for y.

For example, if $x = 5$, $y = \pm 4$.

Since some first coordinates produce two second coordinates, this relation is not a function.

EXAMPLE 4

If $h(x) = 3x - 2$, determine each of the following:

a. $h(-2)$ **b.** $h(a)$ **c.** $h\left(\dfrac{1}{x}\right)$ **d.** $h(x + 5)$

SOLUTION

a. $h(-2) = 3(-2) - 2$
$\quad\quad = -8$

b. $h(a) = 3(a) - 2$
$\quad\quad = 3a - 2$

c. $h\left(\dfrac{1}{x}\right) = 3\left(\dfrac{1}{x}\right) - 2$
$\quad\quad = \dfrac{3}{x} - 2$
$\quad\quad = \dfrac{3 - 2x}{x}$

d. $h(x + 5) = 3(x + 5) - 2$
$\quad\quad = 3x + 15 - 2$
$\quad\quad = 3x + 13$

Exercise 1.2

PART A

1. State which of the following relations are functions.

  a. $(0, 0), (1, 2), (2, 4), (3, 9), (4, 16)$ **b.** $(0, 0), (1, 3), (1, -1), (4, 1), (3, -2)$

  c. $(1, 5), (2, 5), (-1, 5), (0, 5), (3, 5)$ **d.** $(3, 1), (3, 2), (3, 3), (3, -2), (3, 0)$

  e. The optical reader at the checkout counter of the supermarket to convert codes to prices.

  f. $y = 3x - 5$ **g.** $x = y^2$

  h.

x	1	1	2	2	3	3
y	2	-2	3	-3	5	-5

2. If $g(x) = 3x - 1$, determine the following:

  a. $g(2)$ **b.** $g(-3)$ **c.** $g\left(\dfrac{1}{3}\right)$ **d.** $g(a)$

3. Which of the following relations are functions?

  a. 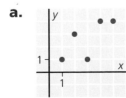   **b.** 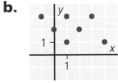   **c.**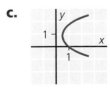

  d. 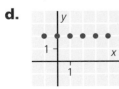   **e.**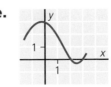

4. Given the function f represented by the given set of ordered pairs
$f = \{(-1, 3), (0, 4), (1, 5), (2, 6)\}$, determine $f(0)$ and $f(2)$.

5. Given the graph of $y = g(x)$, state the value of each
of the following:

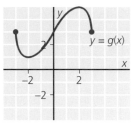

    a. $g(-2)$ **b.** $g(0)$ **c.** $g(3)$

    d. What is the value of x such that $g(x) = 5$?

PART B

6. Determine whether or not each of the following relations represents a function.

    a. $y = 3x - 2$ **b.** $x^2 + y^2 = 25$

    c. $y = x^2 - 2x$ **d.** $x = y^2$

7. Use your graphing calculator or computer to graph each function. Determine the
domain and range of each function.

    a. $f(x) = x^2 - 3$ **b.** $g(x) = \sqrt{x - 3}$ **c.** $h(x) = \dfrac{1}{x + 2}$

8. Given the function $f(x) = -x^2 + 4x - 5$, determine the following:

    a. $f(3)$ **b.** $f(-2)$ **c.** $f(0.25)$ **d.** $f(a)$

9. Given the graph of $y = f(x)$, determine the following:

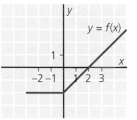

    a. $f(-2)$

    b. the value of x such that $f(x) = 0$

    c. $f(1) + f(-1)$

10. Given $g(x) = 2x - 5$, determine the following:

    a. $g(-1)$ **b.** $g(5)$ **c.** $g\left(-\dfrac{1}{2}\right)$

    d. $g(2x)$ **e.** $g(x + 1)$ **f.** $g(x + a)$

    g. $g\left(\dfrac{3}{x}\right)$ **h.** $g(2x - 5)$ **i.** $g(2x^2 - x)$

11. Given $h(x) = 3x^2 - 2x + 5$, determine the following:

    a. $h(-3)$ **b.** $h(3)$ **c.** $h(3a)$

    d. $h(2x)$ **e.** $h(-x)$ **f.** $h(x + 3)$

    g. $h\left(\dfrac{1}{x}\right)$ **h.** $h(3x + 2)$ **i.** $h(x^2 - 1)$

12. If $f(x) = 3x + 5$ and $g(x) = x^2 + 3x - 5$, determine the following:

a. $f(1) + g(1)$ **b.** $f(-3) - g(-2)$ **c.** $f(x) + g(x)$

d. $3f(x) + 2g(x)$ **e.** $f(2x) - g(x)$ **f.** $g(2x + 1) - f(3x)$

13. Explain the difference between a relation that is a function and one that is not.

PART C

14. If $2f(x) - 3f\left(\dfrac{1}{x}\right) = x^2$, determine $f(2)$.

15. Let f be a function defined by $f(n) = 2f(n - 1) + 3f(n - 2)$, where $f(1) = 1$ and $f(2) = 2$. Determine $f(5)$.

The Transformations of M.C. Escher

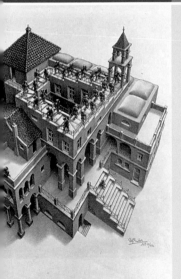

Many artists have used transformations in their work. From the designs on the vases of ancient Egypt to modern art, translations, rotations, and reflections have been used to create beautiful patterns.

Maurits Cornelius Escher (1898–1972) was a master at using transformations. Born in the Netherlands, Escher studied graphic art and woodcutting techniques. In his travels, Escher studied the art forms everywhere he went. He was especially intrigued by the ornamentation of the Moors, which filled all the space available with patterns.

Escher's art involved repeated patterns, regular divisions of the space, and seemingly impossible constructions. Many of his works used tessellation, the fitting of shapes together to form a design or picture. Escher's work is readily recognized for its spatial illusions, impossible buildings, and repeated geometric patterns. Studying Escher's drawing called *Ascending and Descending*, you become lost in climbing endless staircases, only to find that at the next turn, you are again at the bottom of the staircase without ever having descended.

Escher's unique understanding of mathematical concepts enabled him to draw his wondrous creations.

You can learn more about Escher's work and enjoy his marvellous drawings by doing a Web site search using his name.

Photo: M. C. Escher's Ascending and Descending © Cordon Art BV, Baarn, Holland

When we transform something, we change the object into something with a new appearance while maintaining some or all of the characteristics of the original. Mathematically we transform the graph of a function by changing either its appearance or its position or both. Examples of **transformations** are **translations, reflections, rotations,** and **dilatations.** When we apply a transformation, we alter some aspect(s) of the graphical representation while maintaining the characteristics of the original function.

In the following activity, we will examine the effects of applying some transformations to the graph of a given function $y = f(x)$.

ACTIVITY

You will use this graph of $y = f(x)$.

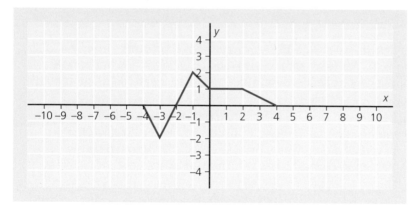

PART 1

1. On graph paper make a copy of $y = f(x)$.

2. Copy and complete the following chart.

x	−4	−3	−2	−1	0	1	2	$\frac{3}{2}$	4
$f(x)$	0	−2	0	2	1	1	1		0
$f(x) - 1$							0		
$f(x) + 2$							3		
$f(x) - 3$							−2		

3. Sketch the graphs for $y = f(x) - 1$, $y = f(x) + 2$, and $y = f(x) - 3$ on the same grid as $y = f(x)$.
Label each graph with its equation.

4. **a.** How is the graph of $y = f(x) - 1$ related to the graph of $y = f(x)$?

   b. How is the graph of $y = f(x) + 2$ related to the graph of $y = f(x)$?

   c. How is the graph of $y = f(x) - 3$ related to the graph of $y = f(x)$?

5. How would the graph of $y = f(x) + c$ be related to the graph of $y = f(x)$? Consider the cases where $c > 0$ and where $c < 0$.

This transformation is called a vertical translation.

PART 2

1. On graph paper make a copy of $y = f(x)$.

2. Copy and complete the following chart.
   Example: When $x = -3$, $f(x - 1) = f(-4) = 0$.
   When $x = -3$, $f(x + 2) = f(-1) = 2$.

   Shaded blocks indicate that we cannot find a value for these values of x. Why not?

x	-6	-5	-4	-3	-2	-1	0	1	2	3	4	5	6	7
$f(x)$			0	-2	0	2	1	1	1	$\frac{1}{2}$	0			
$f(x - 1)$				0										
$f(x + 2)$				2										
$f(x - 3)$														

3. Sketch the graphs for $y = f(x - 1)$, $y = f(x + 2)$, and $y = f(x - 3)$ on the same grid as $y = f(x)$. Label each graph with its equation.

4. **a.** How is the graph of $y = f(x - 1)$ related to the graph of $y = f(x)$?

   b. How is the graph of $y = f(x + 2)$ related to the graph of $y = f(x)$?

   c. How is the graph of $y = f(x - 3)$ related to the graph of $y = f(x)$?

5. How would the graph of $y = f(x + c)$ be related to the graph of $y = f(x)$? Consider the cases where $c > 0$ and where $c < 0$.

This transformation is called a horizontal translation.

PART 3

1. On graph paper make a copy of $y = f(x)$.

2. Copy and complete the following chart.

x	-4	-3	-2	-1	0	1	2	3	4
$f(x)$	0	-2	0	2	1	1	1	$\frac{1}{2}$	0
$2f(x)$				4					
$\frac{1}{2}f(x)$				1					

3. Sketch the graphs for $y = 2f(x)$ and $y = \frac{1}{2}f(x)$ on the same grid as $y = f(x)$. Label each graph with its equation.

4. a. How is the graph of $y = 2f(x)$ related to the graph of $y = f(x)$?

   b. How is the graph of $y = \frac{1}{2}f(x)$ related to the graph of $y = f(x)$?

5. How would the graph of $y = kf(x)$ be related to the graph of $y = f(x)$? Consider the cases where $k > 1$ and $0 < k < 1$.

> This transformation is called a **vertical dilatation**.
> It may be either a **stretch** or a **contraction**.

PART 4

1. On graph paper make a copy of $y = f(x)$.

2. Copy and complete the following chart. Note that it may be necessary to interpolate (read in between values) for some cases.

x	-8	-7	-6	-5	-4	-3	-2	-1	0	1	2	3	4	5	6	7	8
$f(x)$					0	-2	0	2	1	1	1	$\frac{1}{2}$	0				
$f\left(\frac{1}{2}x\right)$					2												
$f(2x)$					0												

3. Sketch the graphs for $y = f\left(\frac{1}{2}x\right)$ and $y = f(2x)$ on the same grid as $y = f(x)$. Label each graph with its equation.

4. a. How is the graph of $y = f\left(\frac{1}{2}x\right)$ related to the graph of $y = f(x)$?

   b. How is the graph of $y = f(2x)$ related to the graph of $y = f(x)$?

5. How would the graph of $y = f(kx)$ be related to the graph of $y = f(x)$? Consider the cases where $k > 1$ and where $0 < k < 1$.

> This transformation is called a **horizontal dilatation**.
> It may be either a **stretch** or a **contraction**.

PART 5

1. On graph paper make a copy of $y = f(x)$.

2. Copy and complete the following chart.

x	−4	−3	−2	−1	0	1	2	3	4
f(x)	0	−2	0	2	1	1	1	2	0
−f(x)				−2					
f(−x)				1					

3. Sketch the graphs for $y = -f(x)$ and $y = f(-x)$ on the same grid as $y = f(x)$. Label each graph with its equation.

4. **a.** How is the graph of $y = -f(x)$ related to the graph of $y = f(x)$?

   b. How is the graph of $y = f(-x)$ related to the graph of $y = f(x)$?

These transformations are referred to as **reflections** in a line.
In the above case, the reflection lines are the x-axis or the y-axis.

Summary

Transformation	Name of Transformation	Effect on Graph
$f(x) + k$	vertical translation	$k > 0$: Shift up k units. $k < 0$: Shift down $\lvert k \rvert$ units. ($\lvert k \rvert = k$ if $k \geq 0$ and $\lvert k \rvert = -k$ if $k < 0$, so, for example, $\lvert -3 \rvert = 3$).
$f(x + k)$	horizontal translation	$k > 0$: Shift left k units. $k < 0$: Shift right $\lvert k \rvert$ units.
$kf(x)$	vertical dilatation	$k > 1$: Stretch vertically by a factor k. $0 < k < 1$: Contract vertically by a factor k.
$f(kx)$	horizontal dilatation	$k > 1$: Contract horizontally by a factor of $\frac{1}{k}$. $0 < k < 1$: Stretch horizontally by a factor of $\frac{1}{k}$.
$-f(x)$	vertical reflection	Reflect in the x-axis.
$f(-x)$	horizontal reflection	Reflect in the y-axis.

1. Copy and complete the following table. Fill in all the blanks for which you have sufficient information.

x	-3	-2	-1	0	1	2	3
$f(x)$	-6	-2	0	3	5	-4	8
$f\left(\tfrac{1}{2}x\right)$						5	
$f(2x)$			-2				

2. Let $y = f(x)$ be represented by the graph on the right. Graph each of the following:

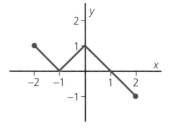

a. $y = f(x) + 1$

b. $y = f(x - 2)$

c. $y = f(2x)$

d. $y = f\left(\tfrac{1}{2}x\right)$

e. $y = 2f(x) - 3$

3. Describe the transformation on the graph of $y = f(x)$ needed to obtain the graph of each of the following:

a. $y = f(x - 4)$

b. $y = 5f(x)$

c. $y = f(x) - 5$

d. $y = f\left(\tfrac{1}{4}x\right)$

e. $y = f(x - 2) - 1$

f. $y = 3f(x + 1)$

g. $y = 2 - f(x)$

h. $y = f(-2x)$

i. $y = \tfrac{1}{4}f(x) + 2$

j. $y = f(1 - x)$

k. $y = f\left(\tfrac{x}{3} - 2\right)$

l. $y = 3 - 3f(2x)$

1.4 The Quadratic Function

Suppose that a baseball is "popped" straight up by a batter. The height of the ball can be modelled by the function $h = f(t) = -4.9t^2 + 21.5t + 1.5$, where t is the time in seconds after the ball leaves the bat and h is the height in metres. The graph of the function, obtained from a graphing calculator, is shown.

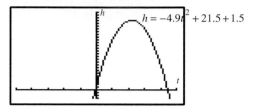

This is an example of a quadratic function. Other examples are $y = x^2$, $y = -3(x - 1)^2 + 2$, and $f(x) = 2x^2 + 4x - 5$. In general, a quadratic function can be written in standard form, $f(x) = ax^2 + bx + c$, $a \neq 0$, or in the vertex form, $f(x) = a(x - p)^2 + q$, $a \neq 0$, where the coordinates of the vertex are (p, q).

In this section, we will use the vertex form so that we can use our knowledge of transformations in graphing the function. Later you will learn how to change from the standard form to the vertex form.

The graph of a quadratic function is a parabola. The graph at the right represents the basic quadratic function $f(x) = x^2$. Basic properties of the function $f(x) = x^2$ are as follows:

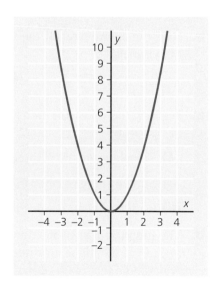

◆ The lowest or minimum point of the curve, in this case $(0, 0)$, is called the **vertex of the parabola.**

◆ The domain is $x \in R$.

◆ The range is $y \geq 0$, $y \in R$.

◆ The graph is symmetric about the y-axis or the line $x = 0$. The vertical line through the vertex of the parabola is the **axis of symmetry.**

◆ The graph opens upward or is concave up.

EXAMPLE 1

a. By using transformations on the graph of the basic quadratic function $y = x^2$, sketch the graph of the function $y = (x - 3)^2 + 2$.

b. State the coordinates of the vertex, the domain, the range, and the equation of the axis of symmetry for $y = (x - 3)^2 + 2$.

SOLUTION

a. To obtain the graph of $y = (x - 3)^2 + 2$, translate the graph of $y = x^2$ three units to the right and two units up.

b. Since the vertex $(0, 0)$ of the basic function is translated three units to the right and two units up, the vertex of the graph of $y = (x - 3)^2 + 2$ is $(3, 2)$.

The domain is $x \in R$.

The range is $y \geq 2$, $y \in R$.

The axis of symmetry is $x = 3$.

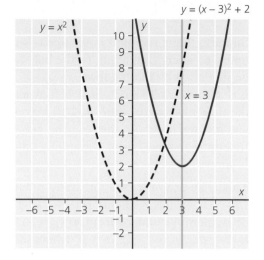

EXAMPLE 2

For the function $f(x) = -2(x + 3)^2$,

a. sketch the graph of $f(x)$;

b. state the coordinates of the vertex, the domain, the range, and the equation of the axis of symmetry.

SOLUTION

a. Perform the following transformations on the graph of $f(x) = x^2$.

   i) Translate $f(x) = x^2$ three units to the left.

   ii) Stretch $f(x) = (x + 3)^2$ vertically by a factor of 2.

   iii) Reflect $f(x) = 2(x + 3)^2$ in the x-axis.

b. The graph of $f(x)$ opens downward, so the vertex $(-3, 0)$ is the highest or maximum point of the curve. A parabola that opens downward is said to be concave down.

The domain is $x \in R$, and the range is $y \leq 0$, $y \in R$.

The equation of the axis of symmetry is $x = -3$.

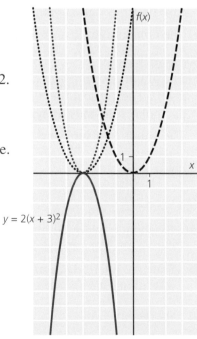

In general, the graph of a quadratic function given by $f(x) = a(x - p)^2 + q$ has vertex (p, q) and axis of symmetry with equation $x = p$.

If $a > 0$, the parabola opens upward. If $a < 0$, the parabola opens downward.

EXAMPLE 3

For the function $y = -3(x + 1)^2 - 2$,

a. sketch the graph;

b. state the coordinates of the vertex, the domain, the range, and the equation of the axis of symmetry.

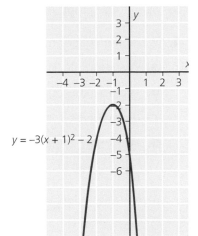

$y = -3(x + 1)^2 - 2$

SOLUTION

The vertex is $(-1, -2)$.

The domain is $x \in R$ and the range is $y \leq -2$, $y \in R$. The equation of the axis of symmetry is $x = -1$.

EXAMPLE 4

Write the equation for a quadratic function with vertex $(-2, 4)$, concave up and congruent to $y = 3x^2$.

SOLUTION

The function equation is $y = 3(x + 2)^2 + 4$.

Exercise 1.4

PART A

1. For each of the following functions state

    i) the direction of opening,   **ii)** the coordinates of the vertex,

    iii) the range,   **iv)** the equation of the axis of symmetry.

    a. $y = x^2 - 3$ **b.** $y = (x - 3)^2$ **c.** $y = -x^2 + 4$

    d. $y = 2(x + 2)^2 + 1$ **e.** $y = -3(x - 4)^2 - 3$ **f.** $f(x) = -(x - 5)^2$

    g. $g(x) = 4x^2 - 3$ **h.** $h(x) = 3 - 2(x + 4)^2$ **i.** $f(x) = \frac{1}{2}(x + 3)^2 - 5$

2. Sketch the graphs of each of the following:

    a. $y = (x - 2)^2$ **b.** $y = x^2 - 2$

    c. $y = -2x^2$ **d.** $y = -(x - 2)^2$

3. Sketch the graph of each of the quadratic functions in Question 1.

4. Write the equation for each of the following quadratic functions.

    a. Vertex $(-1, 2)$, opening down, congruent to $y = 4x^2$.

    b. Vertex $(0, -3)$, opening up, congruent to $y = -2x^2$.

    c. Range $y \leq 3$, axis of symmetry $x = 1$, congruent to $y = x^2$.

    d. Vertex $(-2, -2)$, range $y \geq -2$, congruent to $y = -5x^2$.

Parabolic Trajectory

Galileo discovered in the early 1600s that the trajectory of a projectile fired from a gun is parabolic (assuming that air resistance is ignored).

When great warships of the past were armed for battle, their guns were designed so that the elevation of the muzzle could be changed. If the opposing ship was closer, the shell was fired at a greater angle so that when the shell followed a parabolic path, it would come down closer to the gun. The adjustment of the elevation of the gun was a trial-and-error process.

Now when the military fire a missile at a predetermined target, they take into account the distance to the target and the speed at which the missile will travel. Knowing that the path of the missile will be parabolic, computers calculate the angle of elevation at which to fire the missile.

When NASA launched the space shuttle *Endeavour*, the booster rocket was released shortly after takeoff. The parabolic path of the booster rocket lit up the night sky.

The Space Shuttle Endeavour   *Photo:* © Roger Ressmeyer/CORBIS

1.5 Square Root Function

The time taken for an object to fall to the ground from a given height is given by the function $t = \sqrt{\dfrac{h}{4.9}}$, where t is the time in seconds and h is the height in metres.

This is an example of a square root function. The graph of this function is shown below using a graphing calculator. Note that the time increases slowly, relative to the height.

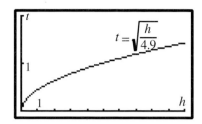

The square root function is $f(x) = \sqrt{x}$. The graph is shown to the right and appears to look like half of a parabola opening to the side (and in fact is exactly that). Basic properties of this function are

◆ the domain is $x \geq 0$, $x \in R$;

◆ the range is $y \geq 0$, $y \in R$.

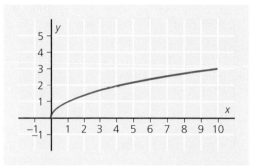

EXAMPLE 1

a. By using transformations on the square root function $y = \sqrt{x}$, sketch the graph of the function $y = \sqrt{x + 2} - 1$.

b. What are the domain and range of the function?

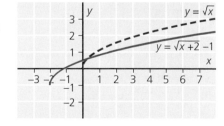

SOLUTION

a. To obtain the graph of $y = \sqrt{x + 2} - 1$, translate the graph of $y = \sqrt{x}$ two units to the left and one unit down.

b. The domain is $x \geq -2$, $x \in R$, and the range is $y \geq -1$, $y \in R$.

EXAMPLE 2

a. By using transformations on the graphs of $y = \sqrt{x}$, sketch the graph of $y = \sqrt{-x}$ and state the domain and range.

b. By using transformations on the graph of $y = \sqrt{x}$, sketch the graph of $y = -\sqrt{x}$ and state the domain and range.

SOLUTION

a. The graph of $y = \sqrt{-x}$ is a reflection of $y = \sqrt{x}$ in the y-axis. The domain is $x \leq 0$, $x \in R$, and the range is $y \geq 0, y \in R$.

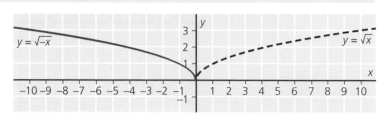

b. The graph of $y = -\sqrt{x}$ is a reflection of $y = \sqrt{x}$ in the x-axis. The domain is $x \geq 0$, $x \in R$, and the range is $y \leq 0, y \in R$.

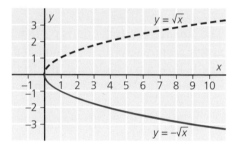

Exercise 1.5

PART A

1. Give the domain and range for each of the following functions.

a. $y = \sqrt{x - 5}$

b. $y = \sqrt{x + 3}$

c. $y = \sqrt{2 - x}$

d. $y = \sqrt{3x}$

e. $y = -\sqrt{3 - x}$

f. $y = 2 + \sqrt{x + 1}$

g. $y = \sqrt{x - 4} - 2$

h. $y = 3 - \sqrt{-x}$

PART B

2. Given that the graphs below represent square root functions, write a possible equation for each.

a.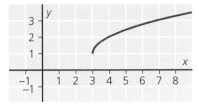

b.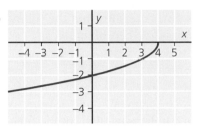

c.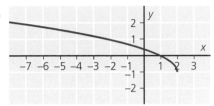

d.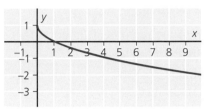

3. Sketch graphs for each of the following functions.

  a. $y = \sqrt{2x}$

  b. $y = \sqrt{2 - x}$

  c. $y = 2\sqrt{x}$

  d. $y = 2 - \sqrt{x}$

  e. $y = \dfrac{\sqrt{x + 3}}{2}$

  f. $y = \sqrt{\dfrac{x + 4}{2}}$

  g. $y = -2\sqrt{1 - x}$

  h. $y = 2 - \sqrt{2 - x}$

4. State the domain for $f(x) = \sqrt{2x - 3}$.

PART C

5. State the domain and range and sketch the graph for $y = 4 - 2\sqrt{2 - 4x}$.

6. **a.** Sketch the graph of $y^2 = x$.

  b. What two functions make up this graph?

  c. Sketch the graph of $y^2 = -x$.

  d. What two functions make up this graph?

The time required for Sammy the Snail to move 1 m is inversely proportional to his speed. Some of his times are shown in the following table where his time in minutes is t and his speed in centimetres per minute is s.

Speed	1	2	4	5	10	20	50
Time	100	50	25	20	10	5	2

This relation can be modelled by the function $t = \dfrac{100}{s}$, which is an example of a reciprocal function.

The graph of the reciprocal function $f(x) = \dfrac{1}{x}$ has some particular features.

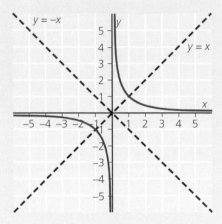

◆ The domain is $x \in R, x \neq 0$, since division by zero is undefined.

◆ The range is $y \in R, y \neq 0$, since $\dfrac{1}{x} \neq 0$ for any value of x.

◆ The graph has no x- or y-intercepts since $x \neq 0$ and $y \neq 0$.

◆ Observe that the curve is symmetrical about the lines $y = x$ and $y = -x$.

Consider the following table of values for $y = \dfrac{1}{x}$, when $x \geq 0$.

x	20	10	5	$\frac{1}{5}$	$\frac{1}{10}$	$\frac{1}{20}$
y	$\frac{1}{20}$	$\frac{1}{10}$	$\frac{1}{5}$	5	10	20

As x increases in value, y decreases in value. As x gets very large, y gets very close to 0. As x gets very close to 0, y gets very large.

The graph of $y = \dfrac{1}{x}$ approaches the x- and y-axis but never touches them.

When a graph approaches a line but never touches it, that line is called an **asymptote.**

Now consider values for $y = \dfrac{1}{x}, x \leq 0$.

x	−20	−10	−5	$-\frac{1}{5}$	$-\frac{1}{10}$	$-\frac{1}{20}$
y	$-\frac{1}{20}$	$-\frac{1}{10}$	$-\frac{1}{5}$	−5	−10	−20

Now we observe that the graph of $y = \dfrac{1}{x}$ approaches the x- and y-axis in quadrant 3.

The axes are asymptotes for the curve defined by $y = \dfrac{1}{x}$.

As with the other functions we have examined, transformations will alter some aspects of the graph, most notably the positions of the asymptotes.

Reflections and Translations

Architects use reflections and translations to create the beautiful symmetry shown in their buildings.

Canada has produced many world-renowned architects, among them Frank Gehry, David Azrieli, and Moshe Safdie.

Frank Gehry is known for his approach to architecture as art. He has designed such internationally influential structures as the Chiat/Day building in Venice, California, and the Guggenheim Museum in Bilbao, Spain.

David Azrieli is renowned for creating some of the Middle East's largest developments, including state-of-the-art skyscrapers and large commercial complexes. Azrieli's interest in building shopping malls grew out of his childhood dream of becoming a fashion designer. He found he could express his ideas in architecture as well as he could have in fashion design.

Frank Gehry

Moshe Safdie designed Habitat '67 for the World Exhibition in Montreal in 1967 and the spectacular National Gallery of Canada in Ottawa. Safdie's work is characterized by the interplay of technology and geometry.

Architects blend their sense of design and artistic ability to create our wonderful surroundings. You can learn more about the work of these famous Canadian architects on the Internet.

Photo: Roger Ressmeyer/CORBIS

EXAMPLE 1

a. By using transformations on the graph of the basic reciprocal function $f(x) = \frac{1}{x}$, sketch the graph of $y = \frac{1}{x-2} + 1$.

b. State the domain, the range, and the equations of the asymptotes.

SOLUTION

a. To obtain the graph of $y = \frac{1}{x-2} + 1$, translate the graph of $y = \frac{1}{x}$ two units to the right and one unit up.

b. The domain is $x \in R, x \neq 2$, and the range is $y \in R, y \neq 1$.
The asymptotes are $x = 2$ and $y = 1$.

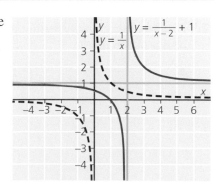

EXAMPLE 2

Sketch the graph of $y = -\frac{2}{x}$.

SOLUTION

To obtain the graph of $y = -\frac{2}{x}$, stretch the function $y = \frac{1}{x}$ vertically by a factor of 2 and then reflect it in the x-axis. Since neither a stretch nor a reflection affects points on the axes, the asymptotes remain unchanged; that is, the vertical asymptote is $x = 0$ and the horizontal asymptote is $y = 0$.

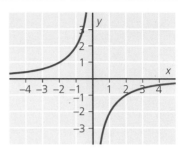

Exercise 1.6

PART B

1. For each of the following,

   i) sketch a graph;

   ii) state the domain, range, and the equations of the asymptotes.

   a. $y = \frac{1}{x + 3}$

   b. $y = \frac{1}{x} - 3$

   c. $y = \frac{2}{x - 4}$

   d. $y = \frac{3}{x + 1} + 1$

   e. $y = \frac{-4}{x - 2}$

   f. $y = 3 + \frac{1}{x + 2}$

   g. $y = \frac{2}{2 - x}$

   h. $y = 1 - \frac{1}{1 - x}$

   i. $y = -\frac{2}{x - 3} + 2$

PART C

2. Describe the transformation on the graph of $y = \frac{1}{x}$ needed to obtain the graph of each of the following:

   a. $y = \frac{x + 3}{x + 1}$

   b. $y = \frac{2x}{x - 1}$

1.7 Inverse Functions

We saw in Section 1.2 that a function is a rule that associates each element in the domain with exactly one element in the range, producing a set of ordered pairs in which no two ordered pairs have the same first element. What happens if we reverse the rule defining the function? An example illustrates the effect. For the function "is the cube of," inputting x-values means that the function produces y-values that are the cube of the x-values. We obtain ordered pairs such as $(3, 27)$, $(4, 64)$, $(1, 1)$, $(-2, -8)$. Reversing or taking the inverse of this gives "is the cube root of," which means that in ordered pairs obtained, the y-value is the cube root of the x-value. Here we obtain ordered pairs $(27, 3)$, $(64, 4)$, $(1, 1)$, $(-8, -2)$. The **inverse of a function,** then, reverses the coordinates of the ordered pairs defined by the function. This will often, but not always, give a new function.

EXAMPLE 1

In each of the following, give the inverse of the function and state whether the inverse is a function.

a. $f = \{(0, 6), (1, 7), (2, 8), (3, 9)\}$

b. $g = \{(3, 5), (-2, 1), (4, 3), (2, 5)\}$

SOLUTION

a. The inverse is $\{(6, 0), (7, 1), (8, 2), (9, 3)\}$. It is a function.

b. The inverse is $\{(5, 3), (1, -2), (3, 4), (5, 2)\}$. Since two pairs have the same first coordinate, it is *not* a function.

> If the inverse of a function f is itself a function,
> we denote the inverse function by f^{-1}.

Be careful to distinguish between f^{-1} and $3^{-1} = \frac{1}{3}$. Recall that $3^{-1} = \frac{1}{3}$, but f^{-1} does *not* mean $\frac{1}{f}$. Rather, f^{-1} means that in $y = f(x)$, we are to interchange x and y. The symbol $y = f^{-1}(x)$ is read as "f inverse of x" or "the inverse of f at x."

EXAMPLE 2

a. Graph $f = \{(-2, 0), (-1, 1), (0, 3), (2, 6)\}$ and its inverse.

b. Find the coordinates of the midpoints of the line segment joining corresponding points on the graph of f and f^{-1}.

SOLUTION

$f^{-1} = \{(0, -2), (1, -1), (3, 0), (6, 2)\}$

The midpoint of $(0, -2)$ and $(-2, 0)$ is $\left(\dfrac{-2 + 0}{2}, \dfrac{0 + (-2)}{2}\right) = (-1, -1)$.

The midpoint of $(-1, 1)$ and $(1, -1)$ is $(0, 0)$.

The midpoints of the other pairs of points are $(1.5, 1.5)$ and $(4, 4)$. Notice that each midpoint lies on the line with equation $y = x$.

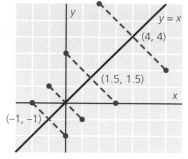

You will be asked to prove this result in general in the exercise.

This example suggests two properties of the graphs of a function and its inverse.

> **1.** If we join a point (p, q) on the graph of $y = f(x)$ and its image (q, p) on the graph of its inverse, the midpoint of the line segment will be on the line with equation $y = x$.
>
> **2.** The graph of the inverse of a function is a reflection of the graph of the function in the line with equation $y = x$.
>
> The second property is useful in graphing the inverse of a function.

EXAMPLE 3

a. Draw the graph of the inverse of the function $y = f(x)$ as shown in the diagram.

b. State the domain and range of both $y = f(x)$ and its inverse.

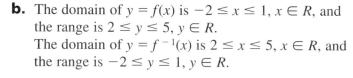

SOLUTION

a. The graph of the inverse is the reflection of $y = f(x)$ in the line $y = x$.

Since the inverse relation is a function, we can use the symbol $y = f^{-1}(x)$ to refer to this relation.

b. The domain of $y = f(x)$ is $-2 \le x \le 1, x \in R$, and the range is $2 \le y \le 5, y \in R$.

The domain of $y = f^{-1}(x)$ is $2 \le x \le 5, x \in R$, and the range is $-2 \le y \le 1, y \in R$.

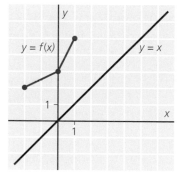

Since the inverse relation or function is obtained by reversing the x- and y-coordinates, the domain and range of the inverse are found by interchanging the domain and range of the original function.

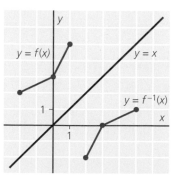

$$\text{If } f^{-1} \text{ is the inverse of } f, \text{ then the domain of } f^{-1} \text{ is the range } f$$
$$\text{and the range of } f^{-1} \text{ is the domain of } f.$$

EXAMPLE 4

If $f(x) = 3x + 2$, determine the inverse of $f(x)$.

SOLUTION

Since the graph of the inverse is obtained by a reflection in the line $y = x$, we can determine the inverse by interchanging x and y.

To determine the inverse, replace $f(x)$ by y.

$$f(x) = 3x + 2$$
$$y = 3x + 2$$

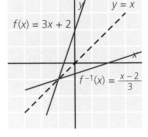

Now interchange x and y and then solve for y in terms of x.

The inverse relation is

$$x = 3y + 2$$
$$3y = x - 2$$
$$y = \frac{x - 2}{3}.$$

This is a function and we can write $f^{-1}(x) = \frac{x - 2}{3}$.

Steps in Finding the Inverse of a Function

1. Replace $f(x)$ by y.
2. Interchange x and y.
3. Solve this equation for y in terms of x, if possible.
4. Check whether or not the inverse is a function.
5. If so, the result is $y = f^{-1}(x)$.

Exercise 1.7

PART A

1. Is the inverse of a function always a function? Explain.

2. State the ordered pairs in the inverses of each of the following functions. Is the inverse a function?

   a. $\{(2, 5), (3, 6), (4, 7), (5, 8)\}$ b. $\{(4, 2), (5, 3), (-3, 2), (6, 3)\}$

   c. $\{(1, 3), (2, 3), (4, 3), (7, 3)\}$ d. $\{(-2, 3), (-1, 4), (2, 7), (4, 9)\}$

3. State the defining equation for the inverse of each of the following relations.

   a. $y = 3x - 2$ **b.** $5x + 2y = 10$

   c. $y = x^2 - 3$ **d.** $4x^2 + 9y^2 = 36$

4. In which of the cases illustrated below are f and g inverse relations?

   a.
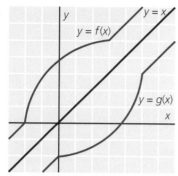
   b.
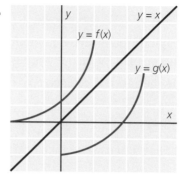

   c.
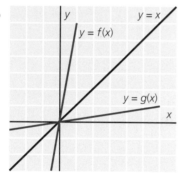
   d.
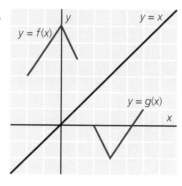

PART B

5. a. Graph the function $f = \{(0, 0), (1, 1), (2, 4), (3, 9)\}$.

   b. On the same grid as in part **a**, graph the inverse of f and the line $y = x$.

   c. Is the inverse a function?

6. a. Graph $y = x^2$.

   b. Is the relation in part **a** a function?

   c. Graph the inverse of the relation in part **a**.

   d. Is the inverse a function?

7. a. What is the inverse of $y = \frac{1}{x}$?

  b. How is the equation of the inverse related to the original function? Explain.

8. For the function $g(x) = 2x + 3$,

  a. evaluate $g(3)$, $g(-2)$, and $g(0)$;

  b. determine the defining equation for $g^{-1}(x)$;

  c. evaluate $g^{-1}(9)$, $g^{-1}(-1)$, and $g^{-1}(3)$;

  d. determine $g\left(\frac{x-3}{2}\right)$;

  e. determine $g^{-1}(2x + 3)$;

  f. determine the value of x if $g^{-1}(x) = 4$.

9. Determine the defining equation for the inverse of each of the following functions. State whether the inverse is also a function. If it is a function, give its domain and range.

  a. $f(x) = 2x - 5$

  b. $g(x) = 3x + 1$

  c. $h(x) = 2x$

  d. $f(x) = x^2 - 4$

  e. $g(x) = \frac{1}{x-3}$

10. Alice makes a base salary of $20 000 plus a commission of 15% of her sales.

  a. Write an equation for her total income y in terms of her sales x.

  b. State the domain and range.

  c. What is the defining equation for the inverse relation?

  d. Is the inverse relation a function? State the domain and range.

11. If $f(x) = 2x - 3$,

  a. determine the defining equation for the inverse of $f(x)$;

  b. graph $f(x)$ and its inverse on the same set of axes;

  c. determine the point of intersection of $f(x)$ and its inverse.

12. If $g(x) = 3x + 4$,

  a. determine $g^{-1}(x)$;

  b. determine n given that $g(-3) = g^{-1}(n)$.

13. Describe how you would graph the inverse of a function given the graph of the function.

14. Under what conditions is the inverse of a function also a function?

PART C

15. a. Graph $f(x) = \dfrac{3}{x + 2}$.

    b. What is the domain and range of $f(x)$?

    c. Determine the defining equation for the inverse of $f(x)$.

    d. Is the inverse a function?

    e. What is the domain and range of the inverse?

16. a. Given two points $P(a, b)$ and $Q(b, a)$, prove that the midpoint of PQ lies on the line with equation $y = x$.

    b. Prove that PQ is perpendicular to the line with equation $y = x$.

    c. Explain the significance of these results in terms of f and f^{-1}.

17. a. Graph the inverse of the function f in the diagram.

    b. Is the inverse a function? Explain.

    c. State the domain and range of both $y = f(x)$ and its inverse.

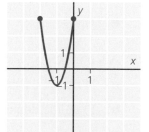

18. a. For each of the following functions, determine the inverse function $f^{-1}(x)$.

    i) $f(x) = 2x + 3$

    ii) $f(x) = ax + b$

    iii) $f(x) = \sqrt{x},\ x \geq 0$

    iv) $f(x) = \dfrac{2x + 3}{x - 1},\ x \neq 1$

    b. For each of the above, evaluate $f(f^{-1}(x))$.

    c. For each part of **a,** evaluate $f^{-1}(f(x))$.

    d. Make a general statement about $f(f^{-1}(x))$ and $f^{-1}(f(x))$ for a properly defined function. (While you are not asked for justification, you might be interested in using the definition of inverse function to do so.)

1. State the domain and range for each of the following relations.

   a. $\{(-2, 1), (-1, 0), (0, -1), (1, 0), (2, 1)\}$

   b. $\left\{(-2, 0), \left(-1, \frac{1}{2}\right), (0, 1), \left(-1, -\frac{1}{2}\right), (0, -1)\right\}$

   c. $x + 2y = 4$

   d. $y = 4 - x^2$

   e. $f(x) = \frac{1}{2}x + 1$

   f. $f(x) = \sqrt{x + 1}$

   Which of the relations above are functions?

2. For *each* part of Question 1, find the inverse relation and its domain and range. Which of these inverses are functions?

3. For each of the following graphed relations,

   a. state the domain and range;   **b.** sketch the inverse relation.

   i)

   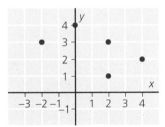

   ii)

   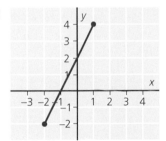

   iii)

   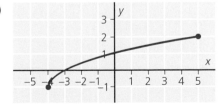

4. The graph at the right gives the position of the elevator in a building over several minutes. Explain what has happened here.

   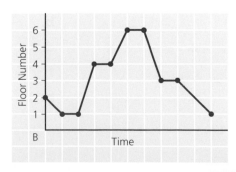

5. The height h, in metres, of a projectile fired vertically is graphed against time t, in seconds, as shown.

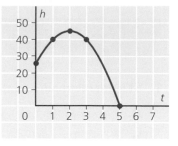

a. What is the maximum height of the projectile?

b. When does it reach this height?

c. What does the h-intercept represent?

d. What does the t-intercept represent?

6. For $f(x) = x^2 - 1$,

a. find $f(-2)$, $f(-1)$, and $f(2)$;

b. state the domain and range for $f(x)$.

7. For $f(x) = \sqrt{x + 3}$,

a. find $f(1)$, $f(-2)$, and $f(-4)$;

b. state the domain and range for $f(x)$.

8. For $f(x) = \dfrac{2}{x - 2}$,

a. find $f(-2)$, $f(0)$, and $f\left(\dfrac{1}{2}\right)$;

b. state the domain and range for $f(x)$.

9. From the graph of $f(x)$ shown,

a. find $f(-2)$, $f\left(\dfrac{1}{2}\right)$, and $\dfrac{1}{f(2)}$;

b. sketch the graph of the inverse of $f(x)$.

10. $f(x) = 2x + 1$ and $g(x) = x^2 + 1$. Determine each of the following:

a. $f(1) + g(1)$

b. $f(-1) \bullet g(-1)$

c. $f(g(1))$

d. $g(f(-1))$

11. Using the graph for $y = f(x)$ as shown, graph the following:

a. $f(x - 2)$

b. $f(x + 2)$

c. $2f(x)$

d. $f(2x)$

e. $-f(x)$

f. $f(-x)$

g. $2f(x - 1)$

h. $2 - f(2x)$

12. Given $f(x) = 2x - 4$, determine the equation of its inverse.

13. Give the vertex and range for each. Sketch the graph of each function.

    a. $y = -2(x - 2)^2 + 1$ **b.** $y = 4(x + 4)^2$ **c.** $y = 16 - x^2$

14. Give the equation for each of the following parabolas.

    a. Vertex $(-1, 1)$, opening up, congruent to $y = -x^2$.

    b. Vertex $(2, 0)$, range $y \leq 0$, congruent to $y = 2x^2$.

    c. Axis $x = -2$, range $y \geq -3$, congruent to $y = \frac{x^2}{2}$.

15. State the domain and range for each of the following:

    a. $y = 2(x - 1)^2 - 3$ **b.** $y = 2\sqrt{x - 1} - 3$ **c.** $y = \frac{2}{x - 1} - 3$

16. Sketch graphs for each of the following:

    a. $y = 2x^2 - 1$ **b.** $y = 2\sqrt{x} - 1$ **c.** $y = \frac{2}{x} - 1$

17. For each of the following, sketch the graph and state the equations of the asymptotes.

    a. $y = \frac{3}{x - 2} + 1$ **b.** $y = 3 - \frac{2}{x + 1}$

Problem Solving

At the end of each chapter we include a problems page. Why a special page, and why do we call it a problems page? What, indeed, is a problem?

As you know, there are relatively routine exercises that you are asked to do following the teaching of a new concept. These are designed to ensure that you become familiar with the new ideas, and they do not usually require that you use creative thinking. Every once in a while we meet a question that requires original thought. The skills we have practised don't tell us how to begin to attack the situation posed. The **key** to solving the question is not obvious. Here we have what we term a mathematical **problem.** A problem requires that we puzzle out an approach without being given detailed instructions.

If we are to solve problems, we must develop a sense of inquiry, a feeling of curiosity, and a use of imagination. We'll have to "try something," and we may have to conclude that an approach we take is not going to succeed. Is it hard? Yes. But that is what makes it worthwhile. The development of the senses of inquiry, curiosity, and imagination, together with some hard work, is precisely what we need for success in life, whether we are to become mathematicians, lawyers, artists, or anything else.

How does one develop problem-solving skills? By trying various ideas out; by doodling; by chatting with friends, parents, or teachers. Above all, remember that these problems are presented for enjoyment, and through working on them we will develop new skills and, most important, our mental discipline.

Each of these problems has appeared on the Canadian Mathematics Competition or was posed for students at special seminars.

1. If we start with the number 5 and put the digit 1 in front of it, we get the number 15, which is exactly three times the original number.

   a. Can you find other numbers with this same property?

   b. Find a number so that when the digit 1 is put in front of it, the new number is five times the original number. Can you find more than one number?

   c. Find a number so that when the digit 1 is put in front of it, the new number is eight times the original number. Explain your answer.

   d. Find a number so that when the digit 3 is put in front of it, the new number is six times the original number. Can you find more than one number?

2. What is the smallest perfect square that is greater than 20 000 and is divisible by 392?

3. **a.** You have a row of six discs, three blue and three green. They are lined up with the blue discs first. The only move allowed is the interchange of two neighbouring discs, and it is desired to get all the green discs to the left and the blue discs to the right. What is the minimum number of moves required? Write a step-by-step solution.

   b. How many moves are required if you start with four discs of each type? How many if you start with five of each type? How many if you start with k discs of each type?

4. How many different three-digit numbers can you form from three 4's, two 7's, and one 3?

5. **a.** A rectangle 6 × 4 units is divided into squares 1 × 1 unit. How many squares of all sizes are there?

   b. Repeat part **a** if the rectangle is 7 × 5 units.

   c. If in part **b** the one-unit square directly in the centre of the rectangle is blacked out, how many squares of all sizes are there?

   d. Repeat part **a** for a rectangle 10 × 6 units.

6. A length of rope is cut into two pieces at a randomly selected point. What is the probability that the longer piece is at least three times as long as the shorter piece?

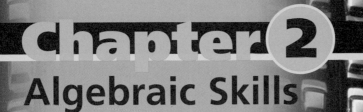

Chapter 2
Algebraic Skills

When would you use algebraic skills to solve real-life problems? In both your personal life and career, you will need to know algebra to make accurate calculations. Techniques used in algebra will not only give you the tools you need to interpret data you encounter in everyday life, but also enable you to plot a visual image of data connections. Whether you are monitoring a patient's vital signs in a hospital, creating a financial plan for a client as a financial adviser, or buying your first car, algebra can give you formulas to develop, interpret, and describe a range of strategies.

As your algebraic skills become more advanced throughout this section, you will develop methods for exploring new problems or unfamiliar situations.

IN THIS CHAPTER, YOU CAN . . .

- simplify and evaluate expressions;
- multiply, divide, add, and subtract rational expressions;
- solve inequalities and represent the solutions on number lines;
- state the restrictions on rational expressions;
- simplify radicals;
- factor polynomials;
- add, subtract, multiply, and divide polynomials.

2.1 Operations With Polynomials

To ensure that we can work efficiently with algebraic expressions, we first review and consolidate our algebraic skills from earlier courses. The first of these is the simplifying of expressions to make them easier to read and to work with. For example, if we wish to evaluate the expression $3x^2 - 4x - 2x^2 + 4x$ when $x = 7$, it is easier if we simplify the expression to x^2 before replacing the variable.

Just as we can add, subtract, multiply, and divide numbers, we can also perform these operations with polynomial expressions. These basic skills are illustrated in the following examples.

EXAMPLE 1

Simplify each of the following by performing the indicated operation.

a. $3x^2 + 4x^2$ **b.** $7x^3 - 2x^3$ **c.** $3x^4 \times 4x^2$ **d.** $\frac{8x^3}{4x}$

SOLUTION

a. $3x^2 + 4x^2 = 7x^2$ **b.** $7x^3 - 2x^3 = 5x^3$

c. $(3x^4)(4x^2) = 12x^6$ **d.** $\frac{8x^3}{4x} = 2x^2$

To add, subtract, or multiply polynomial expressions, multiply each term from one polynomial by each term from the other polynomial and collect any resulting like terms.

EXAMPLE 2

Simplify each of the following:

a. $2x^2 - 5x + 4 - (5x - 4)$

b. $(x + 2)(x^2 - 2x + 3)$

c. $3(x - 5)^2 - (x - 4)(x + 4)$

SOLUTION

a. $2x^2 - 5x + 4 - (5x - 4) = 2x^2 - 5x + 4 - 5x + 4$
$$= 2x^2 - 10x + 8$$

b. $(x + 2)(x^2 - 2x + 3) = x^3 - 2x^2 + 3x + 2x^2 - 4x + 6$
$$= x^3 - x + 6$$

c. $3(x - 5)^2 - (x - 4)(x + 4) = 3(x^2 - 10x + 25) - (x^2 - 16)$
$$= 3x^2 - 30x + 75 - x^2 + 16$$
$$= 2x^2 - 30x + 91$$

Exercise 2.1

PART A

1. Simplify each of the following:

   a. $4x^3 - 5x^3 - 6x^3$

   b. $(4x^3)(-5x^3)(-6x^3)$

   c. $x^2y - 2x^2y + 3x^2y$

   d. $(x^2y)(-2x^2y)(3x^2y)$

   e. $\frac{8x^4}{4x^3}$

   f. $1 - x + 2 - x$

   g. $2x^2 - 3x - x^2 + 3$

   h. $\frac{24x^4y^3}{8xy}$

   i. $2x^2y - xy^2 + 3xy - xy^2$

2. Simplify each of the following:

   a. $4x^3(5x^3 - 6x^3)$

   b. $4x^3 - (5x^3 - 6x^3)$

   c. $\frac{4x^3}{5x^3 - 6x^3}$

   d. $3x(-2x)^3$

   e. $-2x^2(-2x)^2$

   f. $\frac{(-3x)^4}{3x}$

   g. $2x(3x + 4)$

   h. $x^2(2x - 3)$

   i. $3xy(x^2 - 2xy + 3y^2)$

   j. $x^2 - (x^2 - 2x)$

   k. $(4x - 8) \div 4$

   l. $\frac{8x^2 - 2x}{2x}$

3. Simplify each of the following:

   a. $(x + 1)(x + 2)$

   b. $(2x - 3)(x - 4)$

   c. $(3x - 2)(x + 6)$

   d. $(x - 3)(x + 3)$

   e. $(x + 3)^2$

   f. $(4 - x)^2$

   g. $(5 - 2x)(5 + 2x)$

   h. $(x - y)(x + y)$

   i. $(x + 3y)^2$

4. Simplify each of the following and evaluate for $x = 6$.

   a. $7x^2 - 11x^2 + 3x^2$

   b. $4x^2 - 6x - 3x^2 + 5x$

   c. $\frac{15x^6}{5x^5}$

PART B

5. Simplify each of the following:

   a. $(x^2 - 2x + 3) - (x^2 - 3x + 2)$

   b. $(1 - 3x + x^3) + (x^2 + 2x - 2)$

   c. $3x(2x - 1) + 3(x - 2)$

   d. $x(x + 1) - 2x(x - 2) + 3x(x + 3)$

   e. $(4x - 5)(6x - 7)$

   f. $(2x - 1)(x^2 - 1)$

   g. $(x - 2)(x^2 + 2x + 4)$

   h. $(3x^2 - 4)(3x^2 + 4)$

   i. $(5x - 6)^2$

   j. $4x(2x - 1)(3x + 1)$

6. Perform the indicated operations in each of the following:

a. $(2x - 1)(x + 2) + (x - 2)(x + 3)$ **b.** $2x(3x - 1) - (x - 2)(x + 1)$

c. $(2x - 3)^2 - 2x(x - 3)$ **d.** $(x - 4)^2 + (x - 4)(x + 4)$

e. $(2x - 1)(2x - 3)(2x - 5)$ **f.** $(2x - 1)^3$

g. $(x^2 - 2x + 1)^2$ **h.** $4(5x - 1)^2 - 5(4x - 1)^2$

7. Give an expression for the perimeter and area of each of the following:

a.  x cm

$2x$ cm

b.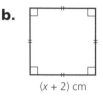

$(x + 2)$ cm

c. 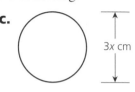 $3x$ cm

8. Evaluate $f(x) = x^2 + 3x + 7$ for each of the following:

a. $x = 2$ **b.** $x = -3$ **c.** $x = \dfrac{2}{3}$

d. $x = a + 2$ **e.** $x = a^2$ **f.** $x = 1 - 2a$

9. If $f(x) = 8x + 10$ and $g(x) = 2x$, determine an expression for each of the following:

a. $f(x) - g(x)$ **b.** $f(x)g(x)$ **c.** $2f(x) - 3g(x)$ **d.** $[f(x)]^2 + [g(x)]^2$

10. If $f(x) = x^2 - 4x$, determine an expression for $g(x)$ in each of the following:

a. $g(x) = f(x) + 4$ **b.** $g(x) = 3f(x)$ **c.** $g(x) = f(x + 2)$

d. $g(x) = f(3x)$ **e.** $g(x) = f(x - 1)$ **f.** $g(x) = f(x + 2) + 4$

g. $g(x) = -f(x)$ **h.** $g(x) = f(-x)$ **i.** $g(x) = f(2x) - 4$

PART C

11. Expand each of the following. Look for any patterns that form.

a. $(x + y)^2$ **b.** $(x + y)^3$ **c.** $(x + y)^4$ **d.** $(x + y)^5$

12. If $f(x) = 4x - 3$ and $g(x) = x + 1$, determine an expression for each of the following:

a. $f(g(x))$ **b.** $g(f(x))$

13. If $m(x) = x^2 - 1$ and $n(x) = 2x - 1$, determine an expression for each of the following:

a. $m(n(x))$ **b.** $n(m(x))$

14. If $f(x) = 3x - 4$ and $g(x) = \dfrac{x + 4}{3}$, determine an expression for each of the following:

a. $f(g(x))$ **b.** $g(f(x))$

15. What connection is there between the functions $f(x)$ and $g(x)$ in Question 14?

Numbers such as $\sqrt{10}$, $\sqrt{25}$, $\sqrt{17}$, and $\sqrt{72}$ are called **radicals.** Recalling that a radical represents the principal (positive) square root of a number, we make the following definitions:

$$\sqrt{a} = b \text{ if and only if } b \geq 0 \text{ and } b^2 = a.$$
$$\sqrt{a^2} = |a|, \text{ where } |a| = a \text{ if } a \geq 0$$
$$\text{and } |a| = -a \text{ if } a < 0 \text{ (e.g., } |-5| = -(-5) = 5 \text{).}$$

Most radicals are irrational numbers and we cannot determine exact values for them. We can, however, combine them as we do algebraic terms, and in fact the same rules apply. Just as $x + x = 2x$, $\sqrt{3} + \sqrt{3} = 2\sqrt{3}$. Just as $p \times q = pq$, $\sqrt{3} \times \sqrt{7} = \sqrt{21}$. In general,

$$\sqrt{a} + \sqrt{a} = 2\sqrt{a}, \sqrt{a} \times \sqrt{b} = \sqrt{ab}, \text{ and } \frac{\sqrt{a}}{\sqrt{b}} = \sqrt{\frac{a}{b}}.$$

EXAMPLE 1

Simplify each of the following:

a. $\sqrt{2} \times \sqrt{8}$ **b.** $3\sqrt{2} + 2\sqrt{3} - \sqrt{2}$ **c.** $3\sqrt{5} \times 4\sqrt{3}$ **d.** $\frac{6\sqrt{15}}{2\sqrt{3}}$

SOLUTION

a. $\sqrt{2} \times \sqrt{8} = \sqrt{16} = 4$ **b.** $3\sqrt{2} + 2\sqrt{3} - \sqrt{2} = 2\sqrt{2} + 2\sqrt{3}$

c. $3\sqrt{5} \times 4\sqrt{3} = 12\sqrt{15}$ **d.** $\frac{6\sqrt{15}}{2\sqrt{3}} = 3\sqrt{5}$

A number such as $\sqrt{12}$ is known as an **entire radical** and the quantity under the root sign is the **radicand.** Recognizing that $\sqrt{12} = \sqrt{4} \times \sqrt{3}$ allows us to simplify this to $2\sqrt{3}$, which we call a **mixed radical.**

EXAMPLE 2

Express the following as mixed radicals in simplest form.

a. $\sqrt{48}$ **b.** $\sqrt{54}$ **c.** $5\sqrt{28}$

SOLUTION

a. $\sqrt{48} = \sqrt{16 \times 3}$ **b.** $\sqrt{54} = \sqrt{9 \times 6}$ **c.** $5\sqrt{28} = 5(\sqrt{4 \times 7})$
$\quad\quad = \sqrt{16} \times \sqrt{3}$ $\quad\quad = \sqrt{9} \times \sqrt{6}$ $\quad\quad = 5(2\sqrt{7})$
$\quad\quad = 4\sqrt{3}$ $\quad\quad = 3\sqrt{6}$ $\quad\quad = 10\sqrt{7}$

Note that $\sqrt{48}$ might also have been changed to $\sqrt{4} \times \sqrt{12} = 2\sqrt{12}$, but this would not be the simplest form since $\sqrt{12}$ could still be changed to $2\sqrt{3}$.

Changing to mixed radicals allows us to add or subtract in cases when the entire radicals have different radicands as in the next example, thereby simplifying expressions.

EXAMPLE 3

Simplify each of the following:

a. $\sqrt{12} + \sqrt{27}$

b. $\sqrt{5} + \sqrt{10} + \sqrt{20}$

SOLUTION

a. $\sqrt{12} + \sqrt{27} = 2\sqrt{3} + 3\sqrt{3}$
$$= 5\sqrt{3}$$

b. $\sqrt{5} + \sqrt{10} + \sqrt{20} = \sqrt{5} + \sqrt{10} + 2\sqrt{5}$
$$= 3\sqrt{5} + \sqrt{10}$$

As mentioned earlier, normal rules for algebraic operations apply, so combining expressions involving radicals involves only one new concern. We must always be alert for radicals that can be simplified.

EXAMPLE 4

Simplify each of the following:

a. $\sqrt{2}(\sqrt{3} + \sqrt{18})$

b. $(\sqrt{3} - 2)(\sqrt{3} + 1)$

SOLUTION

a. $\sqrt{2}(\sqrt{3} + \sqrt{18}) = \sqrt{6} + \sqrt{36}$
$$= \sqrt{6} + 6$$

b. $(\sqrt{3} - 2)(\sqrt{3} + 1) = 3 + \sqrt{3} - 2\sqrt{3} - 2$
$$= 1 - \sqrt{3}$$

Exercise 2.2

PART A

1. Evaluate the following:

a. $\sqrt{121}$ **b.** $\sqrt{625}$ **c.** $\sqrt{\frac{4}{9}}$ **d.** $\sqrt{0.01}$

e. $\sqrt{2\frac{1}{4}}$ **f.** $\sqrt{25 - 16}$ **g.** $\sqrt{100} - \sqrt{36}$ **h.** $(\sqrt{17})^2$

2. Simplify the following:

    a. $\sqrt{5} \times \sqrt{7}$ **b.** $2\sqrt{3} \times 3\sqrt{5}$ **c.** $3\sqrt{5} + \sqrt{5}$ **d.** $2\sqrt{7} - 5\sqrt{7}$

    e. $2(\sqrt{3} + 3)$ **f.** $\sqrt{2}(1 + \sqrt{2})$ **g.** $3\sqrt{2}(\sqrt{3} - \sqrt{5})$ **h.** $\sqrt{\frac{1}{2}} \times 2\sqrt{6}$

PART B

3. Change the following to mixed radicals in simplest form.

    a. $\sqrt{12}$ **b.** $\sqrt{18}$ **c.** $\sqrt{45}$ **d.** $\sqrt{75}$

    e. $\sqrt{200}$ **f.** $\sqrt{512}$ **g.** $3\sqrt{24}$ **h.** $5\sqrt{63}$

    i. $9\sqrt{72}$ **j.** $3\sqrt{125}$

4. Simplify the following:

    a. $\sqrt{2} \times \sqrt{10}$ **b.** $\sqrt{6} \times \sqrt{8}$

    c. $\sqrt{5} \times 2\sqrt{15}$ **d.** $2\sqrt{3} \times 3\sqrt{6}$

    e. $\sqrt{6} \times \sqrt{10} \times \sqrt{15}$ **f.** $2\sqrt{5} \times \sqrt{10} \times 3\sqrt{5}$

    g. $3\sqrt{6} \times 6\sqrt{3} \times \sqrt{2}$ **h.** $3\sqrt{7} \times 2\sqrt{3} \times 2\sqrt{7}$

    i. $4\sqrt{5} \times 3\sqrt{6} \times 7\sqrt{5}$

5. Simplify the following:

    a. $2\sqrt{3} - \sqrt{5} - \sqrt{3} + 4\sqrt{5}$ **b.** $4\sqrt{7} - 7 - 2\sqrt{7} - 7$

    c. $\sqrt{3} + \sqrt{12}$ **d.** $4\sqrt{2} - \sqrt{32}$

    e. $\sqrt{75} - \sqrt{48}$ **f.** $\sqrt{2} + \sqrt{8} + \sqrt{18}$

    g. $3\sqrt{12} - 2\sqrt{27}$ **h.** $\sqrt{2} + \sqrt{4} + \sqrt{8} + \sqrt{16}$

    i. $2\sqrt{18} - \sqrt{24} - 3\sqrt{8}$

6. Simplify the following:

    a. $\sqrt{5}(\sqrt{5} - \sqrt{10})$ **b.** $2\sqrt{3}(3\sqrt{2} + \sqrt{3})$

    c. $3\sqrt{2}(\sqrt{6} + \sqrt{18})$ **d.** $(2 + \sqrt{3})(3 + \sqrt{3})$

    e. $(3 - 2\sqrt{6})(4 + \sqrt{6})$ **f.** $(2\sqrt{5} + 3\sqrt{2})(2\sqrt{5} - 3\sqrt{2})$

What is the domain of the function $f(x) = \sqrt{4x - 2}$? In Chapter 1 we examined the square root function. In the last section we looked at operations with radicals. Here we have a combination of these items and conclude that if $f(x)$ is to have real values, $4x - 2$ must be non-negative. To determine the domain of $f(x)$, we must consider the **linear inequality** $4x - 2 \geq 0$.

$$4x - 2 \geq 0$$
$$4x - 2 + 2 \geq 2$$
$$4x \geq 2$$
$$x \geq \frac{1}{2}$$

The domain of $f(x)$ is $x \geq \frac{1}{2}$.

From this example, we see that linear inequalities are solved in the same way as linear equations. We add or subtract like amounts to each side and then determine a value for the variable. There is one significant difference, which occurs when both sides of an inequality are multiplied by a negative number. For example, $3 > 2$ is a true statement. If we multiply both sides of the inequality by -1, we obtain $-3 > -2$, which is false. This becomes a true statement by reversing the sense of the inequality, writing $-3 < -2$.

> A **linear inequality** is solved in the same way as a linear equation with the exception that if both sides of an inequation are multiplied or divided by a negative quantity, the sense of the inequality is reversed.

EXAMPLE 1

Solve each of the following:

a. $3x - 4 > 5$

b. $3 - 4x \geq 7$

SOLUTION

a. $3x - 4 > 5$

$$3x > 9$$
$$x > 3$$

b. $3 - 4x \geq 7$

$$-4x \geq 4$$
$$x \leq -1$$

To better visualize this solution, we can graph the solution set on a number line by indicating which numbers satisfy the statement. Thus, the solution sets for the previous example could be graphed as follows:

a.

b.

Note the exclusion of $x = 3$ in the first case and the inclusion of $x = -1$ in the second.

We frequently encounter compound statements of linear inequalities joined by the connectives *and* or *or*. A statement containing two inequations connected by *and* is true only for values of the variable satisfying both parts. A statement containing two inequalities connected by *or* is true for values of the variable satisfying either part.

EXAMPLE 2

Solve and graph each of the following:

a. $-3 < 2x - 1 < 5$ **b.** $2 - x \geq 5$ or $2 - x \leq 1$

SOLUTION

a. $-3 < 2x - 1 < 5$ can be written as

$$2x - 1 > -3 \text{ and } 2x - 1 < 5.$$
$$2x > -2 \text{ and } 2x < 6$$
$$x > -1 \text{ and } x < 3$$

The solution is $-1 < x < 3$.

b. $2 - x \geq 5$ or $2 - x \leq 1$

$$-x \geq 3 \text{ or } -x \leq -1$$
$$x \leq -3 \text{ or } x \geq 1$$

The solution is $x \leq -3$ or $x \geq 1$.

Exercise 2.3

PART A

1. Solve each of the following:

  a. $x - 3 < 2$ **b.** $5 - x \geq 4$ **c.** $4x \leq 2$ **d.** $-3x > -6$

  e. $2 > 3 - x$ **f.** $\frac{2x}{3} \geq 6$ **g.** $4x < \frac{1}{2}$ **h.** $1 - \frac{x}{2} \geq 0$

2. Graph each of the solutions to Question 1 on a number line.

3. Write an inequality represented by each of the following:

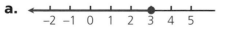

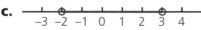

4. Solve and graph each of the following on a number line.

a. $4x - 9 < 11$ b. $6 - 3x > 21$ c. $2x + 5 > 2$ d. $3 - 4x < 5$

e. $3x + 7 \leq 1$ f. $5 - 4x \geq 1$ g. $\frac{x}{2} - 3 \leq 3$ h. $2 + \frac{x}{3} \geq \frac{1}{3}$

5. Solve and graph each of the following on a number line.

a. $3x - 4 \leq x$ b. $2x + 5 \geq 3x$

c. $x + 4 < 2x - 3$ d. $3x - 1 > 5x + 3$

e. $2 - 4x > 4 - 2x$ f. $3 - 4x < 3 - 2x$

g. $2(3x - 1) < 4x$ h. $3(2 - x) - 1 > 1 - 2(3 - x)$

6. Solve and graph each of the following on a number line.

a. $x + 6 > 2x$ and $x + 6 < 4x$

b. $2x - 3 > 3 - x$ or $2x - 3 < x - 3$

c. $-5 \leq 2x - 1 \leq 5$

d. $1 + x \geq 3 - x \geq 5 + x$

e. $x > \frac{x - 2}{3}$ or $x < \frac{2x - 1}{3}$

f. $2(x - 1) \leq 3(x - 2)$ or $2(x + 1) \geq 3(x + 2)$

7. Solve each of the following:

a. $(x - 3)(x + 2) > 0$ b. $(x + 4)(x + 7) \leq 0$

8. Given that $a^2 + b^2 + c^2 = 6$, determine the minimum value of $ab + bc + ca$.

9. Solve: $(2x - 3)(3x + 4)(x - 1) > 0$.

2.4 Factoring Polynomials

The **divisors** of a number are those numbers that divide exactly into the given number. For example, the divisors of 12 are 1, 2, 3, 4, 6, and 12.

When we are asked to **factor** a number or expression, we are required to write it as a product of two or more of its divisors, and often there are options. Thus, if we are asked to factor 12, we might write 1×12, 2×6, 3×4, or $2 \times 2 \times 3$. Generally, we want the completely factored form where none of the factors can be factored again. Similarly, we can factor a polynomial such as $3x^2 - 12$ into $3(x^2 - 4)$ and then further into $3(x - 2)(x + 2)$. Extensive use is made of factoring in working with rational expressions and solving of polynomial equations. It is, for example, much easier to solve the equation $x^2 - x - 42 = 0$ if we first factor $x^2 - x - 42$ to obtain $(x - 7)(x + 6)$.

Factoring of large numbers is sometimes easy (if the number contains easily recognized factors such as 2, 3, or 5) and sometimes very difficult. In fact, the use of large numbers that cannot be factored easily is an important process in the creation of secret codes for the business world and for military messages.

The following is a summary of the methods learned in previous courses for factoring polynomials.

Common Factors: $ax + ay = a(x + y)$

EXAMPLE 1

Factor each of the following:

a. $3x^2 + 6x$

b. $x(x + 3) - 2(x + 3)$

c. $5x^3 - 10x^2 + 4x - 8$

SOLUTION

a. $3x^2 + 6x = 3x(x + 2)$

b. $x(x + 3) - 2(x + 3) = (x + 3)(x - 2)$

c. $5x^3 - 10x^2 + 4x - 8 = (5x^3 - 10x^2) + (4x - 8)$
$$= 5x^2(x - 2) + 4(x - 2)$$
$$= (x - 2)(5x^2 + 4)$$

Difference of Squares: $x^2 - y^2 = (x - y)(x + y)$

EXAMPLE 2

Factor each of the following:

a. $x^2 - 81$ **b.** $4a^2 - 9b^2$ **c.** $(x - 2)^2 - 9$

SOLUTION

a. $x^2 - 81 = (x + 9)(x - 9)$

b. $4a^2 - 9b^2 = (2a - 3b)(2a + 3b)$

c. $(x - 2)^2 - 9 = [(x - 2) + 3][(x - 2) - 3]$
$$= (x - 2 + 3)(x - 2 - 3)$$
$$= (x + 1)(x - 5)$$

Quadratics of the Form $x^2 + bx + c$

Find the factors of "c" whose total is "b".

EXAMPLE 3

Factor each of the following:

a. $x^2 + 20x + 75$ **b.** $x^2 - 10x - 24$ **c.** $x^4 - 6x^2 + 9$

SOLUTION

a. 75 is 1×75 or 3×25 or 5×15. Of these, $5 + 15 = 20$.
$$x^2 + 20x + 75 = (x + 5)(x + 15)$$

b. $-24 = 2 \times (-12)$ and $2 + (-12) = -10$
$$x^2 - 10x - 24 = (x - 12)(x + 2)$$

c. $9 = (-3) \times (-3)$ and $(-3) + (-3) = -6$
$$x^4 - 6x^2 + 9 = (x^2 - 3)(x^2 - 3) \text{ or } (x^2 - 3)^2$$

Quadratics of the Form $ax^2 + bx + c$, $a \neq 1$

Use an organized trial-and-error method involving the factors of "a" and "c" as illustrated in the next two examples.

EXAMPLE 4

Factor $6x^2 - 5x - 4$.

SOLUTION

We know that if the factors exist, they must have the form $(ax + b)(cx + d)$ where $ac = 6$, $bd = -4$, and $(ad + bc) = -5$.

We therefore put the factors down in a chart form and check "cross-multiples" for $(ad + bc)$.

$$
\begin{array}{cc|cccccc}
2 & 1 & 1 & -1 & 4 & -4 & 2 & -2 \\
\hline
3 & 6 & -4 & 4 & -1 & 1 & -2 & 2
\end{array}
$$

Therefore, $6x^2 - 5x - 4 = (2x + 1)(3x - 4)$.

Note that we might have eliminated many of the cases immediately by recognizing that since there is no common factor in the original expression, we cannot have a common factor between a and b or between c and d.

EXAMPLE 5

Factor $12x^2 - 28xy + 15y^2$.

SOLUTION

$$2(-5) + 6(-3) = -28$$

$$
\begin{array}{cc|cc|cc}
3 & 2 & 1 & -1 & -15 & -3 & -5 \\
\hline
4 & 6 & 12 & -15 & -1 & -5 & -3
\end{array}
$$

$$12x^2 - 28xy + 15y^2 = (2x - 3)(6x - 5)$$

These methods may need to be used in combination for some expressions but should be looked for in the sequence given, remembering that we want the expression "factored completely." Always look for a common factor first.

EXAMPLE 6

Factor completely each of the following:

a. $9x^2 - 18x - 27$ **b.** $x^4 - 8x^2 - 9$

SOLUTION

a. $9x^2 - 18x - 27 = 9(x^2 - 2x - 3)$
$$= 9(x - 3)(x + 1)$$

b. $x^4 - 8x^2 - 9 = (x^2 + 1)(x^2 - 9)$
$$= (x^2 + 1)(x + 3)(x - 3)$$

PART A

1. Factor each of the following:

    a. $x^2 + x$ **b.** $2x - 8$ **c.** $ab - bc$ **d.** $wxy + xyz$ **e.** $m^6 + m^2$

2. Factor each of the following:

    a. $y^2 - 1$ **b.** $m^2 - 16$ **c.** $4 - a^2$ **d.** $x^2y^2 - 9$ **e.** $25 - t^4$

3. Factor each of the following:

    a. $x^2 - 3x + 2$ **b.** $y^2 + 6y + 5$ **c.** $x^2 - 5x - 6$ **d.** $t^2 + 7t + 12$

PART B

4. Factor completely each of the following:

    a. $9a^3 - 12a$ **b.** $5a^2b + ab^2$ **c.** $4a^3b^4 - 6a^2b^2 + 2ab$

    d. $\pi r^2 + \pi rh$ **e.** $x(2x - 1) + 2(2x - 1)$ **f.** $(x + 3)^2 - 3(x + 3)$

5. Factor completely each of the following:

    a. $x^2 - 49$ **b.** $100 - y^2$ **c.** $81x^2 - 4y^2$

    d. $25 - x^2y^2$ **e.** $(x + 3)^2 - 16$ **f.** $(2x + 3)^2 - (x - 2)^2$

    g. $x^3 - x$ **h.** $\pi R^2 - \pi r^2$ **i.** $x^4 - y^4$

6. Factor completely each of the following:

    a. $x^2 - 11x + 18$ **b.** $x^2 + 11x - 42$ **c.** $x^2 + 15x + 54$

    d. $x^2 - 21x + 54$ **e.** $x^2 - 16x + 64$ **f.** $x^2 - 16x - 80$

    g. $x^4 + 15x^2 + 50$ **h.** $x^4 - 18x^2 + 81$ **i.** $x^2 + 6xy - 7y^2$

7. Factor completely each of the following:

    a. $6x^2 - 23x - 18$ **b.** $12x^2 - 5x - 2$ **c.** $4x^2 + 17x + 4$

    d. $9x^2 - 30x + 25$ **e.** $4x^4 - 3x^2 - 1$ **f.** $6x^2 - 12x - 18$

    g. $2x^3 - 3x^2 + x$ **h.** $4x^4 - 13x^2 + 9$ **i.** $8x^4 - 31x^2 - 4$

    j. $15x^2 - 29x - 14$ **k.** $21x^2 - 29x + 10$ **l.** $22x^2 + 43x - 2$

8. Factor completely each of the following:

a. $x^3 - 4x^2 + 3x - 12$
b. $2x^3 - 6x^2 - 3x + 9$
c. $4x^3 + 8x^2 - x - 2$
d. $2x^3 - 6x^2 + 10x - 30$
e. $3x^5 - 12x^3 - x^2 + 4$
f. $2x^4 - 4x^3 - 8x^2 + 16x$

9. Factor completely each of the following:

a. $x^2 - 13x + 22$
b. $3x^2 - 9x$
c. $6x^2 + 11x - 7$
d. $10x^3 - 21x^2 + 8x$
e. $x^3 - 3x^2 + 4x - 12$
f. $25x^2 - 49y^2$
g. $4x^2 + 20x + 25$
h. $5x^2 - 30x + 45$
i. $4x^2 + 19x - 5$
j. $100a^2 - 36b^2$
k. $4x^4 + 28x^2 + 49$
l. $12x^2 + 8x + 28$

PART C

10. Use difference of squares factoring to evaluate each of the following:

a. $51^2 - 49^2$
b. $27^2 - 23^2$
c. $121^2 - 111^2$
d. $10\,000^2 - 9\,999^2$

11. Factor completely each of the following:

a. $(x + 2)^2 - 3(x + 2) + 2$
b. $(x - 1)^4 - 1$
c. $x^2 - 8x + 16 - 4y^2$

12. By considering factors, find the smallest value of x such that $120x$ will be a perfect square.

13. It was mentioned at the beginning of this section that there are important uses of large numbers that cannot be factored easily. Compare the following techniques for efficiency in factoring a given number.

i) Divide the given number by 2, 3, 5, 7, 11, and so on, using consecutive primes. (Why would you *not* use composite numbers?)

ii) Determine the square root of the given number and divide the given number by the largest integer smaller than this square root, then the next largest (sensible) integer, and so on. (Why can we be confident that this is a sensible starting point, and why do we say "sensible integer"?)

Try the two techniques on the following:

a. 703
b. 399
c. 6497
d. 31 897
e. 52 371

2.5   Rational Expressions

We mentioned earlier that polynomial expressions are the algebraic equivalent of numbers in our numerical system. It is natural then to consider the question of an algebraic fraction. If $f(x) = 2x - 3$ and $g(x) = 3x + 5$ are functions, or algebraic numbers, then $\frac{2x - 3}{3x + 5}$ can be considered to be an algebraic fraction, or, as it is commonly called, a **rational expression.**

> A **rational expression** is of the form $\frac{f(x)}{g(x)}$, where
> $f(x)$ and $g(x)$ are polynomials, $g(x) \neq 0$.

The following expressions are examples of rational expressions.

$$\frac{2}{x}, \quad \frac{x - 2}{5}, \quad \frac{x - 2}{2x - 1}, \quad \frac{x + 3}{x^2 - x - 6}$$

The reciprocal function that we examined in Section 1.6 is an example of a rational expression. Since division is the indicated operation, and division by zero gives an undefined result, there are implied restrictions on the variable so that the denominator will not equal zero.

EXAMPLE 1

What restriction(s) must be placed on the variable in each of the following?

a. $\frac{2}{x}$ **b.** $\frac{x}{x - 3}$ **c.** $\frac{2x - 3}{2x + 1}$

SOLUTION

a. $x \neq 0$ **b.** $x - 3 \neq 0$ **c.** $2x + 1 \neq 0$

$x \neq 3$ $x \neq -\frac{1}{2}$

If the denominator is of greater degree than one, then we make use of the factoring skills and note that if one of the factors equals zero, the denominator is zero.

EXAMPLE 2

Determine the restriction(s) on the variable in each of the following:

a. $\frac{2}{x^2 - 2x}$ **b.** $\frac{x + 2}{x^2 - 4}$ **c.** $\frac{x^2 - x - 2}{2x^2 + x - 1}$

a. $x^2 - 2x \neq 0$

$x(x - 2) \neq 0$

$x \neq 0$ and $x - 2 \neq 0$

$x \neq 0$ and $x \neq 2$

b. $x^2 - 4 \neq 0$

$(x - 2)(x + 2) \neq 0$

$x - 2 \neq 0$ and $x + 2 \neq 0$

$x \neq 2$ and $x \neq -2$

c. $2x^2 + x - 1 \neq 0$

$(2x - 1)(x + 1) \neq 0$

$2x - 1 \neq 0$ and $x + 1 \neq 0$

$x \neq \dfrac{1}{2}$ and $x \neq -1$

Just as we can reduce rational numbers, we can reduce rational expressions by dividing both numerator and denominator by the same factor(s). When reducing, we must remember to take note of any restrictions on the variable. In simplifying rational expressions, we will list any restriction on the variable.

EXAMPLE 3

Simplify each of the following:

a. $\dfrac{2x - 4}{3x - 6}$

b. $\dfrac{2x^2 + 3x}{4x^2 - 9}$

SOLUTION

a. $\dfrac{2x - 4}{3x - 6} = \dfrac{2(x - 2)}{3(x - 2)}$

$= \dfrac{2}{3}, x \neq 2$

b. $\dfrac{2x^2 + 3x}{4x^2 - 9} = \dfrac{x(2x + 3)}{(2x - 3)(2x + 3)}$

$= \dfrac{x}{2x - 3}, x \neq -\dfrac{3}{2}, \dfrac{3}{2}$

Exercise 2.5

PART A

1. State the restrictions on the variable in each of the following:

a. $\dfrac{x + 2}{2x}$

b. $\dfrac{2x}{x + 2}$

c. $\dfrac{3}{2x - 5}$

d. $\dfrac{x - 2}{x(x + 2)}$

e. $\dfrac{x - 1}{x^2(x + 1)}$

f. $\dfrac{2}{x(x - 1)(x + 1)}$

g. $\dfrac{x}{x - y}$

h. $\dfrac{1}{x^2 + 1}$

2. Simplify each of the following. State any restrictions.

a. $\dfrac{2x}{x^2}$

b. $\dfrac{x - y}{2(x - y)}$

c. $\dfrac{x(x - 1)}{2x(x + 1)}$

d. $\dfrac{x - 2}{2 - x}$

PART B

3. Simplify each of the following. State any restrictions.

a. $\dfrac{5x + 10}{20}$

b. $\dfrac{3x}{6x - 12}$

c. $\dfrac{2x + 8}{3x + 12}$

d. $\dfrac{x + 3}{x^2 + 3x}$

e. $\dfrac{x^2 - 4x}{x^2 + 4x}$

f. $\dfrac{(x - 1)^2}{5x - 5}$

g. $\dfrac{4x - 8}{4 - 2x}$

h. $\dfrac{x^3 + x}{2x^2 + 2}$

4. Simplify each of the following. State any restrictions.

a. $\dfrac{x + 2}{x^2 - 4}$

b. $\dfrac{9 - x^2}{x - 3}$

c. $\dfrac{4x - 8}{2x^2 - 8}$

d. $\dfrac{4x^2 + 16x}{x^3 - 16x}$

e. $\dfrac{x^2 - 1}{x^2 - 2x + 1}$

f. $\dfrac{x^2 - 5x + 6}{2x - 6}$

g. $\dfrac{x^2 + 3x - 10}{x^2 - 3x + 2}$

h. $\dfrac{3x^2 - 6x}{x^3 - 4x^2 + 4x}$

5. Simplify each of the following. State any restrictions.

a. $\dfrac{2x^2 - 14x + 20}{3x^2 - 9x - 30}$

b. $\dfrac{x^3 + 7x^2 + 12x}{3x^4 - 27x^2}$

c. $\dfrac{2x^2 + 9x + 9}{4x^2 - 9}$

d. $\dfrac{x^4 - 5x^2 + 4}{x^2 - x - 2}$

e. $\dfrac{4x^2 - 4x - 15}{4x^2 + 16x + 15}$

f. $\dfrac{16 - 9x^2}{3x^2 - x - 4}$

PART C

6. Determine all integer values of x so that y will always be an integer in the given relations.

a. $y = 3 + \dfrac{5}{x - 2}$

b. $y = 2 - \dfrac{6}{x + 3}$

7. For each of the following, determine all integer points (x, y) on its graph. (*Hint:* In part **a**, note that $x + 4 = (x - 2) + 6$.)

a. $y = \dfrac{x + 4}{x - 2}$

b. $\dfrac{3x + 4}{x - 2}$

c. $y = \dfrac{4x - 9}{x - 3}$

d. $y = \dfrac{6x + 3}{3x - 4}$

2.6 Multiplying and Dividing Rational Expressions

In multiplying or dividing rational numbers, it is always best to simplify by dividing out common factors. For example, in evaluating $\frac{4}{15} \times \frac{33}{32}$, we could simply multiply the numerators and denominators to obtain $\frac{132}{480}$, but we must then simplify the resulting fraction. It is easier to divide all possible common factors. Our solution then is $\frac{\cancel{4}^1}{\cancel{15}_5} \times \frac{\cancel{33}^{11}}{\cancel{32}_8} = \frac{11}{40}$.

EXAMPLE 1

Simplify $\frac{x^2 - 9}{2x + 4} \times \frac{x + 2}{x + 3}$. State any restrictions.

SOLUTION

$$\frac{x^2 - 9}{2x + 4} \times \frac{x + 2}{x + 3} = \frac{(x - 3)(x + 3)}{2(x + 2)} \times \frac{x + 2}{x + 3}$$

$$= \frac{x - 3}{2}, x \neq -2, -3$$

EXAMPLE 2

Simplify $\frac{x^2 - 4x + 4}{2x + 4} \div \frac{x^2 - x - 2}{4x^2 + 8x}$. State any restrictions.

SOLUTION

$$\frac{x^2 - 4x + 4}{2x + 4} \div \frac{x^2 - x - 2}{4x^2 + 8x} = \frac{(x - 2)(x - 2)}{2(x + 2)} \times \frac{4x(x + 2)}{(x - 2)(x + 1)}$$

$$= \frac{2x(x - 2)}{x + 1}, x \neq 2, -2$$

Exercise 2.6

For all questions, state any restrictions.

PART A

1. Simplify each of the following:

a. $\dfrac{x - 2}{4x} \times \dfrac{2x^2}{x - 2}$

b. $\dfrac{(x + 2)^3}{x - 2} \times \dfrac{3x - 6}{(x + 2)^2}$

c. $\dfrac{x - 1}{x + 3} \times \dfrac{2x + 6}{1 - x}$

d. $\dfrac{2x^3}{2x - 3} \div \dfrac{6x}{3 - 2x}$

e. $\dfrac{x + 1}{x + 2} \div \dfrac{(x + 1)^2}{x(x + 2)}$

f. $\dfrac{x^2(x - 2)}{2x(x + 2)} \div \dfrac{x(x - 2)}{x^2(x + 2)}$

PART B

2. Simplify each of the following:

a. $\dfrac{x^2 + 6x}{x^2 - 4} \times \dfrac{x^2 + x - 2}{x^2 + 6x + 9}$

b. $\dfrac{2x^2 - 2}{4x^2 + 4} \times \dfrac{x^3 + x}{x^2 - x}$

c. $\dfrac{x^2 - 3x + 2}{x^2 + 2x - 3} \times \dfrac{x^2 - 9}{x^2 - 5x + 6}$

d. $\dfrac{4x^2 - 1}{2x^2 - 3x - 2} \times \dfrac{2x^2 - 5x + 2}{4x - 2}$

e. $\dfrac{x^4 - 1}{2x^2 - 8x + 6} \times \dfrac{4x^2 - 12x}{3x^2 + 3}$

f. $\dfrac{6x^2 - x - 1}{6x^2 - 11x + 4} \times \dfrac{9x^2 - 16}{9x^2 + 6x + 1}$

3. Simplify each of the following:

a. $\dfrac{2x + 2}{x^2 + 2x + 1} \div \dfrac{4x^2 + 8x}{x^2 + 3x + 2}$

b. $\dfrac{9x^2 + 3x - 2}{4x^2 - 1} \div \dfrac{6x^2 + 7x + 2}{4x^2 + 4x + 1}$

c. $\dfrac{2x^2 - 14x + 24}{4x^2 - 64} \div \dfrac{x^2 - 8x + 15}{x^2 - x - 20}$

d. $\dfrac{x^4 - 5x^3 + 6x^2}{2x^2 - 8} \div \dfrac{9x - x^3}{x^2 + 4x + 4}$

4. Simplify each of the following:

a. $\dfrac{x^2 - 5x + 6}{2x} \div \dfrac{x - 3}{4x^2} \times \dfrac{4x^2 - 4x - 8}{8x - 16}$

b. $\dfrac{4x^2 - 16x + 15}{2x^2 + 3x + 1} \times \dfrac{x^2 - 6x - 7}{2x^2 - 17x + 21} \div \dfrac{4x^2 - 20x + 25}{4x^2 - 1}$

5. If $f(x) = \dfrac{x^2 - 3x + 2}{2x^2 + 5x + 2} \div \dfrac{x^2 + 4x - 5}{2x^2 + 11x + 5}$, evaluate each of the following:

a. $f(2)$

b. $f(0)$

c. $f(10)$

PART C

6. For the function $f(x) = \dfrac{(6x + 12)(x^2 - 3x)}{(x^2 - 4x + 3)(3x^2 + 6x)}$, do the following:

a. Simplify the function.

b. Graph the simplified function using your graphing calculator.

c. From your graph, determine $f(0)$.

d. Graph the function inputting the original expression.

e. From the graph, determine $f(0)$. Explain your result.

f. Are there any other values of x that have similar inconsistencies? Explain.

7. For the function $f(x) = \dfrac{(x^2 - 3x + 2)(2x^2 + 11x + 5)}{(2x^2 + 5x + 2)(x^2 + 4x - 5)}$, determine any values of x for which there are inconsistencies between the graph of the given function and the graph of the function when simplified.

For addition and subtraction of rational expressions, we use the same rules as with rational numbers. Remember that any simplification possible should be done as soon as possible. Although variable restrictions apply here as discussed earlier, these problems are presented for the purpose of developing algebraic skills. It is not expected that restrictions of the variable be stated unless asked for.

If the denominators are the same, we can combine the numerators.

EXAMPLE 1

Simplify $\dfrac{2x}{x-2} - \dfrac{x+4}{x-2}$.

SOLUTION

$$\dfrac{2x}{x-2} - \dfrac{x+4}{x-2} = \dfrac{2x - (x+4)}{x-2}$$
$$= \dfrac{x-4}{x-2}$$

If the denominators are not the same, we need to find equivalent expressions having a common denominator before we combine the numerators. This involves finding the least common multiple (LCM) of denominators. If the denominators cannot be factored, the LCM is the product of the expressions. If the denominators can be factored, the LCM does not repeat factors that occur in both denominators.

EXAMPLE 2

Simplify each of the following:

a. $\dfrac{x+3}{6} + \dfrac{2x-3}{4}$

b. $\dfrac{2x+5}{5x} - \dfrac{x-7}{3x^2}$

SOLUTION

a. $\dfrac{x+3}{6} + \dfrac{2x-3}{4} = \dfrac{2(x+3)}{12} + \dfrac{3(2x-3)}{12}$
$$= \dfrac{2x+6+6x-9}{12}$$
$$= \dfrac{8x-3}{12}$$

b. $\dfrac{2x+5}{5x} - \dfrac{x-7}{3x^2} = \dfrac{3x(2x+5)}{15x^2} - \dfrac{5(x-7)}{15x^2}$
$$= \dfrac{6x^2 + 15x - 5x + 35}{15x^2}$$
$$= \dfrac{6x^2 + 10x + 35}{15x^2}$$

EXAMPLE 3

Simplify each of the following:

a. $\dfrac{x}{x-1} + \dfrac{3x}{x+2}$ **b.** $\dfrac{x-1}{2x^2} - \dfrac{2x+1}{4x^3}$ **c.** $\dfrac{2}{x^2+x} + \dfrac{x}{x^2-1}$

SOLUTION

a. $\dfrac{x}{x-1} + \dfrac{3x}{x+2} = \dfrac{x(x+2)+(3x)(x-1)}{(x-1)(x+2)}$ **b.** $\dfrac{x-1}{2x^2} - \dfrac{2x+1}{4x^3} = \dfrac{2x(x-1)-(2x+1)}{4x^3}$

$$= \dfrac{x^2+2x+3x^2-3x}{(x-1)(x+2)} \qquad\qquad = \dfrac{2x^2-4x-1}{4x^3}$$

$$= \dfrac{4x^2-1}{(x-1)(x+2)}$$

c. $\dfrac{2}{x^2+x} + \dfrac{x}{x^2-1} = \dfrac{2}{x(x+1)} + \dfrac{x}{(x+1)(x-1)}$

$$= \dfrac{2(x-1)+x}{x(x+1)(x-1)}$$

$$= \dfrac{x^2+2x-2}{x(x+1)(x-1)}$$

Note that the denominator may be left in factored form. In fact, it is advisable to do so to see if the resulting rational expression can be reduced. As well, it is advisable to reduce any of the original rational expressions before adding or subtracting, if possible.

Exercise 2.7

PART A

1. Add or subtract as indicated in the following:

a. $\dfrac{x}{3} + \dfrac{5x}{3}$ **b.** $\dfrac{9}{x} - \dfrac{1}{x} - \dfrac{2}{x}$ **c.** $\dfrac{4}{x+1} - \dfrac{2}{x+1}$

d. $\dfrac{5m}{3x} + \dfrac{2m}{3x}$ **e.** $\dfrac{8y}{2x+1} - \dfrac{5y}{2x+1}$ **f.** $\dfrac{3x}{x-3} - \dfrac{5x}{x-3}$

2. Simplify each of the following:

a. $\dfrac{2x}{3} + \dfrac{x}{4}$ **b.** $\dfrac{x}{8} - \dfrac{x}{12}$ **c.** $\dfrac{3}{2x} + \dfrac{2}{x}$

d. $\dfrac{5}{4x^2} - \dfrac{1}{6x^2}$ **e.** $\dfrac{x+2}{3} + \dfrac{x-2}{6}$ **f.** $\dfrac{2}{x-1} + \dfrac{4}{x-1}$

g. $4 - \dfrac{2}{x}$ **h.** $2x - \dfrac{x}{3}$ **i.** $1 - \dfrac{1}{x+1}$

PART B

3. Simplify each of the following:

a. $\dfrac{3}{4x^2} - \dfrac{4}{x^3}$

b. $\dfrac{x+1}{2x} - \dfrac{2}{x^2}$

c. $\dfrac{x}{15} - \dfrac{x}{21} + \dfrac{x}{35}$

d. $4 - \dfrac{3x}{x+1}$

e. $2x + \dfrac{x-2}{2x}$

f. $x - 1 - \dfrac{1}{x}$

4. Simplify each of the following:

a. $\dfrac{x}{x+1} + \dfrac{2x}{x+2}$

b. $\dfrac{x-1}{x} - \dfrac{x}{x+1}$

c. $\dfrac{2x}{2x-3} + \dfrac{1}{x-3}$

d. $\dfrac{x+2}{x-2} - \dfrac{x}{x+2}$

e. $\dfrac{2x-1}{x-1} + \dfrac{x+1}{2x+1}$

f. $\dfrac{x+2}{x-2} - \dfrac{x}{(x-2)^2}$

5. Simplify each of the following:

a. $\dfrac{x}{x^2-1} - \dfrac{1}{x+1}$

b. $\dfrac{2x}{x^2+2x} + \dfrac{x}{x^2-2x}$

c. $\dfrac{x-1}{x^2+2x} + \dfrac{x-2}{2x+4}$

d. $\dfrac{3}{x^2-3x+2} - \dfrac{2}{x^2-4}$

e. $\dfrac{x-1}{x^2-2x+1} - \dfrac{x+2}{x^2-4}$

f. $\dfrac{2}{x^3-9x} - \dfrac{x-2}{x^2+3x}$

6. Simplify each of the following:

a. $\dfrac{5}{x} + \dfrac{2x+1}{x+4} - \dfrac{5x}{2x-1}$

b. $\dfrac{2}{x} - \dfrac{x+2}{x+1} + \dfrac{x-1}{x-3}$

7. Simplify $\dfrac{3x}{x^2+3x+2} - \dfrac{4x}{x^2+5x+6} + \dfrac{5x}{x^2+4x+3}.$

PART C

8. Simplify each of the following:

a. $\dfrac{3 - \dfrac{3}{x+1}}{2 - \dfrac{2}{x+1}}$

b. $\dfrac{3 - \dfrac{2}{x^2-4}}{5 + \dfrac{7}{2-x}}$

9. As you are aware, fractions can be taken apart as well as put together. For example, $\dfrac{11}{12} = \dfrac{3}{12} + \dfrac{8}{12} = \dfrac{1}{4} + \dfrac{2}{3}$. In similar fashion, we can separate rational expressions, provided the denominator has factors. Observe that $\dfrac{3x+2}{x^2-4} = \dfrac{2}{x-2} + \dfrac{1}{x+2}$. How does one obtain the numerators? We simply let $\dfrac{3x+2}{x^2-4} = \dfrac{3x+2}{(x-2)(x+2)} = \dfrac{a}{x-2} + \dfrac{b}{x+2}$ and then attempt to determine values for a and b.

Suggest a method whereby values for a and b can be determined.

10. Rewrite each of the following as the sum or difference of two rational expressions.

a. $\dfrac{2x-1}{x^2+3x+2}$

b. $\dfrac{5x}{x^2+x-6}$

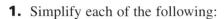

1. Simplify each of the following:

   a. $(4x^3 - 5x + 6) - (2x^2 - x + 3)$ **b.** $(4x^2)(5x)(6x^3)$

   c. $4(2x^2)^3$ **d.** $(3x)^2 - 3x^2$

   e. $2x^2(3x^2 - 2x + 1)$ **f.** $25x^4 \div 5x$

   g. $3x(4x - 5x - 6x)$ **h.** $(3x^2 - 4x^2)^2$

   i. $(7x^2 - 8x^2)(7x - 8x)$

2. Simplify each of the following:

   a. $3(x^2 - 2x + 1) - x(1 - x)$ **b.** $(2x - 1)(x^2 + x + 2)$

   c. $3x(2x - 1)(x + 2)$ **d.** $4(3x - 2)^2$

   e. $(3x + 1)^3$ **f.** $(x - 2)(x + 3) - (x + 2)(x - 3)$

   g. $(2x + 1)^2 - (x + 2)^2$ **h.** $1 - x(1 - x)^2$

3. If $f(x) = 2x + 1$ and $g(x) = x + 2$, find expressions for the following:

   a. $3f(x) - 2g(x)$ **b.** $6f(x)g(x)$

   c. $[f(x)]^2 + [g(x)]^2$ **d.** $[f(x) - g(x)]^2$

4. If $f(x) = 2x^2 - 3x + 1$, find the simplified expression for $g(x)$ in each of the following:

   a. $g(x) = f(x + 1)$ **b.** $g(x) = f(2x)$

   c. $g(x) = f(-x)$ **d.** $g(x) = 2f(x - 1) - 12$

5. Find the simplified expression for the perimeter and area of the given figure.

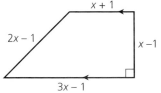

6. Change the following to mixed radicals in simplest form.

   a. $\sqrt{128}$ **b.** $3\sqrt{72}$ **c.** $\sqrt{800}$ **d.** $5\sqrt{32}$ **e.** $3\sqrt{175}$

7. Simplify each of the following:

   a. $3\sqrt{5} \times 2\sqrt{7}$ **b.** $4\sqrt{3} \times 2\sqrt{6}$ **c.** $3\sqrt{5} - 5\sqrt{3} + \sqrt{5} - \sqrt{3}$

   d. $\sqrt{20} + \sqrt{40} + \sqrt{80}$ **e.** $3\sqrt{12} + 2\sqrt{48}$ **f.** $\sqrt{3}(2\sqrt{3} - 3\sqrt{2})$

   g. $(3 + \sqrt{2})^2$ **h.** $(2\sqrt{3} + 3)(2\sqrt{3} - 3)$ **i.** $(3\sqrt{2} + 4\sqrt{5})(3\sqrt{2} - 4\sqrt{5})$

8. Factor completely each of the following:

a. $9x^2 - 18x$

b. $x^2 - 18x - 19$

c. $9x^2 - 18x + 9$

d. $x^4 + 5x^2 + 4$

e. $x^4 - 10x^2 + 9$

f. $4x^2 + 15x - 4$

g. $(2x + 3)^2 - 9$

h. $(x + 1)^2 - 3(x + 1)$

i. $x^2(2x - 1) - 4(2x - 1)$

j. $6x^2 + 7x - 10$

k. $4x^3 - 12x^2 + 9x$

l. $5x^4 - 10x^3 - 3x^2 + 6x$

9. Solve each of the following for x and graph the solution on a number line.

a. $1 - 2x < 3$

b. $2x - 4 < x$

c. $3x + 5 \leq x - 3$

d. $2x - 3 > 4x - 6$

e. $3(4 - x) \geq 4(3 - x)$

f. $1 - 2(3 - x) > x - 6$

10. Solve and graph each of the following on a number line.

a. $-3 \leq x + 2 \leq 3$

b. $2x - 1 < 3$ or $2x - 1 > 7$

11. State any restrictions on the variable in each of the following:

a. $\dfrac{2x + 4}{6}$

b. $\dfrac{4}{4x - 8}$

c. $\dfrac{3}{2x - 1}$

d. $\dfrac{2}{x^2 - 4}$

e. $\dfrac{x}{x^2 + x}$

f. $\dfrac{1}{x^2 - 2x + 1}$

g. $\dfrac{3x}{x^2 + 3x - 4}$

h. $\dfrac{1}{x^3 - x^2 - 2x}$

i. $\dfrac{x - 3}{x^2 - 8x + 15}$

12. Simplify each of the following:

a. $\dfrac{3}{6x - 9}$

b. $\dfrac{2x}{x^2 + x}$

c. $\dfrac{x + 1}{x^2 - 1}$

d. $\dfrac{2x + 4}{3x + 6}$

e. $\dfrac{x^2 - 4}{2x - 4}$

f. $\dfrac{3x^2 - 3}{x^3 - x}$

g. $\dfrac{x^2 + 2x - 3}{x^2 - 2x + 1}$

h. $\dfrac{x^2 - 4}{x^2 - 4x + 4}$

i. $\dfrac{2x^2 + 5x - 12}{2x^2 + 11x + 12}$

13. Simplify each of the following:

a. $\dfrac{3x}{4} \div \dfrac{6}{x}$

b. $\dfrac{x - 2}{3} \times \dfrac{6}{5x - 10}$

c. $\dfrac{(x - 2)^2}{x^2 - 4} \times \dfrac{x + 2}{x}$

d. $\dfrac{3 - x}{4x} \div \dfrac{x^2 - 9}{8}$

e. $\dfrac{2x - 6}{3x + 6} \times \dfrac{x^2 + 5x + 6}{4x^2 - 36}$

f. $\dfrac{x - x^3}{x^2 - 2x - 3} \div \dfrac{x^2 - x}{2x - 6}$

14. Simplify each of the following:

a. $\dfrac{3}{x} + \dfrac{2}{x - 1}$

b. $\dfrac{2}{x - 1} - \dfrac{1}{x + 2}$

c. $\dfrac{x + 1}{x} - \dfrac{x}{x + 1}$

d. $3 + \dfrac{3}{x - 1}$

e. $\dfrac{4}{x - 1} + \dfrac{5}{x^2 - 1}$

f. $\dfrac{x + 1}{x^2 - x} + \dfrac{x - 1}{x^2 + x}$

15. Simplify each of the following:

a. $\dfrac{x^2 - x}{x^2 - x - 2} \times \dfrac{x^2 - 1}{x^2 - 2x} \div \dfrac{x^2 - 2x + 1}{x^2 - 4x + 4}$

b. $\dfrac{x}{x^2 - 4} - \dfrac{x - 2}{x^2 + 2x} + \dfrac{x + 2}{x^2 - 2x}$

The Problems Page

1. Recall the problem from Chapter 1 in which you were asked to move blue and green discs so that the positioning of the discs was reversed. Here is a variation of the game. What's different? Two things. First, we have three blue and three green discs, but also a one-disc space between them. Second, there are two allowable moves; you can move a disc one position if there is a blank space beside it, and you can jump one disc over another if there is an empty space to jump into. What is the minimum number of moves required to get the green discs to the three positions on the left and the blue discs to the three positions on the right? Can you give a general solution for k discs of each colour?

2. This is a set of questions about prime numbers. Your task is to answer the question posed and to draw conclusions where you can.

   a. The numbers 2 and 3 differ by one. Can you find another such pair of primes?

   b. The numbers 3 and 5 differ by two. Can you find other such pairs of primes? How many?

   c. The numbers 3, 5, and 7 are a triple set of primes differing by two. Can you find another such triple?

   d. Can you find a prime that is one less than a perfect square? Can you find more than one such prime?

   e. Can you find a prime that is one more than a perfect square? Can you find more than one such prime?

   f. For each part, explain your answers.

3. When $f(x)$ is divided by $3x + 1$, a quotient of $2x - 3$ and a remainder of 5 are obtained. What is $f(x)$?

4. Bus A and bus B leave the same terminal at 09:00. Bus A requires 45 min to complete its route while bus B requires 54 min for its route. If the buses continue running, what is the next time the two leave the terminal together?

5. If two poles 10 m and 15 m high are 25 m apart, what is the height of the point of intersection of the lines that run from the top of each pole to the foot of the other pole?

6. We define a sequence of numbers by using the symbol t_k where $k = 1, 2, 3, 4, \ldots$. This means that t_1 represents the first number in the sequence, t_2 the second, and so on. If $t_1 = 1$, $t_2 = 2$, and every number thereafter is obtained from the formula
$$t_{k+1} = \frac{t_k + 1}{t_{k-1}} \text{ (for example, } t_3 = \frac{t_2 + 1}{t_1} = \frac{2 + 1}{1} = 3), \text{ what is } t_{63}? \text{ Can you}$$
determine the sum of the first 100 numbers in the sequence?

7. (If you are keen!) Show that for the set of numbers in Problem 6, $t_6 = t_1$ and $t_7 = t_2$.

Chapter 3

Quadratic Functions

With quadratic functions, you can represent the path of an object and the relationship between two variables. Let's say you are playing ball hockey in your driveway, and there is a sports car parked near the end of your driveway on the road. A quadratic function will describe the distance-time relationship for the ball that got away, missed the net, and is spinning through the air toward the car. You can track its path on graph paper or on a calculator and determine information about how long before the ball reaches a certain height as it rises and again as it falls, how long the ball is above the height of the car, and how high the ball is after a length of time. In many other fields, quadratic functions can be used to model situations. For example, in business, you can calculate the maximum profit and the quantity sold to earn this profit. Quadratic functions can be used to describe situations and determine information about maximum values and minimum values.

IN THIS CHAPTER, YOU CAN . . .

- solve quadratic equations algebraically and graphically;
- solve problems represented by quadratic functions;
- classify the roots of quadratic functions to determine whether the roots of the equation are real numbers;
- determine the maximum or minimum value of a quadratic function, and interpret the meaning of the maximum or minimum value;
- identify the structure of the complex number system and express complex numbers.

The height h, in metres, above the ground of a projectile at any time t, in seconds, after it is fired from the top of a building 10 m high is given by the **quadratic function** $h(t) = 10 + 30t - 5t^2$.

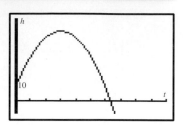

We can determine its height at any time by finding the value of the function for a given value of t; for example, $h(1) = 35$ m. Finding the time when it is at a certain height requires the solving of a quadratic equation. If we want to know when the projectile hits the ground, we must solve the equation $10 + 30t - 5t^2 = 0$. This would be equivalent to asking for the x-intercept of the function $h(t)$. Using the zoom and trace functions of your graphing calculator, the t-intercept for $t > 0$ is approximately 6.32. In this section, we will look at the algebraic solution to this type of question.

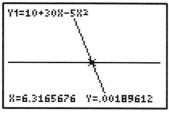

A **quadratic equation** is of the form $ax^2 + bx + c = 0$, $a \neq 0$. When we solve such an equation, we are finding the **roots of the equation.** In terms of the function $f(x) = ax^2 + bx + c$, we are finding the x-intercepts or **zeros of the function.** The graph of the function $f(x) = x^2 - 4x + 3$ graphed below has x-intercepts at 1 and 3, and the solutions to the equation $x^2 - 4x + 3 = 0$ are $x = 1$ and $x = 3$.

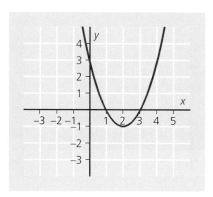

EXAMPLE 1

Solve each of the following:

  a. $x^2 - 15 = 0$ **b.** $(x - 3)^2 = 4$

a. $x^2 - 15 = 0$

$\quad\quad x^2 = 15$

$\quad\quad x = \pm\sqrt{15}$

b. $(x - 3)^2 = 4$

$\quad\quad x - 3 = \pm 2$

$\quad\quad x - 3 = 2 \text{ or } x - 3 = -2$

$\quad\quad\quad x = 5 \text{ or } \quad\quad x = 1$

If the expression $ax^2 + bx + c$ can be factored, then we can arrive at two linear equations to solve, as illustrated in the next example.

EXAMPLE 2

Solve each of the following:

a. $x^2 + 7x + 6 = 0$

b. $2x^2 - 3x + 1 = 0$

SOLUTION

a. $\quad x^2 + 7x + 6 = 0$

$\quad (x + 1)(x + 6) = 0$

$\quad x + 1 = 0 \quad \text{or } x + 6 = 0$

$\quad\quad x = -1 \text{ or } \quad\quad x = -6$

b. $\quad 2x^2 - 3x + 1 = 0$

$\quad (2x - 1)(x - 1) = 0$

$\quad 2x - 1 = 0 \text{ or } x - 1 = 0$

$\quad\quad x = \dfrac{1}{2} \text{ or } \quad\quad x = 1$

If the expression $ax^2 + bx + c$ cannot be factored, we use the **quadratic formula** that we establish below.

Specific Case

$2x^2 + 3x - 1 - 0$

$\quad 2x^2 + 3x = 1$

$\quad x^2 + \dfrac{3x}{2} = \dfrac{1}{2}$

$\quad x^2 + \dfrac{3x}{2} + \dfrac{9}{16} = \dfrac{1}{2} + \dfrac{9}{16}$

$\quad \left(x + \dfrac{3}{4}\right)^2 = \dfrac{17}{16}$

$\quad x + \dfrac{3}{4} = \pm\dfrac{\sqrt{17}}{4}$

$\quad\quad x = \dfrac{3 \pm \sqrt{17}}{4}$

General Case

$ax^2 + bx + c = 0$

$\quad ax^2 + bx = -c$

$\quad x^2 + \dfrac{bx}{a} = \dfrac{-c}{a}$

$\quad x^2 + \dfrac{bx}{a} + \dfrac{b^2}{4a^2} = \dfrac{-c}{a} + \dfrac{b^2}{4a^2}$

$\quad \left(x + \dfrac{b}{2a}\right)^2 = \dfrac{b^2 - 4ac}{4a^2}$

$\quad x + \dfrac{b}{2a} = \pm\dfrac{\sqrt{b^2 - 4ac}}{2a}$

$\quad\quad x = \dfrac{-b \pm \sqrt{b^2 - 4ac}}{2a}$

The roots of the quadratic equation $ax^2 + bx + c = 0$ are

$$x = \frac{-b + \sqrt{b^2 - 4ac}}{2a} \text{ and } x = \frac{-b - \sqrt{b^2 - 4ac}}{2a}.$$

EXAMPLE 3

Use the quadratic formula to solve each of the following:

a. $2x^2 + 3x - 3 = 0$

b. $3x^2 - 8x + 2 = 0$

SOLUTION

a. $2x^2 + 3x - 3 = 0$

$a = 2, b = 3, c = -3$

$x = \dfrac{-3 \pm \sqrt{9 - (-24)}}{4}$

$= \dfrac{-3 \pm \sqrt{33}}{4}$

b. $3x^2 - 8x + 2 = 0$

$a = 3, b = -8, c = 2$

$x = \dfrac{8 \pm \sqrt{64 - 24}}{6}$

$= \dfrac{8 \pm \sqrt{40}}{6}$

$= \dfrac{8 \pm 2\sqrt{10}}{6}$

$= \dfrac{4 \pm \sqrt{10}}{3}$

These roots can be left in this radical form or written in decimal form. If left in the radical form, they should be reduced as in part **b** of Example 3. In part **a,** the roots to two decimal places are $x = 0.69$ and $x = -2.19$.

If we refer back to our original problem of finding the time when the projectile hits the ground, we can now solve the equation $10 + 30t - 5t^2 = 0$ by using the quadratic formula, where $a = -5, b = 30, c = 10$.

$t = \dfrac{-30 \pm \sqrt{900 - (-200)}}{-10}$

$= \dfrac{-30 \pm \sqrt{1100}}{-10}$

$= \dfrac{-30 \pm 10\sqrt{11}}{-10}$

$t = 3 + \sqrt{11}$ or $t = 3 - \sqrt{11}$

Since $t > 0$, the projectile will hit the ground at $t = (3 + \sqrt{11})$ s or at approximately 6.32 s.

In solving the quadratic equation $x^2 + 4 = 0$, we get $x^2 = -4$ or $x = \pm\sqrt{-4}$.

Since $\sqrt{-4}$ is not a real number, there are no real solutions. By extending our number system, we can give meaning to this solution by introducing a number i with the property that $i^2 = -1$ or $i = \sqrt{-1}$. This number i allows us to create an expanded number system called the **Complex Numbers,** which are explained in more detail in Section 3.5.

Using the number i, $\sqrt{-4} = \sqrt{4} \times \sqrt{-1}$

$= 2i.$

The solution to $x^2 = -4$ is $x = \pm 2i$.

EXAMPLE 4

Solve $x^2 - 2x + 5 = 0$ and compare to the graph of $y = x^2 - 2x + 5$.

$x^2 - 2x + 5 = 0$

$$x = \frac{2 \pm \sqrt{4 - 20}}{2}$$

$$= \frac{2 \pm \sqrt{-16}}{2}$$

$$= \frac{2 \pm 4i}{2}$$

$$= -1 \pm 2i$$

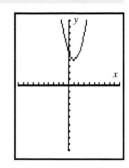

Note that the graph has no x-intercepts.

We can solve quadratic inequalities by using these same methods together with the techniques studied earlier.

If we wish to solve $x^2 - 2x - 3 < 0$, we factor the left side, obtaining $(x - 3)(x + 1) < 0$. Since the product $(x - 3)(x + 1)$ is negative, one of the numbers is negative and one is positive. Clearly, $(x - 3)$ is always smaller than $(x + 1)$, so $x - 3 < 0$ and $x + 1 > 0$. We now solve the double inequality, obtaining $x < 3$ and $x > -1$, or $-1 < x < 3$. We note that values of x satisfying the inequality lie between the roots of the corresponding equation $x^2 - 2x - 3 = 0$.

If we wish to solve $x^2 + 4x - 5 > 0$, we factor the left side, obtaining $(x + 5)(x - 1) > 0$. Now we observe that both $(x + 5)$ and $(x - 1)$ are positive or both are negative. We ensure that both are positive by making $x - 1 > 0$, or $x > 1$. We ensure that both are negative by making $x + 5 < 0$, or $x < -5$. This is equivalent to solving the double inequality $x - 1 > 0$ or $x + 5 < 0$, and we obtain the solution $x < -5$ or $x > 1$. Note that the values of x satisfying the inequality lie outside the roots of the corresponding equation $x^2 + 4x - 5 = 0$.

In solving quadratic inequalities, solve the corresponding quadratic equation $ax^2 + bx + c = 0$, with $a > 0$.

If $ax^2 + bx + c < 0$, values of x are between the roots of the equation.

If $ax^2 + bx + c > 0$, values of x are outside the roots of the equation.

EXAMPLE 5

Solve the inequality $6x^2 - 23x + 20 \leq 0$.

SOLUTION

The corresponding equation is $6x^2 - 23x + 20 = 0$.

Then $(2x - 5)(3x - 4) = 0$

$$2x - 5 = 0 \quad \text{or} \quad 3x - 4 = 0$$

$$x = \frac{5}{2} \quad \text{or} \quad x = \frac{4}{3}$$

The values of x that satisfy the inequality lie between these roots, so $\frac{4}{3} \leq x \leq \frac{5}{2}$.

EXAMPLE 6

For what values of x will $f(x) = x^2 + 4x - 12$ have positive values?

SOLUTION

Graph $f(x)$. Let $f(x) = 0$.

$$x^2 + 4x - 12 = 0$$

$$(x - 2)(x + 6) = 0$$

$$x = 2 \text{ or } x = -6$$

The x-intercepts are 2 and -6.

$f(x)$ is positive if $f(x) > 0$.

$f(x) > 0$ for $x < -6$ or for $x > 2$, $x \in R$.

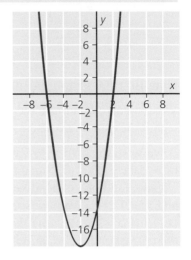

Again, if we refer back to our original projectile problem, we could now be asked to find the time interval during which the height of the projectile would be more than 35 m. This translates into solving the inequality $10 + 30t - 5t^2 > 35$.

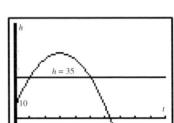

Therefore $\qquad -5t^2 + 30t - 25 > 0$.

Divide by -5. $\qquad t^2 - 6t + 5 < 0$

Consider $\qquad t^2 - 6t + 5 = 0$.

$$(t - 1)(t - 5) = 0$$

$$t = 1 \text{ or } t = 5$$

The solution to $t^2 - 6t + 5 < 0$ is $1 < t < 5$, $t \in R$.

Thus the projectile is higher than 35 m for times between 1 and 5 s.

PART A

1. For each of the quadratic functions graphed below, give the real roots of the corresponding quadratic equation.

a.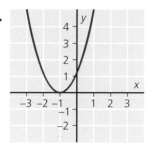

b.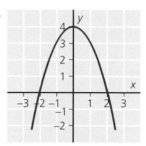

c.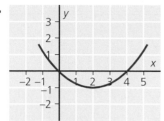

d.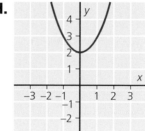

PART B

2. Solve the following equations.

    a. $x^2 = 81$ **b.** $x^2 = 400$ **c.** $x^2 - 10 = 0$

    d. $2x^2 = 72$ **e.** $4x^2 = 25$ **f.** $9x^2 = 27$

    g. $4x^2 - 45 = 0$ **h.** $(x + 5)^2 = 36$ **i.** $(x - 2)^2 = 64$

    j. $4(x - 1)^2 = 25$ **k.** $x^2 + 25 = 0$ **l.** $2x^2 + 36 = 0$

3. Solve the following equations.

    a. $x^2 + 5x = 0$ **b.** $x^2 + 5x + 4 = 0$

    c. $x^2 + 5x - 24 = 0$ **d.** $x^2 - 5x - 6 = 0$

    e. $x^2 = 4x$ **f.** $x^2 = 4x - 4$

    g. $x^2 = 9 - 8x$ **h.** $3x^2 - 18x = 0$

4. Solve the following equations.

   a. $x^2 + 2x - 1 = 0$
         b. $2x^2 - 3x - 4 = 0$

   c. $3x^2 + 2x - 1 = 0$
         d. $3 + x - x^2 = 0$

   e. $2x^2 + 11x = 21$
         f. $x^2 - 10 = 10x$

   g. $4x^2 = 18x + 10$
         h. $x^2 - x - 1 = 0$

   i. $5x^2 - 10x + 10 = 0$
      j. $8 - 6x - 4x^2 = 0$

   k. $2x^2 - 3x + 4 = 0$
        l. $2x^2 - 4x + 2 = 0$

5. Solve the following equations.

   a. $x^2 + (x + 1)^2 = (x + 2)^2$
    b. $3x(x - 3) + 2x(x - 2) = 1$

   c. $(x - 1)(x - 2) + x(x - 1) = x - 1$
   d. $(2x + 1)^2 + (2x + 3)^2 = 26$

   e. $(x - 1)(x + 1)(x + 3) = x^3$

6. Find the x-intercepts for the graphs of each of the following functions.

   a. $f(x) = 2x - x^2$
        b. $g(x) = x^2 - 7x + 6$

   c. $h(x) = 2x^2 - 5$
         d. $m(x) = 3x^2 - x - 2$

   e. $n(x) = 4 - 2x - x^2$

7. Solve the following inequalities.

   a. $x^2 - 4 \geq 0$
         b. $x^2 - 8x < 0$

   c. $x^2 - 4x - 12 < 0$
      d. $2x^2 + 4x - 6 > 0$

   e. $x^2 - 2x + 4 < 0$
       f. $2 - x^2 > x$

8. The safe stopping distance d, in metres, of a motorcycle on grass is given by the function $d = 0.02(3v^2 + 20v + 1000)$ where v is the speed of the motorcycle in metres per second. Determine the speed at which the safe stopping distance is 30 m, correct to one decimal.

Photo: Dick Hemingway

9. Billy throws the baseball back in from the outfield so that the height of the ball is given by the function $h = 1 + 20t - 5t^2$, where h is the height in metres and t is the time in seconds.

    a. If the ball is caught at the same height at which it was thrown, how long is it in the air?

    b. If the ball inadvertently hits a hovering seagull at a height of 16 m, how long was the ball in the air before hitting the bird? (Why are there two answers?)

10. The velocity of a particle acted on by an accelerating force is given by the function $v = t^2 - 6t$, where v is the velocity in centimetres per second and t is the time in seconds. For what values of t is the particle moving forward (i.e., $v > 0$)?

PART C

11. Solve the following equations in terms of k, assuming k is a constant.

    a. $x^2 + 2x + k = 0$ **b.** $kx^2 - 4kx + 4 = 0$

    c. $x^2 + (k - 1)x - k - 0$

12. For each part of Question 11, give the restrictions on k such that the equation will have real roots.

13. Solve the following for x in terms of y.

    a. $x^2 - 4xy + 3y^2 = 0$ **b.** $x^2 - 2x + 1 + y^2 = 0$

    c. $x^2 - 6xy + 6y^2 = 0$

14. Solve the following:

    a. $x^4 - 10x^2 + 9 = 0$ **b.** $x^4 - 14x^2 + 45 = 0$

    c. $x^4 - 9x^2 - 10 = 0$

15. Solve the following:

    a. $\dfrac{3}{x} - \dfrac{4}{x + 1} = 1$ **b.** $x - \dfrac{x}{x + 1} = 2$

    c. $\dfrac{3}{x + 1} + \dfrac{4x}{x^2 - 1} = 3 + \dfrac{2}{x - 1}$

16. a. If p and q are the roots of equation $ax^2 + bx + c = 0$, find expressions for $p + q$ and pq in terms of a, b, and c.

    b. Show that the equation could be written in the form $x^2 - (p + q)x + (pq) = 0$.

    c. Find a quadratic equation whose roots are 2 greater than those of $x^2 - 5x - 10 = 0$.

3.2 Problems Giving Rise to Quadratic Equations

A wide variety of problems can be modelled with a quadratic equation.

EXAMPLE 1

Two squares whose side lengths are consecutive odd integers have a total area of 650 cm^2. What is their total perimeter?

SOLUTION

Since the side lengths are consecutive odd numbers, we can represent them as x and $(x + 2)$ in centimetres.

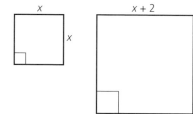

$$x^2 + (x + 2)^2 = 650$$
$$2x^2 + 4x + 4 = 650$$
$$2x^2 + 4x - 646 = 0$$
$$x^2 + 2x - 323 = 0$$
$$(x + 19)(x - 17) = 0$$
$$x = -19 \quad \text{or} \quad x = 17$$

Since $x > 0$, $x = 17$ is the only solution.

Therefore, the sides of the squares are 17 cm and 19 cm and the total perimeter is 144 cm.

EXAMPLE 2

In laying out Dolittle Park, the landscape artist includes a rectangular flowerbed 18 m by 12 m. Within the bed, she plans to have a uniform-width border of annuals around a permanent rose bed. To obtain the best appearance, she wants the rose bed to be exactly half the total area. What is the width of the border, correct to one decimal place?

SOLUTION

Let the width of the border be w metres.

The rose bed is $(18 - 2w)$ by $(12 - 2w)$.

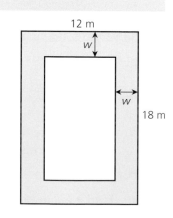

$$(18 - 2w)(12 - 2w) = 18 \times 12 \div 2$$
$$216 - 60w + 4w^2 = 108$$
$$4w^2 - 60w + 108 = 0$$
$$w^2 - 15w + 27 = 0$$
$$w = \frac{15 \pm \sqrt{225 - 108}}{2}$$
$$w \doteq 12.91 \quad \text{or} \quad w \doteq 2.09$$

But, $w \leq 6$ since the whole bed is only 12 m wide.

The width of the border is 2.1 m.

PART B

1. The sum of the squares of two consecutive odd integers is 290. Find the integers.

2. Find two integers whose sum is 96 and whose product is 1728.

3. Find three consecutive odd integers such that the sum of the squares of the first two is 15 less than the square of the third.

4. The sum of a number and its reciprocal is 6.41. What is the number?

5. A rectangular solar-heat collecting panel is 2.5 m longer than it is wide. If its area is 21 m², what are its dimensions?

6. A right triangle has a perimeter of 3 m. Its hypotenuse is 130 cm. What are the lengths of the other sides?

7. A rectangular box is 20 cm high and twice as long as it is wide. If it has a surface area of 1600 cm², what is its volume?

8. A rectangular nuclear-waste holding facility is 100 m long and 70 m wide. A safety zone of a uniform strip of more than 20 m in width must be constructed around the facility as shown.

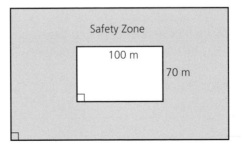

Determine the width of the strip of land to the nearest metre, if the area of the facility and the strip is 90 000 m².

9. What is the length of the diagonal of a square with area 40 cm²?

10. A matte of uniform width is to be placed around a painting so that the area of the matted surface is twice the area of the picture. If the outside dimensions of the matte are to be 20 cm by 30 cm, find the width of the matte.

11. A television screen is 40 cm high and 60 cm wide. The picture is compressed to 62.5% of its original area, leaving a uniform dark strip around the outside. What are the dimensions of the reduced picture?

12. If the average speed of a light plane had been 40 km/h less, the plane would have taken one hour longer to fly 1200 km. What was the speed of the plane?

13. The stopping distance d, in metres, of a car travelling at a velocity of v km/h is given by the formula $d = 0.007v^2 + 0.015v$.

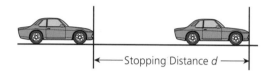

Stopping Distance d

How fast, to the nearest whole number, is a car travelling if it takes 30 m to stop?

14. A number of students charter a bus to go to a school football game at a total cost of $80. Eight of the students are ill and cannot go. Each of the remaining students then has to pay an extra 50¢. How many students go on the bus?

PART C

15. Show that it is impossible to form a 20 cm length of wire into a rectangle with area 30 cm^2.

16. It took a crew 80 min to row 3 km upstream and back again. If the rate of flow of the stream was 3 km/h, what was the rowing rate of the crew?

17. Express y in terms of x for the relation $4x^2 - 4xy + y^2 - 2x + y - 6 = 0$.

3.3   Classifying the Roots of Quadratic Equations

We have seen that the roots of the quadratic equation $ax^2 + bx + c = 0$ can be found by factoring the expression, if this is possible, or by using the formula developed, which gives the roots

$$x = \frac{-b + \sqrt{b^2 - 4ac}}{2a} \text{ and } x = \frac{-b - \sqrt{b^2 - 4ac}}{2a}.$$

It is possible to save time when dealing with quadratic equations if we are sure that the roots we will obtain are going to satisfy the given conditions. For example, if an equation represents a real situation, it is beneficial to know that the roots will be real numbers at the earliest opportunity. In the activity below, we examine a method for determining the nature of the roots of quadratic equations.

ACTIVITY

1. Copy and complete the following chart.

Equation	Roots	Value of $b^2 - 4ac$
$x^2 + 4x - 5 = 0$		
$2x^2 - 3x + 1 - 0$		
$4x^2 - 4x + 1 = 0$		
$4x^2 - 5 = 0$		
$x^2 + 9 = 0$		
$2x^2 + 5x + 1 = 0$		
$9x^2 - 12x + 4 = 0$		
$5x^2 + 6x + 5 = 0$		

2. Complete the following statements in your notebook.

  a. The equation $ax^2 + bx + c = 0$ has two real, unequal roots
     if $b^2 - 4ac$ is ▮▮▮▮▮▮ .

  b. The equation $ax^2 + bx + c = 0$ has two real, equal roots
     if $b^2 - 4ac$ is ▮▮▮▮▮▮ .

  c. The equation $ax^2 + bx + c = 0$ has two non-real roots
     if $b^2 - 4ac$ is ▮▮▮▮▮▮ .

The quantity $b^2 - 4ac$ is called the **discriminant** of the quadratic equation $ax^2 + bx + c = 0$. From it we can determine the nature of the roots of the equation. It can also be used in establishing conditions so that the roots have desired properties.

EXAMPLE 1

Determine the nature of the roots for each of the following quadratic equations.

a. $x^2 - 6x + 3 = 0$

b. $4x^2 - 20x + 25 = 0$

c. $5x^2 + 6x + 12 = 0$

SOLUTION

a. Since $b^2 - 4ac = (-6)^2 - 4(1)(3) = 24$, there are two real, unequal roots.

b. Since $b^2 - 4ac = (-20)^2 - 4(4)(25) = 0$, there are two real, equal roots.

c. Since $b^2 - 4ac = 6^2 - 4(5)(12) = -204$, there are two non-real roots.

EXAMPLE 2

For what values of k will $x^2 + 6x + k = 0$ have real roots?

SOLUTION

For the roots to be real, $b^2 - 4ac \geq 0$.

$$a = 1, b = 6, c = k$$
$$b^2 - 4ac = 6^2 - 4(1)(k)$$
$$= 36 - 4k$$

Then $36 - 4k \geq 0$,

$$-4k \geq -36$$
$$k \leq 9.$$

The equation has real roots if $k \leq 9$.

We have seen in earlier sections that the roots of the quadratic equation $ax^2 + bx + c = 0$ are the zeros of the corresponding function $y = ax^2 + bx + c$. The zeros of the function represent the x-intercepts of the graph of the function. From our work here, we can make the following conclusions.

a. If $b^2 - 4ac > 0$, the graph of the corresponding quadratic function has two distinct x-intercepts.

b. If $b^2 - 4ac = 0$, the graph of the corresponding quadratic function touches the x-axis at one point.

c. If $b^2 - 4ac < 0$, the graph of the corresponding quadratic function does not intersect the x-axis.

PART A

1. State the value of the discriminant for each of the following equations.

    a. $x^2 - 4x + 1 = 0$ **b.** $x^2 + x - 2 = 0$ **c.** $x^2 - 5x = 0$

    d. $x^2 + 4x + 4 = 0$ **e.** $x^2 - 10 = 0$ **f.** $2x^2 - x - 3 = 0$

    g. $3x^2 + 6x + 3 = 0$ **h.** $1 - x - 2x^2 = 0$ **i.** $4x^2 + 3x - 2 = 0$

2. For each of the graphs below, state whether the discriminant of the corresponding quadratic equation is greater than 0, equal to 0, or less than 0.

a.

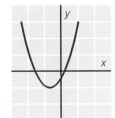

b.

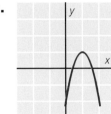

c.

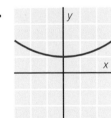

d.

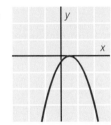

PART B

3. Determine the nature of the roots for each of the following quadratic equations.

    a. $x^2 - 8x + 12 = 0$ **b.** $x^2 + 4x + 5 = 0$

    c. $x^2 - 10x + 5 = 0$ **d.** $4x^2 - 4x + 1 = 0$

    e. $1 - 2x - 3x^2 = 0$ **f.** $2x^2 + 2 = 0$

    g. $3x^2 - 4x = 0$ **h.** $4 - 5x = x^2$

    i. $5 - x^2 = 4x$ **j.** $x^2 - \sqrt{8x} + 2 = 0$

    k. $2 - 2x - 0.25x^2 = 0$ **l.** $\sqrt{2}x^2 - 2x - \sqrt{2} = 0$

4. Determine the value(s) of k that will give the indicated type of roots.

    a. $x^2 - 6x + k = 0$;   equal roots

    b. $x^2 + kx - 16 = 0$;   real distinct roots

    c. $kx^2 - 2x + 1 = 0$;   non-real roots

    d. $2x^2 - kx + k = 0$;   equal roots

    e. $2kx^2 + 3x + 2k = 0$;   real distinct roots

5. For what values of k will the graph of $y = 9x^2 + 3kx + k$ have no x-intercepts?

6. For what values of $b^2 - 4ac$ will a quadratic equation have rational roots? How could you use this information when factoring quadratic expressions?

7. What can be said about the position of the vertex of the graph of $y = ax^2 + bx + c$ if $b^2 - 4ac = 0$?

PART C

8. For a, b, and h real numbers, show that the roots of $(x - a)(x - b) = h^2$ are always real.

9. For what values of m will $\dfrac{x^2 - 2}{2x - 3} = \dfrac{m - 1}{m + 1}$ have

    a. real roots equal in magnitude but opposite in sign?

    b. two real, non-equal roots?

10. Determine k such that the equation $(1 - 3k)x^2 + 3x - 4 = 0$ has real distinct roots.

11. The graph of the quadratic function $f(x) = ax^2 + bx + 1$ passes through the point $(1, 2)$. For what values of a does the graph of $f(x)$ intersect the x-axis at two distinct points?

Maximum-Minimum Values of Quadratic Functions

At the beginning of Section 3.1, we considered a projectile whose height was given by the quadratic function $h(t) = 10 + 30t - 5t^2$. To determine the time at which the projectile reached a specific height, we solved the corresponding quadratic equation. How could we determine the greatest height reached by the projectile and the time at which this occurs? From the graph the greatest height appears to be at approximately 55 m, and this occurs at 3 s.

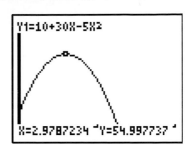

As we have already seen, the graph of a quadratic function always takes the shape of a parabola. The vertex of the parabola is the highest or lowest point on the parabola depending on the concavity of the graph.

We have worked with two forms of the quadratic function: the vertex form $y = a(x - p)^2 + q$ and the standard form $y = ax^2 + bx + c$. From the vertex form, the coordinates of the vertex are (p, q). To find the coordinates of the vertex from the standard form, we rewrite the function in the vertex form.

If $a > 0$, the graph opens upward, the vertex (p, q) represents the minimum point of the graph, and q is the minimum value of the function occurring when $x = p$.

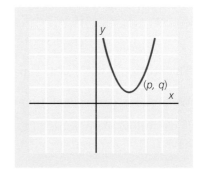

If $a < 0$, the graph opens downward, the vertex (p, q) represents the maximum point of the graph, and q is the maximum value of the function occurring when $x = p$.

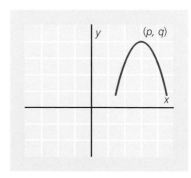

EXAMPLE 1

Find the maximum or minimum value of $f(x) = 3x^2 - 6x + 7$.

SOLUTION

$$f(x) = 3x^2 - 6x + 7$$
$$= 3(x^2 - 2x) + 7$$
$$= 3(x^2 - 2x + 1 - 1) + 7$$
$$= 3(x - 1)^2 - 3 + 7$$
$$= 3(x - 1)^2 + 4$$

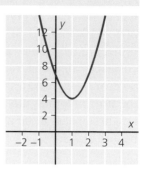

The vertex is at $(1, 4)$ and since $a > 0$, this is the minimum point of the graph.

The function has a minimum value of 4 when $x = 1$.

We can determine algebraically the maximum height reached by the projectile in our problem at the beginning of this section.

$$h(t) = -5t^2 + 30t + 10$$
$$= -5(t^2 - 6t + 9 - 9) + 10$$
$$= -5(t - 3)^2 + 45 + 10$$
$$= -5(t - 3)^2 + 55$$

The projectile will reach a maximum height of 55 m at a time of 3 s.

We obtain a general solution by completing the square on the quadratic.

If we follow the process of completing the square with the general quadratic function, we have $f(x) = ax^2 + bx + c$.

$$f(x) = a\left(x^2 + \frac{b}{a}x\right) + c$$
$$f(x) = a\left(x^2 + \frac{b}{a}x + \frac{b^2}{4a^2} - \frac{b^2}{4a^2}\right) + c$$
$$f(x) = a\left(x + \frac{b}{2a}\right)^2 - \frac{b^2}{4a} + c$$
$$f(x) = a\left(x + \frac{b}{2a}\right)^2 + \frac{4ac - b^2}{4a}$$

The vertex is $\left(-\frac{b}{2a}, \frac{4ac - b^2}{4a}\right)$. Therefore, the maximum or minimum value of a quadratic function occurs at $x = \frac{-b}{2a}$. This formula is useful when solving maximum or minimum problems.

EXAMPLE 2

The outward power P, in watts, of a 100V electric generator is given by the relation $P = 100I - 5I^2$, where I is the current in amperes.

a. For what value of I will the power be a maximum?

b. What is the maximum power?

SOLUTION

By comparing $P = 100I - 5I^2$ to the quadratic function $f(x) = ax^2 + bx + c$, $a = -5$, $b = 100$, $c = 0$.

a. The I value of the vertex is $I = \dfrac{-b}{2a}$

$$= \dfrac{-100}{2(-5)}$$

$$= 10.$$

b. Substitute $I = 10$ in $P = 100I - 5I^2$.

$$P = 100(10) - 5(10)^2$$

$$= 1000 - 500$$

$$= 500$$

The maximum power is 500W when the current I is 10A.

EXAMPLE 3

Determine the maximum product of two integers whose sum is 25.

SOLUTION

Let the numbers be n and $(25 - n)$. Their product is $P = 25n - n^2$.
$a = -1$, $b = 25$, and $c = 0$.

$$n = -\dfrac{b}{2a} = \dfrac{25}{2}$$

$$= 12.5$$

Since $a < 0$, this function has a maximum value when $n = 12.5$.

However, the numbers were required to be integers. Therefore, we must examine the integers on either side of 12.5 for the value.

When $n = 12$, $P = 25 \times 12 - 12^2$ or 156.

When $n = 13$, $P = 25 \times 13 - 13^2$ or 156.

The maximum product is 156 and the two integers are 12 and 13.

EXAMPLE 4

A magazine producer can sell 600 of her magazines at $6.00 each. A marketing survey shows her that for every $0.50 she increases the price, she will lose 30 sales. What price should she set to obtain the greatest income?

SOLUTION

If she increases her price x times, her new price will be $\$(6 + 0.5x)$ and her sales will be $(600 - 30x)$.

Then her income will be given by

$$I = (6 + 0.5x)(600 - 30x)$$
$$= 3600 + 120x - 15x^2$$
$$a = -15, b = 120, c = 3600$$
$$x = \frac{-b}{2a}$$
$$= \frac{-120}{2(-15)}$$
$$= 4$$

When $x = 4$, the price per magazine is $[6 + 0.5(4)]$ or 8.00.

Exercise 3.4

PART A

1. Determine the vertex of each of the following parabolas.

a. $y = x^2 - 9$

b. $y = 4x - x^2$

c. $y = x^2 - 4x + 4$

d. $y = x^2 - 6x + 5$

e. $y = 2x^2 + 4$

f. $y = x^2 + 4x - 5$

g. $y = 3x^2 - 12x - 9$

h. $y = -2x^2 + 4x + 5$

PART B

2. Determine the maximum or minimum value of each of the following functions and state the value of x for which this occurs.

a. $y = x^2 - 10x + 20$

b. $y = 2x^2 + 12x$

c. $y = 5 + 20x - x^2$

d. $y = 3x^2 + 24x - 12$

e. $y = 19 - 4x - 2x^2$

f. $y = 9 - 4x^2$

g. $y = 3x^2 - 15x + 1$

h. $y = 5 + 3x - 3x^2$

3. Determine two numbers whose difference is 12 and whose product is a minimum.

4. The sum of two numbers is 20. What is the least possible sum of their squares?

5. What is the smallest possible total of a number and its square?

6. What number exceeds its square by the greatest amount?

7. A rectangular field is to be enclosed with 600 m of fencing. What is the maximum area that can be enclosed and what dimensions will give this area?

8. A rectangular field is to be enclosed and divided into two sections by a fence parallel to one of the sides using a total of 600 m of fencing. What is the maximum area that can be enclosed and what dimensions will give this area?

9. Still using only 600 m of fencing, you wish to fence in a rectangular lot along a straight stretch of river (no fence needed along the river). What are the maximum area and dimensions in this case?

10. A bullet fired vertically at 80 m/s will have its height h, in metres, after time t, in seconds, given by $h = 80t - 5t^2$. What height will the bullet reach?

11. Sammy McWire hits a massive pop-up. The height h, in metres, of the ball is given by $h = 1.2 + 20t - 5t^2$, where t is in seconds. What is the maximum height of the ball? If it is caught at the same height at which it was hit, how long is it in the air?

12. Lemon Motors has been selling an average of 60 new cars per month at $800 over the factory price. They are considering an increase in this markup. A marketing survey indicates that for every $20 increase, they will sell one less car each month. What should their new markup be to maximize income?

13. Maria produces and sells shell necklaces. The material for each necklace costs her $4. She has been selling them for $8 each and averaging sales of 40 per week. She has been told that she could charge more but has found that for each $0.50 increase in price, she would lose 4 sales each week. What selling price should she set and what would her profit per week be at this price?

14. A caterer in quoting the charge for producing a dinner proposes the following terms. For a group of 60 people, he will charge $30 per person. For every extra 10 people, he will lower the price by $1.50 per person for the whole group. What size group does the caterer want to maximize his income?

15. A strip of sheet metal 30 cm wide is to be made into a trough by turning strips up vertically along two sides.

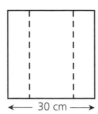

How many centimetres should be turned up at each side to obtain the greatest carrying capacity?

PART C

16. The diagram shows a plan for a deck that is to be built on the corner of a cottage. A railing is to be constructed around the four outer edges of the deck. If $AB = DE$, $BC = CD$, and the length of the railing is 30 m, find the dimensions of the deck which will have maximum area.

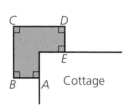

17. A projectile fired vertically with an initial velocity of v and affected only by the force of gravity g will have its height h, in metres per second, after time t, in seconds, is given by the formula $h = vt - \frac{1}{2}gt^2$. Show that the projectile attains a maximum height of $\frac{v^2}{2g}$ at time $\frac{v}{g}$.

18. ABC is a right triangle with $AB \perp BC$, $AB = 50$ cm, $BC = 120$ cm. Find the area of the largest rectangle that can be inscribed in $\triangle ABC$ with one of its corners at B.

19. Prove that the minimum value for $(x - a)^2 + (x - b)^2$ occurs when $x = \frac{a + b}{2}$. What is the minimum value?

20. What is the minimum value of the function $f(x) = 2^{x^2 - 2x}$?

3.5   Complex Numbers

In solving quadratic equations, we found that the roots were not always real numbers. To handle this situation, we introduced the symbol i with the property that $i^2 = -1$. Numbers involving this symbol are an extension of our real number system called the **complex number system** (C). In this section, we take a look at the complex numbers and how to operate with them. For our purposes the following definition will suffice.

> A **complex number** is an expression of the form $a + bi$ where a and b are real numbers and i is a symbol with the property that $i^2 = -1$.

The real number a is called the real part of $a + bi$ and the real number b is the imaginary part.

For example, $1 + 2i$ is a complex number with the real part 1 and the imaginary part 2.

Every complex number has a **conjugate number.**

> **Conjugate complex numbers** are two complex numbers differing only in the sign of their imaginary components.
> The conjugate of $a + bi$ is $a - bi$.

For example, the conjugate of $2 + 3i$ is $2 - 3i$.

The conjugate is denoted by placing a bar over the number. For example, if $z = 1 + 2i$, $\bar{z} = 1 - 2i$.

Adding, subtracting, and multiplying with complex numbers follow the same rules we have identified in working with polynomial expressions:

$$(a + bi) + (c + di) = (a + c) + (b + d)i$$
$$(a + bi) - (c + di) = (a - c) + (b - d)i$$
$$(a + bi)(c + di) = (ac - bd) + (ad + bc)i$$

In other words, we add or subtract complex numbers by combining the real components and combining the imaginary components. We multiply by treating the numbers as binomials and remembering that $i^2 = -1$.

EXAMPLE 1

If $z = 3 - 4i$ and $w = 2 + 3i$, evaluate the following:

a. $z + w$ **b.** $z - w$ **c.** zw

SOLUTION

a. $z + w = (3 - 4i) + (2 + 3i)$
$$= 5 - i$$

b. $z - w = (3 - 4i) - (2 + 3i)$
$$= 1 - 7i$$

c. $zw = (3 - 4i)(2 + 3i)$
$$= 6 + 9i - 8i - 12i^2$$
$$= 6 + i - 12(-1)$$
$$= 18 + i$$

EXAMPLE 2

Given $z = 3 + 4i$, determine the following:

a. $z + \bar{z}$ **b.** $z\bar{z}$

SOLUTION

a. $\bar{z} = 3 - 4i$
$$z + \bar{z} = (3 + 4i) + (3 - 4i)$$
$$= 6$$

b. $z\bar{z} = (3 + 4i)(3 - 4i)$
$$= 9 - 16i^2$$
$$= 9 - 16(-i)$$
$$= 25$$

In general, we can state the following:

> Both the sum and product of complex conjugates
> produce real numbers.

We use this property of conjugates to find the result of dividing complex numbers as illustrated in the following example.

EXAMPLE 3

Write $\dfrac{2 + 3i}{3 + 2i}$ in the form $a + bi$.

To perform this division, we multiply both numerator and denominator by the conjugate of the denominator. According to our property above, this will leave us with real numbers in the denominator.

$$\frac{2 + 3i}{3 + 2i} \times \frac{3 - 2i}{3 - 2i} = \frac{6 + 5i - 6i^2}{9 - 4i^2}$$

$$= \frac{6 + 5i - 6(-1)}{9 - 4(-1)}$$

$$= \frac{12 + 5i}{13}$$

$$= \frac{12}{13} + \frac{5}{13}i$$

EXAMPLE 4

Determine the reciprocal of $1 - i$, in the form $a + bi$.

The reciprocal of $1 - i$ is $\dfrac{1}{1 - i}$.

$$\frac{1}{1 - i} - \frac{1}{1 - i} \times \frac{1 + i}{1 + i}$$

$$= \frac{1 + i}{1 - i^2}$$

$$= \frac{1 + i}{2}$$

The reciprocal of $1 - i$ is $\dfrac{1}{2} + \dfrac{1}{2}i$. We can check this result by showing that the product of $(1 - i)\left(\dfrac{1}{2} + \dfrac{1}{2}i\right)$ is 1.

PART A

1. Determine the following sums or differences in the form $a + bi$.

   a. $(2 - 3i) + (1 - 4i)$ **b.** $(6 - 2i) - (-1 - 2i)$

   c. $-7i - (1 - i)$ **d.** $(-3 + 2i) + (2 - i)$

   e. $(2 - i) + (2 + i)$ **f.** $-6 - (-3 - i)$

2. Determine the following products in the form $a + bi$.

   a. $(2i)(-3i)$ **b.** $(-i)(-i)$

   c. $2i(2 - 2i)$ **d.** $-3i(-3 - i)$

PART B

3. Determine each of the following products in the form $a + bi$.

   a. $(1 + 2i)(2 + 3i)$ **b.** $(2 - i)(3 + i)$

   c. $(-1 - 3i)(-2 - 4i)$ **d.** $(4 - 5i)(4 + 5i)$

   e. $(1 - \sqrt{2}i)(1 + \sqrt{2}i)$ **f.** $(3 - 2i)^2$

4. Simplify each of the following. Express your answer in the form $a + bi$.

   a. $3(2 - 4i) + 4(1 + 2i)$ **b.** $1 - 2(3 - i) + i(2 - i)$

   c. $(1 - i)(2 - i) - (2 - i)(3 - i)$ **d.** $(2 - i)^2 - (1 - i)^2$

   e. $(1 + i)(2 + i)(3 + i)$ **f.** $(2 - i)^3$

5. If $z = 3 - 4i$, determine each of the following in the form $a + bi$.

   a. $\bar{z}$ **b.** $z + \bar{z}$ **c.** $z - \bar{z}$ **d.** $z\bar{z}$

6. Determine each of the following. Express your answers in the form $a + bi$.

   a. $\dfrac{6}{3 + i}$ **b.** $\dfrac{3}{2 + i}$ **c.** $\dfrac{4 + i}{2i}$ **d.** $\dfrac{2 - i}{2 + i}$

   e. $\dfrac{6 - 3i}{2 - i}$ **f.** $\dfrac{2}{2 - \sqrt{3}i}$ **g.** $\dfrac{3 + 2i}{-1 - i}$ **h.** $\dfrac{5i}{1 + 2i}$

7. If $z = 4 - 3i$ and $w = 3 + i$, determine the following in the form $a + bi$.

   a. $z + w$ **b.** $2z - 3w$ **c.** $(z)(w)$ **d.** $\dfrac{1}{zw}$

8. Solve the following equations for z. Write your final answer in the form $a + bi$.

    a. $z + (4 - 2i) = i$

    b. $(2 - 3i) - z = 1 - 2i$

    c. $2z - (1 - i) = 3 - i$

    d. $2iz = 3 - 2i$

    e. $(1 + i)z = 2 + i$

    f. $z^2 - 2z + 2 = 0$

PART C

9. a. For $z = 1 + i$, find z^2, z^4, $\bar{z}^2$, and $\bar{z}^4$.

    b. Repeat for $z = -1 + i$.

    c. Solve $z^4 + 4 = 0$.

Golfing Gravity

Any object that is projected upward at an angle between 0 and 90 degrees to the horizontal will return to ground following a parabolic path because of gravitational acceleration.

Athletes learn to anticipate this parabolic path when they throw a football or catch a baseball.

Golf clubs are designed with different angles on their faces. The angle of the face of the club determines the angle of elevation at which the golf ball will travel. This angle of elevation determines how high the ball will go. Golfers call this the loft. The loft governs how far horizontally the ball will travel before falling to earth.

Designers of golf clubs create the clubs so that you swing the same with all the clubs. The differences in the design of the clubs accounts for the different distances covered by the ball. Drivers have angles between 1 and 10 degrees. They are designed for long horizontal drives. Woods are used when you don't want to hit the ball as far. Woods have angles between 13 and 28 degrees. The larger the angle, the less horizontal distance the ball will travel. Wedges are used to get the ball out of a sand trap and have angles between 50 and 60 degrees.

These differences all sound good in theory. Now all the golfer has to do is learn to hit the ball the way that the golf club designers think they should.

1. Solve the following for x.

    a. $2x^2 - 10 = 0$ **b.** $2x^2 - 10x = 0$ **c.** $2x^2 - 10x + 5 = 0$

    d. $2x^2 - 7x + 5 = 0$ **e.** $4 + x^2 = 0$ **f.** $3 - 2x - x^2 = 0$

    g. $4 - 2x = x^2$ **h.** $x^2 + x + 2 = 0$ **i.** $4x^2 + 5x = 6$

2. Solve the following for x.

    a. $3(x - 2)^2 = 4(x + 1)^2$ **b.** $(x + 2)^2 - (x - 1)^2 = x^2$

    c. $(x - 1)^3 = x^3 - 1$

3. Solve the following inequalities.

    a. $4x^2 \geq 16$ **b.** $2x^2 - 6x < 0$ **c.** $x^2 - 6x + 9 \geq 0$

    d. $9 - x^2 \leq 0$ **e.** $x^2 - 4x < 5$ **f.** $6 - 3x > 3x^2$

4. Determine the x-intercepts and vertex for each of the following:

    a. $f(x) = 2x^2 - 4x$ **b.** $g(x) = 2x^2 - 4x + 1$ **c.** $h(x) = 2x^2 - 4x + 2$

    d. $k(x) = 8 - x^2$ **e.** $m(x) = 2x^2 + 6x + 5$ **f.** $n(x) = 1 - 3x - 4x^2$

5. Find the perimeter and area for the triangle shown to the right.

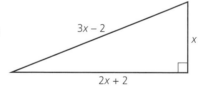

6. Find two consecutive odd integers whose product is 255.

7. A uniform-width boardwalk is built around the inside edge of a rectangular parkland that is 10 m by 15 m. If the boardwalk takes up 20% of the lot, how wide is the boardwalk to the nearest centimetre?

8. The sum S_n of the first n natural numbers $(1 + 2 + 3 + \ldots + n)$ is given by $S_n = \frac{n(n + 1)}{2}$. If Amy has reached a total of 1326 in adding the naturals, how many naturals has she added?

9. If a widget maker increases her production rate by 6 widgets per day, she will finish an order for 12 dozen widgets in four less days. How many widgets does she normally produce per day?

10. Determine the nature of the roots of each of the following quadratic equations.

a. $3x^2 - 4x - 5 = 0$ **b.** $2x^2 + 3x + 7 = 0$ **c.** $9x^2 + 12x + 6 = 0$

11. For what value of k will each of the following have equal roots?

a. $3x^2 - kx + 3 = 0$ **b.** $kx^2 - (k + 2)x + 2 = 0$

12. For what values of m will each of the following have two real roots?

a. $mx^2 - 4x + 2 = 0$ **b.** $4x^2 - mx + 2m = 0$

13. For what values of b will the graph of $y = 3x^2 - bx + 3$ have no x-intercepts?

14. Find the maximum or minimum value for each of the following functions and state the value of x for which this occurs.

a. $y = x^2 - 8x + 6$ **b.** $y = 4x^2 - 6x$

c. $y = 3 - 4x - 2x^2$ **d.** $y = 4x^2 + 20x + 25$

e. $y = -2x^2 + 6x - 5$ **f.** $y = 9 - x^2$

15. A flare is fired vertically from the top of a platform so that its height h, in metres, after time t, in seconds, is given by $h = 4 + 40t - 5t^2$.

a. How high is the platform?

b. What height does the flare reach?

c. How long is it in the air?

16. Max Soso has just hit a towering pop-up. If the height h, in metres, of the ball is given by $h = 1 + 30t - 5t^2$ and t is the time in seconds, what height does the ball reach? If it is caught at the same height at which it was hit, how long is it in the air?

17. A long strip of metal 60 cm wide is to be scored lengthwise in two places equidistant from the edges and folded perpendicular to the centre at the scores to form a trough. What are the dimensions of the trough to allow for maximum flow through the trough?

18. Canada Truck sells an average of 300 spark plug sets each week at a price of $6.40 each. The store decides to reduce the price and estimates that for every $0.10 decrease in price, they will gain 5 sales. What price should they set to maximize the total revenue?

19. If $z = -1 - 2i$, determine each of the following in the form $a + bi$.

a. $2z$ **b.** $\bar{z}$ **c.** z^2 **d.** $z + \bar{z}$ **e.** $z\bar{z}$

1. In the following questions, you are asked to identify some of the properties of consecutive integers.

   a. Choose any positive integer and, starting with it, write down five consecutive integers. Is there (at least) one divisible by 2? by 3? by 4? by 5? Repeat this exercise using different starting integers. If you write down seven consecutive integers, is there at least one in the set divisible by each of 2, 3, 4, 5, 6, and 7 (not the same one for each, of course)? If you write down n consecutive integers, is it reasonable to conclude that at least one is divisible by 2, one (may be the same one) by 3, . . . , and one by n?

   b. What is the largest number that will always be a divisor of the product of two consecutive integers? of three consecutive integers? of four consecutive integers? of k consecutive integers?

   c. Show that $n^3 - n$, $n \ge 2$ and n an integer, is always divisible by 6.

   d. Show that $n^5 - 5n^3 + 4n$, $n \ge 3$ and n an integer, is always divisible by 120.

   e. Show that $n^4 + 2n^3 - n^2 - 2n$, $n \ge 2$ and n an integer, is always divisible by 24.

2. Square $ABCD$ has area 4, and the isosceles triangle BEF has area 1. Determine the lengths of AE and ED.

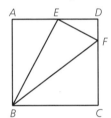

3. Determine all positive integers a and b that satisfy
$$\frac{1}{a} + \frac{a}{b} + \frac{1}{ab} = 1.$$

4. Two boys agree to mow a rectangular lawn 80 m by 60 m, each mowing half. The first boy mows a strip of uniform width around the outside of the lawn. How wide is the strip?

5. a. Determine all integral values of z such that $z^2 + 4z - 3$ is the square of an integer.

   b. Repeat part **a** for $z^2 + 13z + 3$.

6. If $2x = t + \sqrt{t^2 + 4}$ and $3y = t - \sqrt{t^2 + 4}$, determine the value of y when $x = \frac{2}{3}$.

7. If $ab = p$ and $a + b = q$, determine the value of $a^3 + b^3$ in terms of p and q.

8. If $f(x)$ is a quadratic function such that $f(0) = 2$, $f(1) = 4$, and $f(2) = 16$, determine $f(3)$.

1. Simplify each of the following:

    a. $(4x - 5)(2x + 1) - 3(2x - 1)^2$ **b.** $(3x - 2)^2 - 3(x - 2)^2$

    c. $1 - x(1 - x)^2 - (1 - x^2)$

2. If $f(x) = 4x - x^2$, determine each of the following:

    a. $f(3)$ **b.** $f(-5)$ **c.** $f\left(\frac{1}{2}\right)$ **d.** $f(2x)$ **e.** $f(x - 2)$

3. Simplify each of the following:

    a. $(3\sqrt{6})(2\sqrt{5})$ **b.** $\sqrt{3}(\sqrt{48} - \sqrt{12})$

    c. $\sqrt{8} - \sqrt{2}$ **d.** $3\sqrt{45} - 4\sqrt{5}$

    e. $(2\sqrt{3} + \sqrt{2})(2\sqrt{3} - \sqrt{2})$ **f.** $(3\sqrt{2} + 4)(4\sqrt{2} - 3)$

4. Factor each of the following completely.

    a. $4x^2 - 16x$ **b.** $4x^2 - 16x - 20$ **c.** $x^4 - 8x^2 + 16$

    d. $3x^4 - 6x^3 - 12x^2 + 24x$ **e.** $4x^2 - 15x - 4$ **f.** $12x^2 - xy - 6y^2$

5. Simplify each of the following. State any restrictions.

    a. $\dfrac{2x}{2x - 4}$ **b.** $\dfrac{x^2 - 4}{2x - 4}$ **c.** $\dfrac{x^2 - 4x - 5}{1 - x^2}$ **d.** $\dfrac{2x^2 - 7x + 3}{4x^2 - 8x + 3}$

6. Simplify each of the following. State any restrictions.

    a. $\dfrac{x - 3}{3x} \times \dfrac{x^2}{x^2 - 9}$ **b.** $\dfrac{2x - 2}{x^2 - x} \times \dfrac{x^3}{4x - 8}$ **c.** $\dfrac{1 - x^2}{x^2 - 2x + 1} \div \dfrac{x^2 - x - 2}{2x - 4}$

    d. $\dfrac{x}{x - 1} + \dfrac{2}{x}$ **e.** $\dfrac{1}{x - 1} - \dfrac{1}{x + 1}$ **f.** $\dfrac{2x}{x^2 + 2x} - \dfrac{x + 2}{2x}$

7. Solve each of the following for x.

    a. $\dfrac{3}{x - 4} = \dfrac{1}{x}$ **b.** $\dfrac{4}{x - 3} = x$ **c.** $2x^2 = 8x$

    d. $x^2 + 6x - 8 = 0$ **e.** $2x^2 = 8x + 24$ **f.** $6x^2 + 5x - 6 = 0$

8. Solve each of the following for x.

    a. $3(x - 2) \leq 4 - x$ **b.** $-4 \leq 2x + 4 \leq 8$ **c.** $2x^2 > 6x + 20$

9. Determine the domain and range for each of the following:

    a. $h(x) = 1 - x^2$ **b.** $f(x) = \sqrt{x - 4}$ **c.** $g(x) = -2(x - 2)^2 + 4$

10. Determine the x- and y-intercepts for each of the following:

    a. $y = 8 - 4x^2$ **b.** $y = x^2 - 9x - 36$ **c.** $y = 2x^2 - 7x + 6$ **d.** $y = \dfrac{x - 4}{x - 2}$

11. For each of the following, determine the coordinates of the vertex and indicate whether the function has a maximum or minimum value. Sketch the graph.

  a. $y = 4x^2 - 24x$ **b.** $y = 6 - 2x - x^2$ **c.** $y = 2x^2 + 4x + 6$

12. For what value(s) of k will the graph of $y = 4x^2 - kx + k$ have no x-intercepts?

13. The graph of $y = f(x)$ is shown at the right.

  a. State the domain and range of f.

  b. Is f a function?

  c. Sketch the graph for $f^{-1}(x)$.

  d. Is f^{-1} a function?

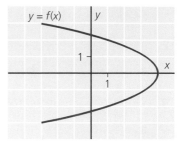

14. For the graph of $y = f(x)$, sketch graphs for each of the following:

  a. $y = f(x - 2)$

  b. $y = \frac{1}{2}f(x)$

  c. $y = f(x - 3) - 2$

  d. $y = f^{-1}(x)$

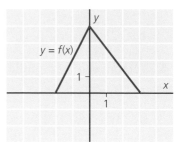

15. Sketch the graph of $y = 1 - \dfrac{2}{x - 3}$. Identify any transformations used.

16. Simplify each of the following. Write your answers in the form $a + bi$.

  a. $3i(2 - i)$ **b.** $(3 - i)(4 + i)$ **c.** $(4 - i)^2$ **d.** $\dfrac{3}{2 - i}$

17. A rectangular warehouse 90 m by 60 m is to be built on a lot so that there is a uniform strip around the building for parking and loading. If the total area of the lot is to be 9000 m², how wide is the uniform strip?

18. If a bullet is fired vertically with an initial speed of 100 m/s, its height h, in metres, after time t, in seconds, is given by $h = 100t - 5t^2$.

  a. What maximum height will the bullet reach?

  b. How long is it in the air (assuming it falls back to the ground)?

19. A small company produces and sells x widgets per week. They find that their cost function in dollars is $C(x) = \$50 + 3x$ and their revenue function in dollars is $R(x) = \left(6x - \dfrac{x^2}{100}\right)$. How many widgets per week should be produced for maximum profit?

20. A rectangular field is to be enclosed by a fence. Two fences, parallel to one side of the field, divide the field into three rectangular fields. If 2400 m of fence are available, find the dimensions giving the maximum area.

Chapter 4

Exponents and the Exponential Function

Where would you use numbers in an exponential form? Finance, medicine, science, and the media are only a few of the many fields where numbers are expressed in exponential form. Naming numbers with exponents can be used to interpret values and to perform calculations. Exponential functions provide a method of modelling situations where one quantity increases or decreases according to the value represented by the exponent, resulting in an exponential change. The growth of a forest, the growth of the population of the earth, and radioactive decay are based on the exponential function. An exponential function could also approximate the growth of cancer in a certain population. What is the probability of a malignant cell reproducing in a given length of time? You might examine how cancer cells grow exponentially by graphing a function to show the results of a simulation.

In this chapter, you will encounter positive and negative exponents. Positive exponents can express very large numbers such as distances in outer space, exponential growth of populations, and compound interest. Negative exponents are applied to name very small numbers such as microscopic measurements and extremely short lengths of time. Use a graphing calculator or software to model exponential functions calculations and consider the results and repercussions of exponential changes.

IN THIS CHAPTER, YOU CAN . . .

- simplify and evaluate expressions containing integer and rational exponents, using the laws of exponents;
- graphically represent exponential functions;
- solve exponential equations;
- evaluate logarithms and solve equations with logarithms.

4.1 The Laws of Exponents

An **exponential expression** is used for indicating the numbers of factors involved in multiplication. For example, 5^4 means that there are four factors of five or $5 \times 5 \times 5 \times 5$. In general, a^m means that there are m factors of a. Using this expression, we can develop laws for operating with exponents.

Multiplying Powers With the Same Base

$$a^m \bullet a^n = a^{m+n}$$

To multiply powers with the same base, add the exponents and leave the bases unchanged.

a. $5^3 \bullet 5^4 = 5^7$ **b.** $a^3 \bullet a^5 = a^8$

Dividing Powers With the Same Base

$$\frac{a^m}{a^n} = a^{m-n}, a \neq 0, \text{ if } m > n$$

$$\frac{a^m}{a^n} = \frac{1}{a^{n-m}}, a \neq 0, \text{ if } m < n$$

To divide powers with the same base, subtract the exponents and leave the bases unchanged.

a. $\frac{7^5}{7^2} = 7^3$ **b.** $\frac{7^3}{7^6} = \frac{1}{7^3}$ **c.** $\frac{a^7}{a^2} = a^5$ **d.** $\frac{a^2}{a^5} = \frac{1}{a^3}$

The Power of a Power

$$(a^m)^4 = a^m \bullet a^m \bullet a^m \bullet a^m$$
$$= a^{4m}$$

In general, $(a^m)^n$ means that there are n factors of a^m, so

$$(a^m)^n = a^{mn}.$$

To find a power of a power, multiply the exponents.

The Power of a Product

$$(a^3b^2)^4 = (a^3b^2)(a^3b^2)(a^3b^2)(a^3b^2)$$
$$= (a^3 \bullet a^3 \bullet a^3 \bullet a^3)(b^2 \bullet b^2 \bullet b^2 \bullet b^2)$$
$$= (a^3)^4(b^2)^4$$
$$= a^{12}b^8$$

In general, $(ab)^m = a^m b^m$.

To find the power of a product, apply the exponent to each factor of the product.

The Power of a Quotient

$$\left(\frac{a}{b}\right)^4 = \left(\frac{a}{b}\right)\left(\frac{a}{b}\right)\left(\frac{a}{b}\right)\left(\frac{a}{b}\right)$$

$$= \frac{a^4}{b^4}$$

In general, $\left(\frac{a}{b}\right)^m = \frac{a^m}{b^m}$.

To find the power of a quotient, apply the exponent to each part of the fraction. This is a very useful property. It makes it easy to simplify powers of fractions.

For example: $\left(\frac{2}{5}\right)^3 = \frac{2^3}{5^3}$

$$= \frac{8}{125}$$

Here is a summary of the exponent laws.

Exponent Laws

$a^m \cdot a^n = a^{m+n}$ $(a^m)^n = a^{mn}$

$\dfrac{a^m}{a^n} = a^{m-n}, a \neq 0, m > n$ $(ab)^m = a^m b^m$

$\dfrac{a^m}{a^n} = \dfrac{1}{a^{n-m}}, a \neq 0, m < n$ $\left(\dfrac{a}{b}\right)^m = \dfrac{a^m}{b^m}, b \neq 0$

We can use the exponent laws to simplify algebraic expressions.

EXAMPLE 1

Write each of the following as a single power.

a. $4^3 \times 4^5 \times 4^2$ **b.** $\dfrac{7^2 \times 7^4}{7^3}$ **c.** $\left(\dfrac{5}{3}\right)^2\left(\dfrac{5}{3}\right)^4$ **d.** $(5^3)^2(5^4)^3$

SOLUTION

a. $4^3 \times 4^5 \times 4^2 = 4^{3+5+2}$

$$= 4^{10}$$

b. $\dfrac{7^2 \times 7^4}{7^3} = \dfrac{7^{2+4}}{7^3}$

$$= \dfrac{7^6}{7^3}$$

$$= 7^{6-3}$$

$$= 7^3$$

c. $\left(\dfrac{5}{3}\right)^2\left(\dfrac{5}{3}\right)^4 = \left(\dfrac{5}{3}\right)^{2+4}$

$$= \left(\dfrac{5}{3}\right)^6 \text{ or } \dfrac{5^6}{3^6}$$

d. $(5^3)^2(5^4)^3 = 5^{3\times2} \times 5^{4\times3}$

$$= 5^6 \times 5^{12}$$

$$= 5^{6+12}$$

$$= 5^{18}$$

EXAMPLE 2

Evaluate $\dfrac{3^6 \times 2^5}{2^2 \times 3^4}$.

SOLUTION

$$\dfrac{3^6 \times 2^5}{2^2 \times 3^4} = 3^{6-4} \times 2^{5-2}$$

$$= 3^2 \times 2^3$$

$$= 9 \times 8$$

$$= 72$$

EXAMPLE 3

Simplify each of the following:

a. $(4b^3)^2$ **b.** $x^{a+b} \bullet x^{2a-b}$ **c.** $\dfrac{28x^{2a+3b}}{4x^{a-2b}}$

SOLUTION

a. $(4b^3)^2 = 4^2(b^3)^2$ **b.** $x^{a+b} \bullet x^{2a-b} = x^{a+b+2a-b}$ **c.** $\dfrac{28x^{2a+3b}}{4x^{a-2b}} = 7x^{(2a+3b)-(a-2b)}$

$\quad\quad = 16b^6$ $\quad\quad\quad\quad\quad\quad = x^{3a}$ $\quad\quad\quad\quad\quad\quad = 7x^{2a+3b-a+2b}$

$\quad\quad\quad\quad\quad\quad\quad\quad\quad\quad\quad\quad\quad\quad\quad\quad\quad\quad = 7x^{a+5b}$

We have developed rules for multiplying and dividing powers with the same base. If the powers have different bases, but the same exponents, we can simplify the expression by combining the bases.

EXAMPLE 4

Evaluate or write as a single power and then evaluate.

a. $8^6\left(\dfrac{1}{4}\right)^6$ **b.** $\dfrac{8^2 \times 4^4}{16^3}$

SOLUTION

a. $8^6\left(\dfrac{1}{4}\right)^6 = \left(8 \times \dfrac{1}{4}\right)^6$

$\quad\quad = 2^6$

$\quad\quad = 64$

b. *The powers do not have the same base or the same exponents.* However, if you look at the numbers carefully, you can see that the bases are all powers of 2, so we can proceed as follows:

$$\dfrac{8^2 \times 4^4}{16^3} = \dfrac{(2^3)^2(2^2)^4}{(2^4)^3}$$

$$= \dfrac{2^6 \times 2^8}{2^{12}}$$

$$= \dfrac{2^{14}}{2^{12}}$$

$$= 2^2$$

$$= 4$$

Exercise 4.1

PART A

1. Evaluate each of the following:

a. 3^4

b. $(\sqrt{3})^2$

c. $(-2)^4$

d. $\left(\frac{1}{2}\right)^3$

e. $\left(-\frac{2}{3}\right)^3$

f. -4^2

g. $\left(-\frac{2}{5}\right)^2$

h. $\left(-\frac{3}{2}\right)^3$

i. $-\frac{2^2}{3}$

2. Use the exponent laws to simplify each of the following expressions.

a. $5^4 \times 5^7$

b. $a^5 \times a^3$

c. $r^2 \times r^3 \times r^5$

d. $\left(\frac{5}{7}\right)^9\left(\frac{5}{7}\right)^6$

e. $(-m)^2(-m)^3$

f. $(2ab)^4(2ab)^9$

3. Simplify each of the following expressions.

a. $\frac{6^9}{6^3}$

b. $t^{15} \div t^9$

c. $\frac{(-2m)^{14}}{(-2m)^7}$

d. $\frac{6q^{11}}{3q^9}$

e. $25k^{15} \div (5k^7)$

f. $\frac{16b^9}{4b^7}$

4. Simplify each of the following expressions.

a. $\frac{m^7 \times m^{15}}{m^{20}}$

b. $(-3)^3(-3)^{10}(-3)^4$

c. $\frac{a^{20}}{a^2 \times a^8}$

d. $\frac{(2c)^5(2c)^9}{(2c)^3(2c)^5}$

e. $2^{10} \div 2^8 \times 2^4$

f. $\left(\frac{3}{4}\right)^{20} \div \left(\frac{3}{4}\right)^7 \div \left(\frac{3}{4}\right)^9$

5. Simplify each of the following:

a. $(4^9)^2$

b. $\left[\left(\frac{3}{2}\right)^3\right]^5$

c. $[(-m)^6]^3$

d. $(3^6)^5(3^5)^3$

e. $(k^3)^9 \div (k^2)^{11}$

f. $\frac{(h^3)^{15}}{(h^2)^{16}}$

PART B

6. Simplify each of the following:

a. $2m^6 \times 3m^7$

b. $\frac{12f^{15}}{6f^7}$

c. $\frac{15y^9}{3y^5}$

d. $\frac{4p^7 \times 6p^9}{12p^{15}}$

e. $\frac{10n^3 \times 3n^5}{5n^8}$

f. $\frac{6x^7 \times 12x^2}{4x^3 \times 9x^4}$

7. Simplify each of the following:

a. $(2x^7)^3$

b. $(-2m^2)^7$

c. $\frac{(4a^3)^2}{a^5}$

d. $(3e^6)(2e^3)^4$

e. $\frac{(5m^3)^4}{(5m^4)^2}$

f. $(-2p^2)^3(-3p)^2$

8. In each of the following, two quantities are given. Are the quantities equivalent or not? If not, why not?

a. $\dfrac{18^5}{6^5}$ and 3

b. $\left(\dfrac{15}{9}\right)^2$ and $\dfrac{25}{9}$

c. $5^2 \times 5^3$ and 25^5

d. $\dfrac{7^2 \times 7^6}{7^4}$ and 7^2

9. Simplify each expression and then evaluate if $a = 3$ and $b = -2$.

a. $(a^3b^2)^2$

b. $\dfrac{a^2}{b^3}$

c. $\dfrac{a^4b^3}{a^3b}$

d. $\left(\dfrac{a^3}{b}\right)^2$

10. Simplify the following:

a. $\dfrac{x^4y^5w^3}{x^2y^4}$

b. $\dfrac{12(a^2b^3)^3}{(2a^2b^2)^2}$

c. $\dfrac{3^4 \times 2^3 \times 3 \times 2^5}{2^4 \times 3^2}$

d. $\dfrac{8^{12} \times 6^9}{6^7 \times 8^{11}}$

11. Write each of the following as a power in the form a^b.

a. $5^3 \times 2^3$

b. $\left(\dfrac{1}{2}\right)^4 \times 8^4$

c. $16^3 \times (0.25)^3$

d. $4^9 \times 3^9 \div 6^9$

12. Simplify $[(5^2)^3]^4$.

13. Evaluate each of the following:

a. $\dfrac{27^4}{3^9}$

b. $\dfrac{8^5 \times 4^2}{2^{16}}$

c. $\dfrac{32^5}{4^2 \times 16^4}$

d. $\dfrac{25^3 \times 5^2}{125^2}$

14. Determine all positive integral values for a and b that make each of the following statements true.

a. $27^2 = (3^a)^b$

b. $64^3 = (2^a)^b$

c. $64^3 = (4^a)^b$

d. $625^2 = (5^a)^b$

PART C

15. Simplify each of the following:

a. $x^{p+q} \cdot x^{p-q}$

b. $x^{p+2q} \cdot x^{4p-3q}$

c. $\dfrac{y^{a^2-b^2}}{y^{a-b}}$

d. $(m^{p+q})^{p-q}$

e. $\dfrac{x^{2a} \cdot y^{2b}}{x^{a-b} \cdot y^{a+b}}$

f. $\dfrac{(r^x)^4(r^{2x+3y})^3}{(r^{y-2x})^5}$

16. Simplify each of the following:

a. $\dfrac{6^4 \times 15^3}{3^3 \times 2^4}$

b. $\dfrac{33^3 \times 15^6 \times 21}{25^3 \times 99^3 \times 81}$

4.2 Zero and Negative Exponents

We know that a power is a mathematical abbreviation.

$$2^3 = 2 \times 2 \times 2 = 8$$

We use exponents so that we do not have to write a string of numbers.

We know how to multiply and divide powers.

$$a^5 \bullet a^7 = a^{5+7} = a^{12}$$
$$\frac{a^5}{a^3} = a^{5-3} = a^2, a \neq 0$$

But look what happens here.

$$\frac{a^3}{a^3} = a^{3-3} = a^0, a \neq 0$$

Let's look at this question again.

$$\frac{a^3}{a^3} = \frac{\not{a} \bullet \not{a}^1 \bullet \not{a}}{\not{a} \bullet \not{a}_1 \bullet \not{a}} = 1, a \neq 0$$

Any number divided by itself is 1, so $\frac{a^m}{a^m} = 1, a \neq 0$.

By our exponent laws, $\frac{a^m}{a^m} = a^{m-m} = a^0$.

We define $a^0 = 1, a \neq 0$.

Using the exponent laws $\frac{a^3}{a^5} = a^{3-5} = a^{-2}, a \neq 0$.

We also know that $\frac{a^3}{a^5} = \frac{\not{a} \bullet \not{a}^1 \bullet \not{a}}{\not{a} \bullet \not{a}_1 \bullet \not{a} \bullet a \bullet a} = \frac{1}{a^2}, a \neq 0$.

We have done the question in two different ways. We must get the same answer.

We conclude that $a^{-2} = \frac{1}{a^2}, a \neq 0$.

In the same way, we can show that $a^{-1} = \frac{1}{a}$ and in general $a^{-n} = \frac{1}{a^n}, a \neq 0$.

For example: $5^{-7} = \frac{1}{5^7}$

EXAMPLE 1

Write each of the following with positive exponents.

a. x^{-7} **b.** $5m^{-2}$ **c.** $(3b)^{-3}$

SOLUTION

a. $x^{-7} = \frac{1}{x^7}$

b. $5m^{-2} = 5 \bullet \frac{1}{m^2}$
$$= \frac{5}{m^2}$$

c. $(3b)^{-3} = \frac{1}{(3b)^3}$
$$= \frac{1}{3^3 b^3}$$
$$= \frac{1}{27b^3}$$

EXAMPLE 2

Write each of the following with positive exponents.

a. $\dfrac{1}{w^{-5}}$ 　　　　　　　　　　　　　　　　　**b.** $\left(\dfrac{a}{b}\right)^{-2}$

a. $\dfrac{1}{w^{-5}} = \dfrac{1}{\frac{1}{w^5}}$

$= 1 \div \dfrac{1}{w^5}$

$= 1 \cdot \dfrac{w^5}{1}$

$= w^5$

Notice that $\dfrac{1}{w^{-5}} = w^5$ and in general, $\dfrac{1}{a^{-n}} = a^n$, $a \neq 0$.

b. $\left(\dfrac{a}{b}\right)^{-2} = \dfrac{1}{\left(\frac{a}{b}\right)^2}$

$= \dfrac{1}{\frac{a^2}{b^2}}$

$= 1 \cdot \dfrac{b^2}{a^2}$

$= \left(\dfrac{b}{a}\right)^2$

Notice that $\left(\dfrac{a}{b}\right)^{-2} = \left(\dfrac{b}{a}\right)^2$ and in general, $\left(\dfrac{a}{b}\right)^{-n} = \left(\dfrac{b}{a}\right)^n$, $a, b \neq 0$.

Summary

$$x^0 = 1$$
$$x^{-n} = \dfrac{1}{x^n}$$
$$\dfrac{1}{x^{-n}} = x^n$$
$$\left(\dfrac{a}{b}\right)^{-n} = \left(\dfrac{b}{a}\right)^n$$
$$\text{for } x, a, b \neq 0$$

EXAMPLE 3

Evaluate each of the following:

a. -5^{-3} 　　　**b.** $\left(\dfrac{3}{4}\right)^{-3}$ 　　　**c.** $7^0 - 2^{-2}$ 　　　**d.** $\dfrac{2^{-1} + 2^{-2}}{3^{-1}}$

a. $-5^{-3} = -\dfrac{1}{5^3}$

$= -\dfrac{1}{125}$

b. $\left(\dfrac{3}{4}\right)^{-3} = \left(\dfrac{4}{3}\right)^3$

$= \dfrac{4^3}{3^3}$

$= \dfrac{64}{27}$

c. $7^0 - 2^{-2} = 1 - \dfrac{1}{2^2}$

$\qquad = \dfrac{4}{4} - \dfrac{1}{4}$

$\qquad = \dfrac{3}{4}$

d. $\dfrac{2^{-1} + 2^{-2}}{3^{-1}} = \dfrac{\frac{1}{2} + \frac{1}{2^2}}{\frac{1}{3}}$

$\qquad = \dfrac{\frac{1}{2} + \frac{1}{4}}{\frac{1}{3}}$

$\qquad = \dfrac{\frac{2}{4} + \frac{1}{4}}{\frac{1}{3}}$

$\qquad = \dfrac{\frac{3}{4}}{\frac{1}{3}}$

$\qquad = \dfrac{3}{4} \times \dfrac{3}{1}$

$\qquad = \dfrac{9}{4}$

EXAMPLE 4

Use the exponent laws to simplify the following. Write your answers with positive exponents.

a. $x^{-5} \cdot x^8$

b. $(a^4 b^{-5})(a^{-6} b^{-2})$

c. $(a^{-4})^{-2}$

d. $\dfrac{(3a^{-4})[2a^2(-b)^3]}{12a^5 b^2}$

SOLUTION

a. $x^{-5} \cdot x^8 = x^{-5+8}$

$\qquad = x^3$

b. $(a^4 b^{-5})(a^{-6} b^{-2}) = a^{4-6} b^{-5-2}$

$\qquad = a^{-2} b^{-7}$

$\qquad = \dfrac{1}{a^2 b^7}$

c. $(a^{-4})^{-2} = a^{(-4)(-2)}$

$\qquad = a^8$

d. $\dfrac{(3a^{-4})[2a^2(-b)^3]}{12a^5 b^2} = \dfrac{-6a^{-2} b^3}{12a^5 b^2}$

$\qquad = -\dfrac{b^{3-2}}{2a^{5-(-2)}}$

$\qquad = -\dfrac{b}{2a^7}$

PART **A**

1. Write each of the following with positive exponents.

   a. 5^{-3}

   b. $\dfrac{1}{4^{-3}}$

   c. $\dfrac{x^3}{y^{-2}}$

   d. $\left(\dfrac{3}{5}\right)^{-4}$

   e. $\dfrac{2x^{-2}y^3}{3w^{-4}}$

   f. $\dfrac{1}{a^3b^{-2}}$

2. Evaluate each of the following:

   a. 2^{-3}

   b. $(-5)^{-2}$

   c. $\left(\dfrac{5}{3}\right)^{-3}$

   d. $(-7)^0$

   e. $\dfrac{2}{3^{-2}}$

   f. $\dfrac{5^{-1}}{2^3}$

   g. $\left(\dfrac{3^3}{5}\right)^{-2}$

   h. $\left(\dfrac{2^{-2}}{3^{-1}}\right)^{-2}$

   i. $\left(-12\dfrac{3}{4}\right)^0$

3. Simplify each of the following expressions. Write the answer with positive exponents.

   a. $m^6 \bullet m^{-3}$

   b. $a^{-3} \div a^{-5}$

   c. $(x^2y^{-1})^{-1}$

   d. $\left(\dfrac{b^{-2}}{b^{-1}}\right)^{-3}$

   e. $(x^{-1}y^0)^{-3}$

   f. $(15r^{-4})(2r^{-3})$

   g. $12a^4 \div (3a^{-3})$

   h. $\left[\dfrac{(3a)^{-1}}{2a^{-2}}\right]^{-1}$

PART **B**

4. Evaluate each of the following:

   a. $(3^{-1} + 3^0)^{-1}$

   b. $\left(\dfrac{1}{3}\right)^{-2} + 3^3$

   c. $5^{-2}(5^2 - 5^0)$

   d. $(2^3 - 2^2)(2^{-1} - 2^2)$

   e. $(4 - 2^3) \div (4^{-1} - 2^{-3})^0$

   f. $\dfrac{2^{-1} + 2^{-3}}{2^{-2}}$

   g. $\dfrac{3^{-2} + 3^{-3}}{3^{-2} - 3^{-3}}$

   h. $\dfrac{3^{-6} \times 3^{-5}}{3^{-9}}$

5. Evaluate each of the following:

   a. $3^9 \times 27^{-4}$

   b. $2^{16} \times 4^2 \times (-8)^{-7}$

   c. $\dfrac{32^3 \times 4^{-2}}{16^4}$

   d. $25^4 \times \left(\dfrac{1}{5}\right)^{-2} \div 125^2$

6. Simplify each of the following:

   a. $(8^{4+2a})(16^{a-1}) \div 4^{3a+2}$

   b. $[(27^{m+1}) \div (9^{m-2})]^{-2}$

7. If $x = 2$ and $y = -1$, evaluate each of the following:

   a. $\dfrac{25x^{-2}y^{-1}}{15x^{-1}y^2}$

   b. $(2x^{-2}y^{-4})^{-2}$

4.3 Rational Exponents

We know the meaning of integral exponents. Do fractional exponents have any meaning? What does $3^{\frac{1}{2}}$ mean?

Look at what happens when we multiply.

i) $3^{\frac{1}{2}} \times 3^{\frac{1}{2}} = 3^{\frac{1}{2}+\frac{1}{2}}$

$\qquad = 3$

But $\sqrt{3} \times \sqrt{3} = \sqrt{9}$

$\qquad = 3.$

This suggests that $3^{\frac{1}{2}} = \sqrt{3}$.

ii) For $a \geq 0$,

$\qquad a^{\frac{1}{2}} \times a^{\frac{1}{2}} = a^{\frac{1}{2}+\frac{1}{2}}$

$\qquad\qquad = a$

But $\sqrt{a} \times \sqrt{a} = a.$

This suggests that $a^{\frac{1}{2}} = \sqrt{a}$.

Similarly $3^{\frac{1}{3}} = \sqrt[3]{3}$, which means the cube root of 3.

In general, $a^{\frac{1}{n}} = \sqrt[n]{a}$, for $a \geq 0$.
$\sqrt[n]{a}$ is called a **radical** and means the nth root of a.

We can write an expression in exponential form or radical form.

$$2^{\frac{1}{5}} \longleftrightarrow \sqrt[5]{2}$$

Exponential Form   Radical Form

Consider the following examples for $a < 0$ in the definition $a^{\frac{1}{n}} = \sqrt[n]{a}$.

EXAMPLE 1

Evaluate each of the following, if possible.

a. $\sqrt[3]{-8}$ 　　　 **b.** $\sqrt[5]{-243}$ 　　　 **c.** $\sqrt{-16}$

SOLUTION

a. $\sqrt[3]{-8} = -2$ since $(-2)^3 = -8$.

b. $\sqrt[5]{-243} = -3$ since $(-3)^5 = -243$.

c. We cannot find an even root of a negative number in the real number system. For example, $\sqrt[4]{-625}$ does not exist.

EXAMPLE 2

Evaluate each of the following:

a. $27^{\frac{1}{3}}$ **b.** $(-32)^{\frac{1}{5}}$ **c.** $\left(\frac{1}{8}\right)^{\frac{1}{3}}$

SOLUTION

a. $27^{\frac{1}{3}} = \sqrt[3]{27}$ **b.** $\left(-32\right)^{\frac{1}{5}} = \sqrt[5]{-32}$

$= 3$ $= -2$

c. $\left(\frac{1}{8}\right)^{\frac{1}{3}} = \sqrt[3]{\frac{1}{8}}$

$= \frac{\sqrt[3]{1}}{\sqrt[3]{8}}$

$= \frac{1}{2}$

We have learned that we can write powers in different ways.

$$a^{xy} = (a^x)^y = (a^y)^x$$

This will help us evaluate powers with more complicated radical exponents.

EXAMPLE 3

a. Evaluate $32^{\frac{3}{5}}$ by considering the following:

i) $\left(32^3\right)^{\frac{1}{5}}$ **ii)** $\left(32^{\frac{1}{5}}\right)^3$

b. Which expression is easier to evaluate?

SOLUTION

We can write $32^{\frac{3}{5}}$ as $\left(32^3\right)^{\frac{1}{5}}$ or $\left(32^{\frac{1}{5}}\right)^3$.

a. i) $(32^3)^{\frac{1}{5}} = \sqrt[5]{32^3}$ **ii)** $\left(32^{\frac{1}{5}}\right)^3 = \left(\sqrt[5]{32}\right)^3$

$= \sqrt[5]{32\ 768}$ $= 2^3$

$= 8$ $= 8$

b. $\left(32^{\frac{1}{5}}\right)^3$ is easier to evaluate because we are finding the fifth root of a smaller number.

In general, $a^{\frac{p}{q}} = \left(\sqrt[q]{a}\right)^p$ or $\sqrt[q]{a^p}$.

Alternatively, $a^{\frac{p}{q}} = \left(a^{\frac{1}{q}}\right)^p$ or $(a^p)^{\frac{1}{q}}$.

EXAMPLE 4

Evaluate the following:

a. $16^{-\frac{3}{4}}$

b. $-25^{-\frac{3}{2}}$

a. $16^{-\frac{3}{4}} = \dfrac{1}{16^{\frac{3}{4}}}$

$= \dfrac{1}{(\sqrt[4]{16})^3}$

$= \dfrac{1}{2^3}$

$= \dfrac{1}{8}$

b. $-25^{-\frac{3}{2}} = -\dfrac{1}{25^{\frac{3}{2}}}$

$= -\dfrac{1}{(\sqrt{25})^3}$

$= -\dfrac{1}{5^3}$

$= -\dfrac{1}{125}$

We can use the exponent laws with rational exponents.

EXAMPLE 5

Simplify each of the following:

a. $\left(x^{\frac{1}{2}}\right)\left(x^{\frac{2}{3}}\right)$

b. $\left(8x^6\right)^{\frac{2}{3}}$

a. $\left(x^{\frac{1}{2}}\right)\left(x^{\frac{2}{3}}\right) = x^{\frac{1}{2}+\frac{2}{3}}$

$= x^{\frac{3}{6}+\frac{4}{6}}$

$= x^{\frac{7}{6}}$

b. $\left(8x^6\right)^{\frac{2}{3}} = 8^{\frac{2}{3}}(x^6)^{\frac{2}{3}}$

$= \left(8^{\frac{1}{3}}\right)^2(x^4)$

$= (2^2)(x^4)$

$= 4x^4$

PART A

1. Write each of the following in exponential form.

a. $\sqrt{14}$ **b.** $\sqrt[3]{5}$ **c.** $\sqrt[4]{9}$

d. $\sqrt[5]{-8}$ **e.** $\sqrt{35}$ **f.** $\sqrt[3]{-4^2}$

g. $(\sqrt{5})^3$ **h.** $\sqrt[5]{(-6)^3}$ **i.** $(\sqrt[4]{3})^5$

2. Write each of the following in radical form.

a. $12^{\frac{1}{2}}$ **b.** $5^{\frac{1}{3}}$ **c.** $7^{\frac{1}{5}}$

d. $(-6)^{\frac{1}{3}}$ **e.** $-9^{\frac{1}{4}}$ **f.** $7^{\frac{3}{2}}$

g. $8^{\frac{2}{5}}$ **h.** $19^{\frac{5}{7}}$ **i.** $(-13)^{\frac{2}{5}}$

3. Use your calculator to evaluate each of the following. Write your answer correct to three decimals.

a. $(4.217)^{\frac{7}{3}}$ **b.** $(-293)^{\frac{2}{5}}$ **c.** $(975.4)^{-\frac{4}{5}}$

PART B

4. Evaluate each of the following:

a. $16^{\frac{1}{4}}$ **b.** $27^{\frac{1}{3}}$ **c.** $25^{\frac{1}{2}}$

d. $(-8)^{\frac{2}{3}}$ **e.** $49^{-\frac{1}{2}}$ **f.** $-8^{\frac{4}{3}}$

g. $16^{-\frac{3}{4}}$ **h.** $64^{-\frac{5}{6}}$ **i.** $27^{-\frac{2}{3}}$

5. Evaluate each of the following:

a. $(\sqrt[3]{27})^{-1}$ **b.** $-\sqrt[5]{-32}$ **c.** $\sqrt[5]{\frac{1}{32}}$

d. $(-\sqrt{36})^2$ **e.** $\left(\frac{25}{36}\right)^{\frac{1}{2}}$ **f.** $\left(16^{\frac{1}{2}}\right)^0$

g. $\sqrt[3]{\frac{27}{64}}$ **h.** $\left(\frac{98}{72}\right)^{\frac{1}{2}}$

6. Evaluate each of the following:

a. $-16^{\frac{3}{4}}$

b. $-32^{-\frac{4}{5}}$

c. $16^{\frac{1}{2}} + 16^{\frac{3}{4}}$

d. $27^{\frac{2}{3}} - 49^{\frac{1}{2}}$

e. $25^{0.5} - 16^{0.25}$

f. $8^{-\frac{1}{3}} + 27^{-\frac{2}{3}}$

7. Evaluate each of the following:

a. $25^{-0.5}$

b. $(27 + 9)^{\frac{1}{2}}$

c. $(-4)^3 - 8^2$

d. $9^{-1} + \left(\dfrac{8}{27}\right)^{\frac{2}{3}}$

e. $\left(\dfrac{1}{4} - \dfrac{9}{100}\right)^{-\frac{3}{2}}$

f. $\left(\dfrac{5^{\frac{1}{3}} \times 5^{\frac{1}{2}}}{5^{\frac{5}{6}}}\right)^4$

g. $\dfrac{4^{-2} - 3^{-2}}{4^{-1} - 3^{-1}}$

h. $\dfrac{2^{-3} + 4^{-1}}{3^{-2}}$

8. Evaluate each of the following:

a. $\dfrac{5^{-1}}{3^{-1}}$

b. $\left(8^{-\frac{2}{3}}\right) \div \left(16^{\frac{3}{4}}\right)$

c. $(4^{-1}) \div (3^{-2})$

d. $\dfrac{4^{\frac{1}{2}} - 8^{\frac{2}{3}}}{16^{\frac{1}{4}} \times 27^{\frac{1}{3}}}$

e. $\dfrac{64^{\frac{5}{6}} \times 16^{-\frac{3}{4}}}{9^{\frac{3}{2}}}$

9. Simplify each of the following:

a. $\dfrac{a^{\frac{1}{2}} \cdot a^{\frac{1}{3}}}{a^{\frac{1}{6}}}$

b. $\dfrac{m^{\frac{1}{4}} \cdot m^{\frac{2}{5}}}{m^{\frac{4}{5}}}$

c. $\left(\dfrac{\sqrt[5]{p}}{\sqrt[3]{p^2}}\right)^4$

d. $\sqrt{\dfrac{(25x)^{\frac{1}{2}}(8x^2)^{\frac{1}{3}}}{40x^{\frac{1}{6}}}}$

e. $\sqrt[3]{\dfrac{\sqrt{x}\sqrt{x^5}}{x^{\frac{3}{4}}}}$

f. $\left(\sqrt[5]{\dfrac{x^{\frac{1}{2}}\sqrt{x^3}}{\sqrt{x}}}\right)^2$

Photo: Dick Hemingway

Emily contracted with her grandfather to cut his lawn for 12 weeks in the summer. Her grandfather offered two methods of payment. The first choice was that he would pay her $10 per week. The second choice was that he would pay 3¢ the first week, 9¢ the second week, 27¢ the third week, and so on, each week's pay being 3 times that of the previous week. Which method of payment would you choose?

Not wanting to look like a fool, Emily had no intention of cutting the lawn for 3¢, so she opted to be paid $10 per week. Her grandfather chuckled at her choice of payment and said that to bind her to the agreement, he would give her a bonus of $10 if she took the first choice and 1¢ if she took the second. She cheerfully cut her grandfather's lawn for the 12 weeks that summer, and with the $10 for week zero safely tucked in her pocket, she earned $130 for the summer. However, her grandfather smiled every time he paid her. Finally, Emily got out a pencil and paper and figured out what his second proposal would have paid her.

Here is a table showing Emily's potential weekly wages for cutting the lawn using the second method.

Before she completed the table for all 12 weeks, she knew why her grandfather had offered her two methods of payment. Emily earned $130 that summer cutting his grass. How much would she have earned had she been smart enough to use the second method of payment?

Let's look at a graph of these values.

Week Number (x)	Wages for the Week (y)
0	$0.01
1	$0.03
2	$0.09
3	$0.27
4	$0.81
5	$2.43
6	$7.29
7	$21.87
8	$65.61
⋮	⋮

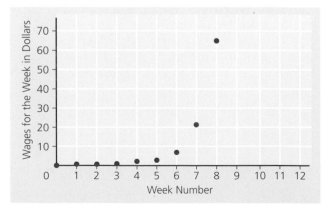

Here is our table of values again. We have added a column showing the finite differences. Recall that finite differences are differences between consecutive y-values determined from evenly spaced x-coordinates. (These even spaces can be any size; here we use differences of 1.) If the first finite differences are constant, the data will produce the graph of a line, and if the first differences are not constant but the second differences are, the data will produce the graph of a parabola.

Week Number (x)	Wages for Week (in ¢) (y)	First Difference	Second Difference
0	1	3 − 1 = 2	6 − 2 = 4
1	3	9 − 3 = 6	18 − 6 = 12
2	9	27 − 9 = 18	54 − 18 = 36
3	27	54	108
4	81	162	324
5	243	486	
6	729		

We do not have a constant first difference or a constant second difference. The data do not fit anything we have met so far. Let's examine the pattern in the sequence of first differences.

Week Number (x)	Wages for Week (in ¢) (y)	First Difference
0	1	2
1	3	6
2	9	18
3	27	54
4	81	162
5	243	486
6	729	

Each first difference in the sequence is 3 times the previous one. We also notice that all the entries in the wages column are powers of 3; that is, the weekly wages are 3^0, 3^1, 3^2, 3^3, ..., 3^{12}.

Generalizing, the wage for week x in cents is 3^x.

When we write $f(x) = 3^x$, we call this an **exponential function.** Other examples of exponential functions are $g(x) = 2^x$, $h(x) = 5^x$, and in general $f(x) = b^x$, where $b > 0$. Note if $b = 1$, $f(x) = 1$, a constant function.

PART A

1. **a.** Using a graphing calculator or graphing software, draw the graphs of $y = 2^x$, $y = 5^x$, and $y = 10^x$. Sketch the graphs in your notebook.

   b. Where do you think that the graphs of $y = 3^x$ and $y = 7^x$ would lie relative to the graphs in part **a**?

2. **a.** With a graphing calculator or graphing software, draw the graphs of $y = \left(\frac{1}{2}\right)^x$, $y = \left(\frac{1}{5}\right)^x$, and $y = \left(\frac{1}{10}\right)^x$. Sketch the graphs in your notebook.

   b. Where do you think that the graphs of $y = \left(\frac{1}{4}\right)^x$ and $y = \left(\frac{1}{12}\right)^x$ would lie relative to the graphs in part **a**?

3. Write a description of the graphs of the exponential function $y = b^x$, $b > 0$. Make sure you describe what happens when $b > 1$ and when $b < 1$. Include sketches to illustrate your description.

PART B

4. In this question we want to examine the first differences for the function $y = 5^x$.

   a. Our first step is to find the sequence of first differences. You may use your calculator for this. In your notebook, write the sequence of first differences.

x	y	First Difference
0	1	
1	5	
2	25	
3	125	
4	625	
5	3125	

   b. By what number is each first difference in the sequence multiplied to get the next term in the sequence?

   c. How does this number relate to the base of the exponential function?

5. **a.** Find the first differences for $y = \left(\frac{1}{3}\right)^x$, $-3 \le x \le 3$, $x \in I$.

   b. By what number is each first difference in the sequence multiplied to get the next term in the sequence?

   c. How does this number relate to the base of the exponential function?

6. Repeat the three steps in Question 4 with $y = 7^x$, $y = \left(\frac{1}{4}\right)^x$, $y = 6^x$, and $y = \left(\frac{1}{5}\right)^x$.

7. a. For the exponential function $y = 2^x$, determine the values of the function for $0 \le x \le 7$, $x \in I$.

   b. Determine the sequence of first differences.

   c. Determine the sequence of second differences.

   d. Predict the sequence of third differences.

8. Recall that if we draw the graph of $y = f(x)$ and $y = f(-x)$, the one is the reflection in the y-axis of the other. Refer back to the graphs of $y = 2^x$ and $y = \left(\frac{1}{2}\right)^x$, and state an optional way of writing each of the functions in Question 2.

PART C

9. For the function $f(x) = 5^x$, determine $f(n)$, $f(n + 1)$, and the first differences $f(n + 1) - f(n)$.

10. For the function $g(x) = b^x$, determine $g(n)$, $g(n + 1)$, and the first differences $g(n + 1) - g(n)$.

11. Graph $y = (-2)^x$ on your calculator using the following **domain** and **range.**

   a. Domain $-10 \le x \le 10$ and range $-10 \le y \le 10$.

   b. Domain $-4.7 \le x \le 4.7$ and range $-3.1 \le y \le 3.1$.

   c. Why in the definition for the exponential function is $b > 0$?

12. a. Use your graphing calculator to draw the graphs of $f(x) = x^2 + 1$ and $g(x) = 2^x$ in the domain $0 \le x \le 5$.

   b. Determine the coordinates of their intersection points.

   c. Determine the domain when

      i) $f(x) > g(x)$ and

      ii) $f(x) < g(x)$.

   d. Sketch these graphs in your notebook clearly marking their intersection points.

4.5 Investigating the Exponential Function $y = b^x$

In this section, you will be investigating the exponential function $f(x) = b^x$. Since you will be drawing several curves on each grid, remember to label each curve with its equation.

EXAMPLE

a. Using a graphing calculator or graphing software, draw the graph of $y = 2^x$.

b. What is the value of the y-intercept?

c. Starting with the trace function on a graphing calculator at $x = -10$, examine the y-values as the cursor moves to the right to $x = 10$.

d. What happens to the y-values for values of $x > 10$? Use the trace function to verify your answer.

e. What happens to the y-values for values of $x < -10$? Use the trace function to verify your answer. Will the function ever be exactly 0?

f. What is the domain and range of $y = 2^x$?

SOLUTION

a.

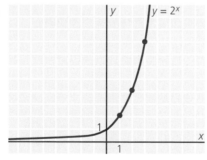

b. To find the y-intercept let $x = 0$.
$$y = 2^0$$
$$= 1$$
The y-intercept is 1.

c. As the x-values increase from -10 to 10, the y-values also increase.

d. As the x-values get larger than 10, the y-values get larger and larger.

e. As the x-values get smaller than -10, the y-values get smaller and smaller but will never reach 0.

f. The domain is $x \in R$ and the range is $y > 0$, $y \in R$.

Note that, no matter how small we make x, y is never 0. The graph of $y = 2^x$ gets closer and closer to the x-axis but never touches it. When a graph approaches a line but never touches it, this line is called an **asymptote.** For the graph of $y = 2^x$, the x-axis or $y = 0$ is an asymptote.

1. Using your graphing calculator or graphing software, investigate the function $y = 3^x$ as we did for $y = 2^x$. Follow the same six steps as in the example.

2. Use your graphing calculator or graphing software to graph $y = 1.2^x$, $y = (1.8)^x$, and $y = 2.5^x$ on the same set of axes. Sketch, labelling the functions carefully.

   a. What do these curves have in common?

   b. How do these curves differ?

3. Write a general description of the graph of the exponential curve $y = b^x$ where $b > 1$.

   a. State the y-intercept and justify your answer.

   b. State the domain and range and justify your answer.

   c. State the equation of the asymptote and justify your answer.

   d. Draw a sketch to illustrate parts **a** and **c**.

4. Using your graphing calculator or graphing software, investigate the function $y = \left(\frac{1}{2}\right)^x$ as we did for $y = 2^x$. Follow the same six steps as in the example.

5. Use your graphing calculator or graphing software to graph $y = \left(\frac{1}{2}\right)^x$, $y = \left(\frac{1}{3}\right)^x$, and $y = \left(\frac{1}{5}\right)^x$ on the same set of axes. Sketch, labelling the functions carefully.

   a. What do these curves have in common?

   b. How do these curves differ?

6. Write a general description of the graph of the exponential curve $y = b^x$, where $0 < b < 1$.

   a. State the y-intercept and justify your answer.

   b. State the domain and range and justify your answer.

   c. State the equation of the asymptote and justify your answer.

   d. Draw a sketch to illustrate parts **a** to **c**.

7. Graph $y = 2^x$ and $y = \left(\frac{1}{2}\right)^x$. Sketch, labelling the functions carefully.

   a. How are these graphs related?

   b. How are they alike?

   c. How are they different?

8. Graph $y = 2^x$ and $y = -2^x$.

    a. How are the graphs the same?

    b. How are they different?

    c. What transformation changes $y = 2^x$ to $y = -2^x$?

From our work in this section, we can summarize the properties of the exponential function.

Properties of the Exponential Function $f(x) = b^x$, $b > 0$

- The base b is positive.
- $f(0) = 1$. (The y-intercept of the graph is 1.)
- The x-axis is a horizontal asymptote.
- The domain is the set of real numbers R.
- The range is the set of positive real numbers R.
- The exponential function is increasing if $b > 1$.
- The exponential function is decreasing if $0 < b < 1$.

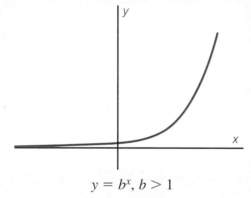

$y = b^x, b > 1$

The function is increasing
as x increases.

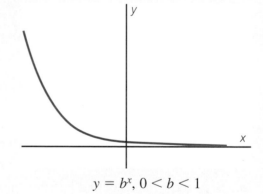

$y = b^x, 0 < b < 1$

The function is decreasing
as x increases.

1. The graphs below represent functions $f(x) = b^x$, $b > 0$.

i) 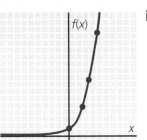 ii) 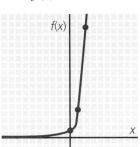 iii)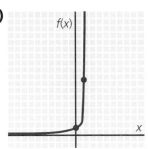

Answer the following questions about each of the functions represented by these graphs. For all graphs, one square represents one unit.

a. What is the value of $f(0)$?

b. What is the value of $f(1)$?

c. What is the value of b?

d. State the equation of the function.

e. What are the values of $f(2), f(3), f(-1)$, and $f(-3)$?

2. The graphs below represent functions $f(x) = b^x$, $0 < b < 1$.

i) 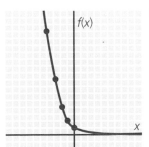 ii) 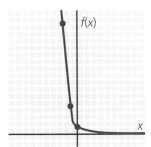 iii)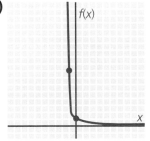

Answer the following questions about each of the functions represented by these graphs. For all graphs, one square represents one unit.

a. What is the value of $f(0)$?

b. What is the value of $f(-1)$?

c. What is the value of b?

d. State the equation of the function.

e. What are the values of $f(2), f(3), f(1)$, and $f(-3)$?

3. Describe how you determine the equation of a function of the form $y = b^x$, $b > 0$, if you are given its graph.

A mathematical model of the size of a bacteria population y, at time x, is given by the function $y = 1000(2^x)$, where time x is measured in hours.

At time $x = 10$ hours, the population would be

$$y = 1000(2^{10})$$
$$= 1000(1024)$$
$$= 1\ 024\ 000.$$

The bacteria population is approximately 1 000 000.

If we wanted to find the time at which there would be 8000 bacteria, we would substitute $y = 8000$ and solve the exponential equation $8000 = 1000(2^x)$.

$$8000 = 1000(2^x)$$
$$8 = 2^x$$

But $2^3 = 8$, therefore $x = 3$.

The population would be 8000 in three hours.

EXAMPLE 1

Let's start with an easy exponential equation.

Solve $5^{3x-2} = 25$.

SOLUTION

$$5^{3x-2} = 25$$
$$5^{3x-2} = 5^2$$

The bases are equal. Therefore, the exponents must be equal.

$$3x - 2 = 2$$
$$3x = 4$$
$$x = \frac{4}{3}.$$

Verification

Substitute $x = \frac{4}{3}$ into both sides of the equation.

L.S. $= 5^{3x-2}$ R.S. $= 25$

$= 5^{3\left(\frac{4}{3}\right)-2}$

$= 5^{4-2}$

$= 5^2$

$= 25$

Since L.S. $=$ R.S., then $x = \frac{4}{3}$ is the correct root.

EXAMPLE 2

Solve $5^{x^2+2x} = 125$.

SOLUTION

$5^{x^2+2x} = 125$

Write the right side of the equation as a power of 5.

$$5^{x^2+2x} = 5^3$$
$$x^2 + 2x = 3$$
$$x^2 + 2x - 3 = 0$$
$$(x + 3)(x - 1) = 0$$
$$x + 3 = 0 \quad \text{or} \quad x - 1 = 0$$
$$x = -3 \qquad\qquad x = 1$$

If we check these values, we can see that they both verify.

If the bases are not the same, we write both sides of the equation with the same base whenever possible.

EXAMPLE 3

Determine the root of $9^{x+1} = 27^{3x-4}$.

SOLUTION

$$9^{x+1} = 27^{3x-4}$$
$$(3^2)^{x+1} = (3^3)^{3x-4}$$
$$3^{2x+2} = 3^{9x-12}$$
$$2x + 2 = 9x - 12$$
$$2x - 9x = -12 - 2$$
$$-7x = -14$$
$$x = 2$$

The root is 2.

EXAMPLE 4

Solve $5\left(\frac{1}{8}\right)^x = 20$.

$5\left(\frac{1}{8}\right)^x = 20$

Divide both sides of the equation by 5.

$$\left(\frac{1}{8}\right)^x = 4$$

Write the equation with the same bases.

$$(2^{-3})^x = 2^2$$
$$2^{-3x} = 2^2$$
$$-3x = 2$$
$$x = -\frac{2}{3}$$

EXAMPLE 5

Solve $2^{2x} - 6(2^x) + 8 = 0$.

$2^{2x} - 6(2^x) + 8 = 0$

We can solve this equation in two different ways.

Method 1

Recalling that the term 2^{2x} can be written as $(2^x)^2$, our equation becomes
$(2^x)^2 - 6(2^x) + 8 = 0$.

$$(2^x)^2 - 6(2^x) + 8 = 0$$
$$(2^x - 4)(2^x - 2) = 0$$
$$2^x = 4 \quad \text{or} \quad 2^x = 2$$
$$2^x = 2^2 \qquad\quad 2^x = 2^1$$
$$x = 2 \qquad\qquad x = 1$$

Method 2

Let $2^x = a$, from which $2^{2x} = a^2$, to get the **quadratic equation** $a^2 - 6a + 8 = 0$.

$$a^2 - 6a + 8 = 0$$
$$(a - 4)(a - 2) = 0$$
$$a = 4 \quad \text{or} \quad a = 2$$
$$2^x = 4 \quad \text{or} \quad 2^x = 2$$
$$x = 2 \qquad\quad x = 1$$

EXAMPLE 6

Use your graphing calculator to solve $5^x = 400$.

SOLUTION

We cannot write 400 as an integral power of 5. However, we can use our graphing calculator to draw the graph of $y = 5^x$, and then use the trace feature of the calculator to approximate the answer to $5^x = 400$.

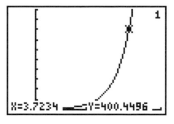

The solution to $5^x = 400$ is approximately $x \doteq 3.72$.

The accuracy of your answer can be improved by using the zoom feature of the calculator.

Alternatively, you can guess and check using the power key on the calculator.

$$5^x = 400$$
$$5^{3.6} = 328$$
$$5^{3.8} = 452$$

The value of x must be between 3.6 and 3.8. Continue to experiment with values to improve the accuracy of your approximation.

Exercise 4.6

PART A

1. Solve each of the following:

    a. $2^x = 64$

    b. $5^{2x+4} = 5^{5x-7}$

    c. $5^{2x+6} = 125$

    d. $5^x = \dfrac{1}{25}$

    e. $4^{x^2+2} = 4^{3x}$

    f. $3^{2x+6} = 9^{1-x}$

    g. $4^{2x+3} = \left(\dfrac{1}{8}\right)^x$

    h. $2^{x^2+5x} = \dfrac{1}{64}$

PART B

2. Solve each of the following:

    a. $3(2^x) = 48$

    b. $5(3^{2x}) = \dfrac{5}{27}$

    c. $4(7^{2x-1}) = 28$

    d. $3(5^{x^2+3x}) = \dfrac{12}{100}$

3. Find the root of each of the following:

a. $\left(\frac{1}{4}\right)^{x+3} = \left(\frac{1}{8}\right)^{x-1}$

b. $\left(\frac{1}{27}\right)^{x+3} = \left(\frac{1}{3}\right)^{x^2-5x}$

c. $\left(\frac{1}{4}\right)^{x^2+2} = 8^{x-2}$

d. $25^{x-1} = \frac{1}{125}$

4. Solve each of the following:

a. $4^{2x} - 20(4^x) + 64 = 0$

b. $2^{2x} + 16 = 10(2^x)$

c. $3^{2x} - 12(3^x) + 27 = 0$

d. $5^{2x} = 30(5^x) - 125$

e. $2^{2x} - 12(2^x) + 32 = 0$

f. $3(3^{2x}) - 10(3^x) + 3 = 0$

g. $3^{2x} - 6(3^x) - 27 = 0$

h. $5^{2x} - 4(5^x) - 5 = 0$

5. Find the root(s) of each of the following:

a. $(3^{x-3})^x = \frac{1}{9}$

b. $2^{x^2+5x} = 64$

c. $3^{3x+1} = 27(9^x)$

d. $2^{2x+4} - 5 = 59$

e. $(2^{x+2})(4^{x-1})(8^{2x-3}) = 256^x$

f. $5^{x+1} + 50 = 175$

g. $3^{2x+4} \cdot 27^{4-x} = \left(\frac{1}{81}\right)^{x-2}$

h. $5^{2x^2+2} = 125^x(25^x)^x$

6. Give an approximate value for x in each of the following:

a. $5^x = 875$

b. $2^x = 923$

c. $3^x = 757$

d. $2^x = 188$

e. $(4.3)^x = 115$

f. $(7.2)^x = 1100$

PART C

7. Solve each of the following:

a. $3^{2x+1} - 10(3^x) + 3 = 0$

b. $2^{2x} - 2^{x+1} - 8 = 0$

c. $2^{2x} - 3 \cdot 2^{x+2} + 32 = 0$

d. $5^{2x+1} - 2 \cdot 5^{x+1} + 5 = 0$

e. $(27 \cdot 3^x)^x = 27^x \cdot 3^{\frac{1}{x}}$

8. Use your graphing calculator to estimate the roots of the following:

a. $\begin{cases} y = 2^x \\ y = x + 4 \end{cases}$

b. $x + 3 = 3^x$

c. $2x + 4 = 5^x$

9. Solve each of the following systems of equations.

a. $\begin{cases} 2^{2x+y} = 32 \\ 2^{x-3y} = \frac{1}{2} \end{cases}$

b. $\begin{cases} 9^{x+2y} = \frac{1}{9} \\ 3^{2x+y} = 81 \end{cases}$

4.7 Logarithms

The inverse of a function is the relation obtained by interchanging the independent and dependent variables. The inverse of the exponential function $y = a^x$ is $x = a^y$. Here, if given the base a, we are to determine the exponent y that, applied to the base, will give us a value for any given x.

EXAMPLE 1

If $x = 3^y$, determine y if $x = 81$.

SOLUTION

Since $3^4 = 81$, the value of y resulting from $x = 81$ is 4.

Is the relation $x = a^y$ a function? We answer this question by considering the graph of $x = a^y$ if given the graph of $y = a^x$. From our previous work, we know that the graph of the inverse of a function is obtained by reflecting the graph of $y = a^x$ in the line $y = x$.

It is clear from the graph that for a given value of x, there is exactly one value for y, so this expression is a function. It is also clear that since $y = a^x$ is defined only for $a > 0$, $x = a^y$ is likewise defined only for $a > 0$, and therefore the domain of $x = a^y$ is $x > 0$, since $a^y > 0$ for $a > 0$.

To write this function in the form $y = f(x)$, we define a new term.

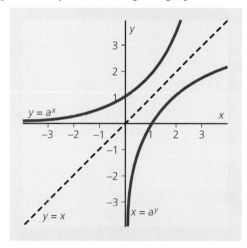

> The **logarithm** of a number to a given base is the **exponent** that must be used with that base to obtain the given number.

The logarithm of 64 to base 2 is 6 since $2^6 = 64$. For notation we write $\log_2 64 = 6$, and in general we write $y = \log_a x$, where a is the given base.

EXAMPLE 2

Determine each of the following:

a. $\log_{10} 1000$ **b.** $\log_5 625$ **c.** $\log_2 1024$ **d.** $\log_6 6$

SOLUTION

a. $\log_{10} 1000 = 3$ because $10^3 = 1000$ **b.** $\log_5 625 = 4$

c. $\log_2 1024 = 10$ **d.** $\log_6 6 = 1$

For convenience in simple numeric calculations, we generally use base 10, and most calculators have a **logarithmic function,** using this base, accessed by the *LOG* key. When the base is omitted in using logarithmic notation, it is understood to be base 10. Now we write, for example, log 100 = log 10^2 = 2.

Two special cases of logarithmic calculation should be noted.

$$\log_a a = 1, \text{ because } a^1 = a, a > 0.$$
$$\log_a 1 = 0, \text{ because } a^0 = 1, a > 0.$$

For calculations we note that $a = 10^{\log a}$ from the definition. Then

$$a^x = (10^{\log a})^x$$
$$= 10^{x \log a}$$

Therefore, log a^x = log($10^{x \log a}$)

$$= x \log a, \text{ because } \log 10^p = p.$$

$$\log a^x = x \log a$$

EXAMPLE 3

Determine the value of x in each of the following:

a. $12^x = 400$ **b.** $(1.08)^x = 4.39$

SOLUTION

a. $12^x = 400$

To solve this equation, we first take the logarithm of both sides.

$$\log 12^x = \log 400$$

Since log a^x = x log a,

$$x \log 12 = \log 400$$
$$x = \frac{\log 400}{\log 12}$$
$$x \doteq 2.41.$$

(Check on your calculator that $12^{2.41}$ gives approximately 400.)

b. $(1.08)^x = 4.39$

$$\log (1.08)^x = \log 4.39$$
$$x \log 1.08 = \log 4.39$$
$$x \log 1.08 = \log 4.39$$
$$x = \frac{\log 4.39}{\log 1.08}$$
$$x \doteq 19.22$$

PART A

1. Change the following to logarithmic notation, $y = \log_a x$.

   a. $3^5 = 243$ **b.** $2^7 = 128$

   c. $7^3 = 343$ **d.** $11^4 = 14\ 641$

2. Change the following to exponential notation, $x = a^y$.

   a. $\log_4 1024 = 5$ **b.** $\log_7 7 = 1$

   c. $\log_6 7776 = 5$ **d.** $\log_a p = r$

3. Evaluate each of the following:

   a. $\log_{10} 100$ **b.** $\log_6 216$

   c. $\log_7 1$ **d.** $\log_4 1024$

4. Use your calculator to determine each of the following:

   a. $\log 327$ **b.** $\log (6^3)$

   c. $\log (0.001)$ **d.** $\log (0.039)$

PART B

5. Evaluate each of the following:

   a. 1.03^{15} **b.** 1.02^{20}

   c. 1.06^{12} **d.** 1.03^{24}

6. Solve for x in each of the following:

   a. $3^x = 32$ **b.** $5^x = 66$

   c. $10^x = 157$ **d.** $1.03^x = 1.69$

   e. $4.5^x = 600$ **f.** $1.02^x = 3.65$

   g. $1.05^x = 2.60$ **h.** $1.045^x = 2.87$

   i. $\log 1.07^x = 2$

PART A

1. Simplify each of the following:

   a. $x^5 \bullet x^3$

   b. $(-2ab)^2(-2a)^3$

   c. $\dfrac{27a^2b^3}{3ab}$

   d. $(k^3)^2 \div (k^2)^3$

   e. $(3b^3)^3$

   f. $\dfrac{(2a^2b)^3}{4a^4b^2}$

2. Evaluate each of the following:

   a. 3^{-2}

   b. $\left(\dfrac{1}{2}\right)^{-3}$

   c. $\left(\dfrac{3}{4}\right)^2$

   d. $\dfrac{4}{3^{-2}}$

   e. $2^2 + 3^{-1} + 4^0$

   f. $4(2)^{-2} \times 9(3)^{-3}$

3. Write each of the following in exponential form.

   a. $\sqrt[3]{22}$

   b. $\sqrt[5]{2^4}$

   c. $\sqrt{12}$

4. Write each of the following in radical form.

   a. $3^{\frac{1}{2}}$

   b. $17^{\frac{2}{3}}$

   c. $(-4)^{\frac{1}{3}}$

PART B

5. Evalute each of the following:

   a. $(2^{-1} + 3^{-1})^{-1}$

   b. $\dfrac{2^{-3}}{2^{-1} + 2^{-2}}$

   c. $\dfrac{4^{-8}}{4^{-3} \times 4^{-7}}$

   d. $(3^2 - 3)(3^{-2} - 3^{-1})$

6. Evaluate each of the following:

   a. $\dfrac{4^5}{2^6}$

   b. $\dfrac{27^3 \times 9^{-2}}{81}$

7. Evaluate each of the following:

   a. $16^{-\frac{3}{4}}$

   b. $\sqrt[3]{\dfrac{1}{27}}$

   c. $\left(\dfrac{8}{50}\right)^{\frac{1}{2}}$

   d. $4^{-1} + \left(\dfrac{3}{2}\right)^{-2}$

8. Solve each of the following:

   a. $5^x = 125$

   b. $4^x = \dfrac{1}{8}$

   c. $9^{x^2+x} = 81$

   d. $2^{2x} - 5(2^x) + 4 = 0$

9. Solve for x.

   a. $3^x = 579$

   b. $(1.025)^x = 1.12$

   c. $(1.03)^x = 2$

   d. $4^x = 100$

PART C

10. Simplify the following:

   a. $x^{2a+3b} \div x^{a-b}$

   b. $(y^{a+b})^{a-b}$

   c. $(25^{m+3n})(125^{2m+1})^{-1}$

   d. $\sqrt{\dfrac{x^{\frac{1}{3}} \cdot \sqrt[3]{x^2}}{\sqrt{x}}}$

1. Given $f(x) = 4^x$ and $f(x + 2) - f(x + 1) = k\,f(x)$, determine the value of k.

2. A bag contains 60 jelly beans, 15 of which are red, 15 are green, 15 are yellow, and 15 are black. What is the least number that a blindfolded person must eat to be certain of having eaten at least one of each colour?

3. You are given 15 discs in a row. Each is to be coloured either blue or yellow under the following rules:

   ◆ The seventh disc is to be blue.

   ◆ A disc is to be coloured blue if there are either three or six discs between it and a blue disc, regardless of the colour of in-between discs.

   ◆ All other discs are to be coloured yellow.

   How many discs are yellow?

4. Solve the equation $x + \sqrt{x + 12} = 8$.

5. Determine the minimum value of the function $f(x) = 5^{x^2 - 2x}$.

6. Part of a pasture is to be fenced into a rectangular enclosure. The interior is to be divided into six areas by fencing as shown, where all interior fences are parallel to one of the outer fences. Determine the maximum total area enclosed if 3600 m of fencing are available.

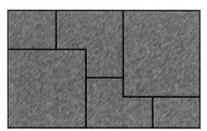

7. If n is chosen from the set of positive integers $\{1, 2, 3, \ldots, 24, 25\}$, for which values is $n^2 + 2n + 4$ divisible by 7? Can you generalize for n any integer?

Chapter 5

Sequence and Series

What is the difference between sequences and series? Although the words *sequences* and *series* are frequently interchanged in conversation, their meanings in mathematics are entirely different. Sequences are sets of numbers that follow a pattern; a series is the sum of terms in a sequence. Expressing a sequence as a formula helps communicate information about a pattern and generates numbers in the pattern. For money invested with simple interest, the value of the investment increases according to an arithmetic sequence where you can see a common difference or pattern between numbers. In this chapter, you will learn sequence and series formulas that are often used in finance.

IN THIS CHAPTER, YOU CAN . . .

- identify sequences as arithmetic, geometric, or neither;
- determine the value of any term in an arithmetic or a geometric sequence;
- determine the sum of the terms of an arithmetic or a geometric sequence;
- write the terms of a sequence, given the formula for the nth term or given a recursion formula.

A **function** generates numbers. Given the function $f(x) = 2x$, we determine $f(1) = 2$, $f(3) = 6, f(\sqrt{3}) = 2\sqrt{3}$, and so on. If the domain of the function is restricted to the natural numbers, patterns frequently result. We use n rather than x as an easy reminder that we wish to limit the domain to the natural numbers.

The function $f(n) = 2n$, $n \in N$, generates the set of values $f(1) = 2, f(2) = 4, f(3) = 6$, and so on, or the set of numbers 2, 4, 6, 8, ...

We call such a set of numbers a **sequence**. Each element in a sequence is a **term**. For convenience we refer to the terms as t_1, t_2, t_3, and so on. The subscript gives the position of the term in the sequence; t_7 is the seventh term, and t_n represents the nth, or general term.

A sequence can be finite, meaning it has a specific number of terms, or it can be infinite, meaning the terms continue without end. The sequence 1, 4, 7, 10, . . . , 31 is finite and has eleven terms. The sequence 1, 4, 9, 16, . . . is infinite and has an unlimited number of terms.

The function $f(n) = n^2 + 3n$, $n \in N$, generates the set of numbers 4, 10, 18, 28, 40, . . . when we replace n by 1, 2, 3, 4, 5, ... Examine this set of function values. Can you describe the pattern? It is not always easy to see the pattern in a set of function values, but it is possible, given a set of numbers forming a pattern, to determine the function that generates the set.

If we know the function or are given an expression for the general term, we can generate terms in the sequence.

EXAMPLE 1

Determine the first 3 terms of the following sequences.

a. $f(n) = 3n - 1$ **b.** $t_n = 2^n$ **c.** $t_n = n^2 + 4$ **d.** $t_n = \frac{1}{2}n^3 - 3$

SOLUTION

a. The function is $f(n) = 3n - 1$.

$$t_1 = f(1)$$
$$= 2$$
$$t_2 = f(2)$$
$$= 5$$
$$t_3 = f(3)$$
$$= 8$$

b. The general term is $t_n = 2^n$.

$$t_1 = 2$$
$$t_2 = 2^2$$
$$= 4$$
$$t_3 = 2^3$$
$$= 8$$

c. The general term is $t_n = n^2 + 4$.

$t_1 = 1^2 + 4$

$\quad = 5$

$t_2 = 2^2 + 4$

$\quad = 8$

$t_3 = 3^2 + 4$

$\quad = 13$

d. The general term is $t_n = \frac{1}{2}n^3 - 3$.

$t_1 = \frac{1}{2}(1^3) - 3$

$\quad = \frac{1}{2} - 3$

$\quad = -\frac{5}{2}$

$t_2 = \frac{1}{2}(2^3) - 3$

$\quad = \frac{8}{2} - 3$

$\quad = 1$

$t_3 = \frac{1}{2}(3^3) - 3$

$\quad = \frac{27}{2} - 3$

$\quad = \frac{21}{2}$

A sequence is defined **recursively** if the first term is given and each term thereafter is given by an expression involving the previous term or terms.

EXAMPLE 2

For the sequence $t_1 = 3$, $t_k = t_{k-1} + k^2$, $k \in N$, $k > 1$, find t_2, t_3, t_4.

SOLUTION

$t_1 = 3$, $t_k = t_{k-1} + k^2$, $k \in N$, $k > 1$

For $k = 2$,

$t_2 = t_1 + 2^2$

$\quad = 3 + 4$

$\quad = 7$.

For $k = 3$,

$t_3 = t_2 + 3^2$

$\quad = 7 + 9$

$\quad = 16$.

For $k = 4$,

$t_4 = t_3 + 4^2$

$\quad = 16 + 16$

$\quad = 32$.

If we have a given sequence, can we determine the function from which it came? Sometimes it is easy and sometimes it is very difficult.

Given the sequence 1, 3, 5, 7, 9, 11, . . . , you can see that it is likely the sequence is generated by the function $f(n) = 2n - 1$, $n \in N$. Given the sequence 2, 10, 30, 68, 130, . . . , it is unlikely that one could determine that $f(n) = n^3 + n$ is the defining function.

We cannot always be sure of the next term in a sequence. For example, the sequence 1, 2, 4, . . . could be generated by the general term $t_n = 2^{n-1}$ or by the recursive definition $t_n = t_{n-1} + (n - 1)$, $t_1 = 1$.

Using the general term $t_n = 2^{n-1}$, the sequence becomes 1, 2, 4, 8, 16, ...

Using the recursive definition, the sequence becomes 1, 2, 4, 7, 11, ...

In the sequence 1, 2, 3, 4, . . . , we are fairly confident that the next term is 5 because we recognize the pattern.

Similarly in the sequence 5, 10, 15, 20, . . . , we can recognize a pattern suggesting that the next term could be 25.

Recognizing patterns enables us to write a function or a general term of a sequence.

EXAMPLE 3

Write a general term t_n for each of the following sequences.

a. 1, 2, 3, 4, . . .

b. 5, 10, 15, 20, . . .

c. 1, 4, 9, 16, . . .

d. $\frac{1}{2}, \frac{2}{3}, \frac{3}{4}, \frac{4}{5}, \ldots$

e. 3, 5, 7, 9, . . .

SOLUTION

a. 1, 2, 3, 4, . . .

$t_1 = 1$

$t_2 = 2$

$t_3 = 3$

Term Number   Value of Term

Notice how the subscript for the term number compares with the value of the term.

The term number is the same as the value of the term.

Thus an nth, or general term, is $t_n = n$.

However, a general term of this sequence could also be

$$t_n = (n - 1)(n - 2)(n - 3)(n - 4) + n,$$

which would give us a different value for t_5, namely 29.

b. 5, 10, 15, 20, . . .

$t_1 = 5$
$t_2 = 10$
$t_3 = 15$

Compare the subscript for the term number with the value of the term.

What operation carried out on 3 produces 15? We would probably guess that $3 \times 5 = 15$. Check whether this operation works for the other terms.

We can multiply a term number by 5 to get the value of the term.

So a general term could be $t_n = 5n$.

c. 1, 4, 9, 16, . . .

$t_1 = 1$

$t_2 = 4$

$t_3 = 9$

$t_4 = 16$

Compare the subscript for the term number with the value of the term.

What operation carried out on 4 produces 16?

You might guess, multiply by 4. But $t_3 = 9$ and $3 \times 4 \neq 9$. So multiplying by 4 does not work.

Guessing again, we consider $4^2 = 16$.

When you check, you can see that squaring the term numbers produces the value of the terms.

So a general term could be $t_n = n^2$.

d. $\dfrac{1}{2}, \dfrac{2}{3}, \dfrac{3}{4}, \dfrac{4}{5}, \ldots$

$t_1 = \dfrac{1}{2}$

$t_2 = \dfrac{2}{3}$

$t_3 = \dfrac{3}{4}$

Compare the term number with the value of the term.

Look for a pattern.

A general term could be $t_n = \dfrac{n}{n + 1}$.

e. 3, 5, 7, 9, . . .

$t_1 = 3$

$t_2 = 5$

$t_3 = 7$

$t_4 = 9$

You can easily recognize the pattern and find the next term in the sequence to be 11. Finding the general term may be a little more difficult.

Compare the term number with the value of the term.

$t_4 = 9$

What operation carried out on 4 produces 9? You might guess, add 5.

But $t_3 = 7$ and $3 + 5 \neq 7$.

Guess again. What can you do to 4 to get 9?

$4 \times 2 + 1 = 9$

Try this with the other terms: $t_3 = 7$ and $3 \times 2 + 1 = 7$; $t_2 = 5$ and $2 \times 2 + 1 = 5$.

A general term could be $t_n = 2n + 1$.

PART A

1. Look at the pattern. Then write the next two terms in each of the following sequences.

   a. 3, 9, 27, . . .

   b. 4, 8, 12, . . .

   c. 9, 3, 1, . . .

   d. 100, 95, 90, . . .

   e. $\frac{1}{2}, \frac{1}{4}, \frac{1}{8}, \dots$

   f. 5, 50, 500, . . .

   g. $-2, -4, -6, \dots$

PART B

2. Determine the first 4 terms for each of the following sequences.

   a. $f(n) = 3n$

   b. $t_n = n - 2$

   c. $t_n = \dfrac{1}{n}$

   d. $f(n) = n^2 - 1$

   e. $t_n = 3 - n$

   f. $t_n = \dfrac{2n}{n + 1}$

   g. $t_n = 3n^2$

   h. $f(n) = 1 - n^3$

   i. $t_n = 2n + 3$

   j. $t_n = 3^{-n}$

3. Determine the first 4 terms for each of the following sequences.

   a. $t_n = t_{n-1} + 2, t_1 = 5$

   b. $t_n = 2t_{n-1}, t_1 = -5$

   c. $t_n = 6 - t_{n-1}, t_1 = 3$

   d. $t_n = (t_{n-1})^2, t_1 = -1$

4. Explain how you find the next term in the sequence.

   a. 5, 10, 15, . . .

   b. 2, 12, 72, . . .

   c. 3, 12, 21, . . .

   d. 5, 10, 20, . . .

5. Determine a general term for each of the following sequences.

   a. 2, 4, 6, 8, . . .

   b. 1, −1, 1, −1, . . .

   c. 2, 4, 8, 16, . . .

   d. 2, 8, 32, 128, . . .

   e. −2, −6, −18, −54, . . .

   f. $1, \frac{1}{4}, \frac{1}{9}, \frac{1}{16}, \dots$

6. Determine the first 6 terms of each of the following sequences.

a. $t_n = t_{n-1} + t_{n-2}, t_1 = 1, t_2 = 1, n \geq 3$

b. $t_n = t_{n-1} + 3t_{n-2}, t_1 = 3, t_2 = -2, n \geq 3$

These sequences are known as Fibonacci sequences. Leonardo Fibonacci (1175–1250) defined the first of these sequences when he was investigating the growth of a rabbit population.

Leonardo Fibonacci (1175–1250)

Leonardo Fibonacci, considered to be the greatest mathematician of the Middle Ages, was born in Pisa where his father was a customs manager. He travelled with his father throughout the Mediterranean. There he learned about the arithmetic systems used by merchants of different lands.

Fibonacci thought that the Hindu-Arabic decimal system (0, 1, 2, 3, . . .) with its positional notation and zero was more useful than the Roman numerals (I, II, III, IV, . . .) that were still used in Europe. His book *Liber Abaci* (Book of Counting), published in 1202, was influential in introducing Arabic numerals into Western culture.

Today Fibonacci is best known for identifying the sequence that we call the Fibonacci numbers.

$$t_1 = 1, t_2 = 1, t_k = t_{k-1} + t_{k-2}, k \in N, k > 2$$

Each number in this sequence, after the second, is the sum of the two preceding numbers. The first few Fibonacci numbers are 1, 1, 2, 3, 5, 8, 13, ... Many naturally occurring patterns in nature, such as the patterns in pine cones, sunflowers, shells, and pineapples follow the numbers in the Fibonacci sequence.

Activity: **Fibonacci Numbers**

1. Collect pine cones or sunflowers to show how they physically fit this pattern.

2. Search the Internet for other applications of the Fibonacci numbers in nature.

5.2 Arithmetic Sequences

The first 5 terms of the sequence generated by the function $f(n) = 4n - 2$ are 2, 6, 10, 14, and 18. Each term after the first one is found by adding the same number, 4, to the preceding term. For the sequence 11, 9, 7, 5, the number -2 is added to determine successive terms of the sequence. Sequences such as these are called **arithmetic sequences.** The fixed quantity added to successive terms is called the **common difference** d and is obtained by subtracting any term from the one following it in the sequence.

To get the next term in an arithmetic sequence, we add a constant.

2, 6, 10, 14, . . . is an arithmetic sequence. To get the next term add 4.

20, 17, 14, . . . is an arithmetic sequence. To get the next term add (-3).

If we represent the first term in the sequence by a, and the common difference by d, then the sequence becomes $a, a + d, a + 2d, \ldots$.

$$t_1 = a$$
$$t_2 = a + d$$
$$t_3 = a + 2d$$
$$t_4 = a + 3d$$

Observe that the coefficient of d in each term is one less than the number of the term. Therefore, the general term is $t_n = a + (n - 1)d$.

> An **arithmetic sequence** with first term a and
> **common difference** d is $a, a + d, a + 2d, \ldots$.
> The general term is $t_n = a + (n - 1)d, n \in N$.

EXAMPLE 1

In each of the following, the first 4 terms of an arithmetic sequence are given. Determine t_n and t_{20}.

a. 3, 7, 11, 15, 19

b. $4, \frac{7}{2}, 3, \frac{5}{2}$

SOLUTION

a. $t_n = a + (n - 1)d$

$a = 3, d = 7 - 3$

$\quad = 4$

$t_n = 3 + (n - 1)4$

$\quad = 3 + 4n - 4$

$\quad = 4n - 1$

b. $t_n = a + (n - 1)d$

$a = 4, d = \frac{7}{2} - 4$

$\quad = -\frac{1}{2}$

$t_n = 4 + (n - 1)\left(-\frac{1}{2}\right)$

$\quad = 4 - \frac{1}{2}n + \frac{1}{2}$

$\quad = -\frac{n}{2} + \frac{9}{2}$

$$t_{20} = 4(20) - 1 \qquad\qquad\qquad t_{20} = -\frac{20}{2} + \frac{9}{2}$$

$$= 79 \qquad\qquad\qquad\qquad\qquad = -\frac{11}{2}$$

EXAMPLE 2

Determine the number of terms in the arithmetic sequence 5, 11, 17, 23, . . . , 119.

SOLUTION

For the given sequence $a = 5$, $d = 6$.

$$t_n = 5 + (n - 1)6$$
$$= 6n - 1$$

To determine the number of terms we equate t_n and 119.

$$6n - 1 = 119$$
$$6n = 120$$
$$n = 20$$

There are 20 terms in the sequence.

EXAMPLE 3

Determine t_{20} in the arithmetic sequence where $t_5 = 27$ and $t_9 = 43$.

SOLUTION

First, we express the given information in terms of a and d.

Since $t_n = a + (n - 1)d$, then $t_5 = a + 4d$ and $t_9 = a + 8d$.

Therefore $a + 4d = 27$ and $a + 8d = 43$.

Now solve the system of linear equations.

$$a + 8d = 43 \qquad ②$$
$$a + 4d = 27 \qquad ①$$
$$4d = 16 \qquad ② - ①$$
$$d = 4$$

Substitute $d = 4$ into $a + 4d = 27$.

$$a + 16 = 27$$
$$a = 11$$

Using $a = 11$, $d = 4$, and $n = 20$ in t_n, we obtain

$$t_{20} = 11 + (20 - 1)(4)$$
$$= 11 + (19)(4)$$
$$= 87.$$

An arithmetic sequence is the function $f(n)$, where

$$f(n) = a + (n - 1)d$$
$$= a + dn - d$$
$$= dn + (a - d), \text{ where } n \in N.$$

An arithmetic sequence is therefore a linear function with the domain $n \in N$. In an arithmetic sequence the difference between consecutive terms is constant, so that the sequence constantly increases or decreases by a fixed amount. Since we are working with a constant interval of 1 in the domain, this fixed increase or decrease corresponds to the slope of the graph of the linear function.

An alternative definition of an arithmetic sequence is the recursive definition

$$t_n = t_{n-1} + d, t_1 = a.$$

Since $t_1 = a$

$$t_2 = t_1 + d \qquad \text{and} \qquad t_3 = t_2 + d$$
$$= a + d, \qquad\qquad\qquad = a + d + d$$
$$= a + 2d.$$

From this definition we obtain the same terms as from $t_n = a + (n - 1)d$.

EXAMPLE 4

Determine the first 4 terms of the arithmetic sequence defined by $t_1 = 8$, $t_k = t_{k-1} + 3$, $k \in N, k > 1$.

SOLUTION

For the given arithmetic sequence, $t_1 = 8$.

$$t_2 = t_1 + 3 \qquad t_3 = t_2 + 3 \qquad t_4 = t_3 + 3$$
$$= 8 + 3 \qquad\quad = 11 + 3 \qquad\quad = 14 + 3$$
$$= 11 \qquad\qquad = 14 \qquad\qquad = 17$$

The first 4 terms are 8, 11, 14, and 17.

The arithmetic sequence has application in financial calculations.

If you put $1 in the bank for a year and the bank pays 5% interest per annum, your $1 earns 5¢ interest so your $1 grows to $1.05.

EXAMPLE 5

On November 1, 2001, Carol invests $100. The investment earns 6% per annum **simple interest,** paid November 1 of each year. The value of the investment on November 1 of four successive years is $100, $106, $112, $118.

Determine the value of the investment on November 1, 2010.

SOLUTION

This is an arithmetic sequence with $a = 100$, $d = 6$. We are looking for t_{10}.

$$t_{10} = a + (n - 1)d$$
$$= 100 + (10 - 1)(6)$$
$$= 100 + (9)(6)$$
$$= 154$$

The investment is worth \$154 on November 1, 2010.

Exercise 5.2

PART A

1. State which of the following could be arithmetic sequences. Give the value of d in each case.

a. $1, 3, 5, 7, \ldots$ **b.** $1, 3, 9, 27, \ldots$ **c.** $5, 10, 15, 20, \ldots$

d. $-1, -4, -7, -10, \ldots$ **e.** $2, 1, \frac{1}{2}, \frac{1}{4}, \ldots$ **f.** $2, 1\frac{1}{2}, 1, \frac{1}{2}, \ldots$

g. $0, 2.5, 5, 7.5, \ldots$ **h.** $x, 3x, 5x, 7x, \ldots$ **i.** $x, x^2, x^3, x^4, \ldots$

j. $2a, 2a - b, 2a - 2b, 2a - 3b, \ldots$

2. State the first 4 terms of the following arithmetic sequences.

a. $a = 3, d = 6$ **b.** $a = 6, d = -2$ **c.** $a = -1, d = 7$

d. $a = 10, d = -4.5$ **e.** $a = 2x, d = 3x$ **f.** $a = 3, d = \frac{1}{4}$

g. $a = \frac{3}{2}, d = \frac{2}{3}$ **h.** $a = x + 1, d = x + 2$ **i.** $t_n = 4n + 3$

j. $t_n = 11 - 4n$

PART B

3. a. Determine the first 5 terms of the arithmetic sequence defined by $f(n) = 2n + 3$, $n \in N$.

b. Graph the ordered pairs represented by the first 5 terms of the arithmetic sequence.

c. Can you join the points graphed in part **b** with a solid line? Explain your answer.

4. Determine t_n and the number of terms in each of the following arithmetic sequences.

a. $1, 3, 5, \ldots, 123$ **b.** $75, 70, 65, \ldots, -25$

c. $1\frac{1}{4}, 1\frac{3}{4}, 2\frac{1}{4}, \ldots, 10\frac{1}{4}$ **d.** $4, 4.5, 5, \ldots, 24$

e. $-45, -42, -39, \ldots, 15$ **f.** $\sqrt{2}, 5\sqrt{2}, 9\sqrt{2}, \ldots, 53\sqrt{2}$

5. Determine the general term t_n for each of the following arithmetic sequences.

   a. $t_3 = 7, t_4 = 10$ **b.** $t_5 = 10, t_8 = 16$

   c. $t_4 = 4, t_7 = 25$ **d.** $t_5 = 5, t_9 = -15$

   e. $t_8 = -8, t_{16} = -40$ **f.** $t_6 = 3, t_{12} = -9$

6. How many multiples of 5 are there from 20 to 200 inclusive?

7. How many multiples of 6 are there between 10 and 1000?

8. In a theatre, each row of seats has 3 seats more than the row in front of it. If there are 14 seats in the first row, how many are there in row 14?

9. A wall of cinder blocks is built up so that each row has 2 less blocks than the one below it. If there are 43 blocks in the bottom row and 11 blocks in the top row, how many rows high is the wall?

10. If Peter Piper picks 4 pecks of pickled peppers on day one, 7 pecks on day two, and each day thereafter picks 3 pecks more than the previous day, how many pecks will Peter pick on day 12?

11. Determine the value of x such that $x + 2, 2x + 3, 4x - 3$ are consecutive terms of an arithmetic sequence.

12. If in an arithmetic sequence $t_{10} = 43$ and $t_{11} = 47$, determine t_{14}.

13. If in an arithmetic sequence $t_{20} = 0$ and $t_{24} = -2$, determine the general term of the sequence.

14. The formula $t_n = n + (n - 1)(n - 2)(n - 3)(n - 4)$, $n \in N$, will produce an arithmetic sequence for $n < 5$ but not for $n \geq 5$. Explain why.

PART C

15. Determine the general term of the sequence defined by $t_1 = 2, t_k = t_{k-1} + 4, k \in N,$ $k > 1$.

16. Given $t_k = t_{k-1} - 3, t_{20} = 17, k \in N, k > 1$, find t_{24}.

17. In the sequence $\ldots, a, b, c, d, 1, 1, 2, 3, 5, 8, \ldots$, each term is the sum of the 2 terms to its left. Find the value of a.

5.3 Geometric Sequences

Consider the sequence 2, 6, 18, 54, 162. Each successive term of the sequence after the first one is found by multiplying the previous term by the constant 3. Sequences that satisfy this property are called **geometric sequences.**

In the geometric sequence defined by the 5 terms $20, -10, 5, -\frac{5}{2}, \frac{5}{4}$, the value of successive terms is obtained by multiplying by $-\frac{1}{2}$. The fixed multiplier is called the **common ratio** r. If 2 consecutive terms are known, the common ratio is determined by dividing the first term into the second. In the example, $r = \frac{-10}{20} = -\frac{1}{2}$.

If we represent the first term in the sequence by a, and the ratio by r, then the sequence becomes $a, ar, ar^2, \ldots$.

$$t_1 = a$$
$$t_2 = ar$$
$$t_3 = ar^2$$
$$t_4 = ar^3$$

Observe that the exponent of r in each term is one less than the number of the term. Therefore, the general term is $t_n = ar^{n-1}$.

> A **geometric sequence** with first term a and **common ratio** r is $a, ar, ar^2, \ldots$.
>
> The general term is $t_n = ar^{n-1}, n \in N, r \in R$.

EXAMPLE 1

In each of these, the first 4 terms of a geometric sequence are given. Determine t_n and t_{10}.

a. 3, 12, 48, 192

b. $5, -\frac{10}{3}, \frac{20}{9}, -\frac{40}{27}$

SOLUTION

a. $t_n = ar^{n-1}$

$a = 3, r = \frac{12}{3}$

$\qquad = 4$

$t_n = 3(4)^{n-1}$

$t_{10} = 3(4)^{10-1}$

$\qquad = 3(4)^9$

$\qquad = 786\ 432$

b. $t_n = ar^{n-1}$

$a = 5, r = \dfrac{-\frac{10}{3}}{5}$

$\qquad = -\frac{10}{3} \times \frac{1}{5}$

$\qquad = -\frac{2}{3}$

$t_n = 5\left(-\frac{2}{3}\right)^{n-1}$

$t_{10} = 5\left(-\frac{2}{3}\right)^9$

$\qquad = 5\left(-\frac{512}{19\ 683}\right)$

$\qquad = -\frac{2560}{19\ 683}$

EXAMPLE 2

Determine the number of terms in the geometric sequence $5, -10, 20, -40, \ldots, 1280$.

SOLUTION

For the given sequence $a = 5$, $r = -2$.

We can represent 1280 by the expression for the general term.

$$1280 = ar^{n-1}$$
$$1280 = 5(-2)^{n-1}$$
$$256 = (-2)^{n-1}$$

Since $256 = (-2)^8$, then $2^8 = (-2)^{n-1}$.

$$n - 1 = 8$$
$$n = 9$$

There are 9 terms in the sequence.

EXAMPLE 3

Find t_9 in the geometric sequence where $t_3 = 12$ and $t_6 = -96$.

SOLUTION

First, we express the given information in terms of a and r.

Since $t_3 = 12$, $\qquad\qquad\qquad$ Since $t_6 = -96$,

$\qquad ar^2 = 12.$ ① $\qquad\qquad\qquad ar^5 = -96.$ ②

The easiest way to solve this system of equations is to divide.

$$② \div ① \qquad \frac{ar^5}{ar^2} = \frac{-96}{12}$$
$$r^3 = -8$$
$$r = -2$$

Substitute $r = -2$ into $ar^2 = 12$

$$a(-2)^2 = 12$$
$$4a = 12$$
$$a = 3.$$

Then, $t_n = 3(-2)^{n-1}$, and

$$t_9 = 3(-2)^8$$
$$= 768.$$

EXAMPLE 4

Find t_{10} in the geometric sequence where $t_3 = 45$ and $t_7 = 3645$.

SOLUTION

First, we express the given information in terms of a and r.

Since $t_3 = 45$,   Since $t_7 = 3645$,

$\quad ar^2 = 45. \; ①$ $\quad ar^6 = 3645. \; ②$

The easiest way to solve this system of equations is to divide.

$$② \div ① \qquad \frac{ar^6}{ar^2} = \frac{3645}{45}$$

$$r^4 = 81$$

$$r^2 = 9 \text{ or } r^2 = -9$$

If $r^2 = 9$, $r = \pm 3$. If $r^2 = -9$, there is no real solution.

Substitute for r in $ar^2 = 45$.

If $r = 3$,   If $r = -3$

$\quad a(3)^2 = 45$ $\quad a(-3)^2 = 45$

$\quad\quad 9a = 45$ $\quad\quad 9a = 45$

$\quad\quad\ a = 5.$ $\quad\quad\ a = 5.$

$\quad\quad t_n = 5(3)^{n-1}$ $\quad\quad t_n = 5(-3)^{n-1}$

$\quad\quad t_{10} = 5(3)^9$ $\quad\quad t_{10} = 5(-3)^9$

$\quad\quad\quad = 98\ 415$ $\quad\quad\quad = -98\ 415$

The tenth term of the geometric sequence is $98\ 415$ or $-98\ 415$.

A geometric sequence is an exponential function with domain $n \in N$.

$$f(n) = ar^{n-1}, \text{ where } n \in N.$$

An alternative definition of a geometric sequence is the **recursive** definition.

$$t_1 = a, \, t_k = rt_{k-1}, \, k \in N, \, k > 1$$

Since $t_1 = a$

$\quad t_2 = rt_1$

$\quad\quad = ar.$

$\quad t_3 = rt_2$

$\quad\quad = r(ar)$

$\quad\quad = ar^2.$

From this definition we obtain the same terms as from $t_n = ar^{n-1}$.

EXAMPLE 5

Determine the first 4 terms of the geometric sequence defined by $t_1 = 5$, $t_k = 7t_{k-1}$, $k \in N$, $k > 1$, and find the general term for this sequence.

SOLUTION

$t_1 = 5$, $t_k = 7t_{k-1}$, $k \in N$, $k > 1$

$t_2 = 7t_1$ $t_3 = 7t_2$ $t_3 = 7t_2$

    $= 7 \times 5$ $= 7 \times 35$ $= 7 \times 245$

    $= 35$ $= 245$ $= 1715$

The first 4 terms are 5, 35, 245, 1715.

Since $a = 5$, $r = \frac{35}{5}$ or 7, $t_n = 7(5)^{n-1}$.

In the last section, we saw that an arithmetic sequence can be used to calculate the amount of a loan or investment with simple interest. In the next chapter, you will see that a geometric sequence can be used in calculating the amount of a loan or an investment with compound interest.

Exercise 5.3

PART A

1. Classify each of the following sequences as arithmetic, geometric, or neither.

    a. 4, 9, 14, 19, . . .   **b.** 2, −2, 2, −2, . . .

    c. 5, 3, 1, −1, . . .   **d.** 1, 4, 9, 16, . . .

    e. $-12, -6, -3, -\frac{3}{2}, \ldots$ **f.** 4, 12, 36, 108, . . .

    g. −16, −13, −10, −7, . . .   **h.** 1, 1, 2, 3, 5, . . .

2. State the first 3 terms of each of the following geometric sequences.

    a. $a = 6$, $r = -2$ **b.** $a = 1$, $r = 5$

    c. $a = -2$, $r = -3$ **d.** $t_n = 5\left(\frac{1}{2}\right)^{n-1}$

    e. $f(n) = -3(2)^{n-1}$ **f.** $t_n = p(2q)^{n-1}$

3. In each of the following, the first 3 terms of a geometric sequence are given. Determine the general term.

    a. −2, −4, −8   **b.** 6, 18, 54

    c. $\frac{1}{2}, -\frac{5}{2}, \frac{25}{2}$ **d.** $\sqrt{3}, \sqrt{6}, 2\sqrt{3}$

    e. $-\frac{2}{3}, -\frac{10}{9}, -\frac{50}{27}$ **f.** $m^2, \frac{m^4}{3p}, \frac{m^6}{9p^2}$

4. In each of the following, the first 3 terms of a geometric sequence are given. Determine t_n and the indicated term.

a. t_7 for 1, 3, 9

b. t_8 for $-16, -8, -4$

c. t_6 for $-\frac{1}{100}, \frac{1}{10}, -1$

d. t_{10} for $1, -1, 1, -1$

e. t_8 for $\sqrt{3}, 3, 3\sqrt{3}$

f. t_9 for $\frac{a}{b^2}, \frac{a^2}{2b^3}, \frac{a^3}{4b^4}$

5. Determine the number of terms in each of the following geometric sequences.

a. $3, 9, 27, \ldots, 59\ 049$

b. $-5, 10, -20, \ldots, -1280$

c. $8, 4, 2, \ldots, \frac{1}{128}$

d. $2, 6, 18, \ldots, 1458$

e. $4, 12, 36, \ldots, 8748$

f. $6, -12, 24, \ldots, -192$

6. Determine the indicated term for the geometric sequences determined by the following pairs of terms.

a. If $t_6 = 486$ and $t_2 = 6$, find t_5.

b. If $t_3 = -3$ and $t_7 = -3$, find t_2.

c. If $t_4 = 40$ and $t_7 = 320$, find t_9.

d. If $t_2 = 20$ and $t_5 = \frac{4}{25}$, find t_3.

e. If $t_6 = 6$ and $t_4 = 24$, find t_9.

f. If $t_7 = 2916$ and $t_5 = 324$, find t_3.

7. Each of the following sequences is either geometric or arithmetic.

a. Determine t_{10} for the sequence 4, 10, 16, ...

b. Determine the number of terms in the sequence 4, 8, 16, . . . , 1024.

c. Determine the general term of the sequence 4, 12, 36, ...

d. Determine the number of terms in the sequence 4, 8, 12, . . . , 120.

8. Determine the value of x such that $x + 2$, $2x + 3$, and $4x + 3$ are consecutive terms in a geometric sequence.

9. Determine the value of x such that $x - 2$, $-2 - x$, and $x + 10$ are consecutive terms in

a. an arithmetic sequence,

b. a geometric sequence.

10. Peter Piper picked a peck of pickled peppers on day one, 2 pecks on day two, 4 pecks on day three, and 8 pecks on day four. If he followed this pattern, how many pecks of pickled peppers should he pick on day ten?

11. A stereo system, costing $1200, depreciates by 30% per year. Find the value of the stereo system after four years.

12. The population of a town grows by 15% per year. If the population was 100 000 in the year 2000, find the projected population in 2020.

13. Determine the first 4 terms defined by the geometric sequence $t_1 = 4$, $t_k = -3t_{k-1}$, $k > 1$, and determine the general term for this sequence.

PART C

14. The following sequences are defined recursively. Determine whether the sequence is arithmetic, geometric, or neither, and write the general term, if possible.

  a. $t_1 = 5$, $t_k = t_{k-1} - 3$, $k \in N$, $k > 1$
  b. $t_1 = 5$, $t_k = -3t_{k-1}$, $k \in N$, $k > 1$

15. Three numbers form an arithmetic sequence, the common difference being 11. If the first number is decreased by 6, the second number is decreased by 1, and the third number doubled, the resulting numbers are in geometric sequence. Determine the numbers that form the arithmetic sequence.

16. Determine all possible values of x such that $x - 2$, $x + 5$, $4x - 8$ are consecutive terms in a geometric sequence.

17. Determine all possible values of x such that $x + 2$, $2x + 6$, $3x + 9$ are consecutive terms in a geometric sequence.

18. a. Find the arithmetic sequence with first term 1 and common difference not equal to 0, whose second, tenth, and thirty-fourth terms are the first 3 terms in a geometric sequence.

  b. The fourth term in the geometric sequence in part **a** appears as the nth term in the arithmetic sequence in part **a**. Determine the value of n.

5.4 Arithmetic Series

A **series** is the indicated sum of the terms of a sequence.

For the sequence $1, 4, 9, 16, \ldots, n^2$, the corresponding series is
$1 + 4 + 9 + 16 + \ldots + n^2$.

For the sequence $t_1, t_2, t_3, \ldots, t_n$, the sum of the corresponding series is
$$S_n = t_1 + t_2 + t_3 + \ldots + t_n.$$

EXAMPLE 1

For the series $1 + 4 + 9 + 16 + \ldots + n^2$, determine S_1, S_2, S_3, and S_4.

SOLUTION

$S_1 = 1$

$S_2 = 1 + 4$

$\quad = 5$

$S_3 = 1 + 4 + 9$

$\quad = 14$

$S_4 = 1 + 4 + 9 + 16$

$\quad = 30$

Each sum S_n can be found by adding the corresponding term to a previous sum.

For example, $S_4 = S_3 + t_4$.

In general, $S_n = S_{n-1} + t_n$.

We rewrite this to isolate the term t_n.

$$t_n = S_n - S_{n-1}$$

We can see that each term of the sequence is the difference between two consecutive sums in the corresponding series.

An **arithmetic series** is one formed from an arithmetic sequence.

For the arithmetic sequence $1, 4, 7, 10, \ldots, (3n - 2)$, the corresponding arithmetic series is $S_n = 1 + 4 + 7 + 10 + \ldots + (3n - 2)$.

When the great German mathematician Carl Friedrich Gauss was only eight years old, he observed that it is possible to find the sum of an arithmetic series by reversing the order of the terms of the series and then adding the two series. We use his method in the next example.

EXAMPLE 2

Determine the sum of the first 50 odd natural numbers $1 + 3 + 5 + 7 + \ldots + 99$.

SOLUTION

The sum of the first 50 terms of the series is
$$S_{50} = 1 + 3 + 5 + 7 + \ldots + 97 + 99.$$
We can write this sum in the reverse order.
$$S_{50} = 99 + 97 + 95 + \ldots + 3 + 1.$$
Adding these, we get
$$2S_{50} = \underbrace{100 + 100 + 100 + \ldots + 100 + 100}_{50 \text{ terms}}$$

$$= 50(100)$$
$$= 5000$$
$$S_{50} = \frac{5000}{2}$$
$$= 2500.$$

We can use Gauss' method of reversing the order of the terms to find the sum of any arithmetic series.

The general arithmetic series has a first term of a and a common difference of d. The sum of the first n terms of the arithmetic series is
$$S_n = a + (a + d) + (a + 2d) + (a + 3d) + \ldots + [a + (n - 2)d] + [a + (n - 1)d].$$
We can write this sum in the reverse order.
$$S_n = [a + (n - 1)d] + [a + (n - 2)d] + \ldots + (a + 3d) + (a + 2d) + (a + d) + a.$$
Adding these two lines we get
$$2S_n = \underbrace{[2a + (n - 1)d] + [2a + (n - 1)d] + \ldots + [2a + (n - 1)d]}_{n \text{ terms}}$$

$$= n[2a + (n - 1)d]$$
$$S_n = \frac{n}{2}[2a + (n - 1)d].$$

Sometimes a different form of this formula is useful.
$$S_n = \frac{n}{2}[2a + (n - 1)d]$$
$$= \frac{n}{2}[a + \{a + (n - 1)d\}]$$
$$= n\left[\frac{t_1 + t_n}{2}\right]$$

Note that $(t_1 + t_n)$ represents the sum of the first term and the last term of the series. Dividing by two gives the average of these 2 terms. To get the sum we multiply the average by n, the number of terms to be summed.

For an arithmetic series with first term a and common difference d, the sum of the first n terms is given by

$$S_n = \frac{n}{2}[2a + (n - 1)d] \text{ or } S_n = n\left[\frac{t_1 + t_n}{2}\right].$$

You can use either form of this formula, as is convenient. Sometimes, one form is more useful than the other in a particular question.

EXAMPLE 3

Determine S_{30} for the arithmetic series $5 + 9 + 13 + 17 + \ldots + (4n + 1)$.

SOLUTION

For the given series, $a = 5$, $d = 4$, and we want S_{30}.

Using $S_n = \frac{n}{2}[2a + (n - 1)d]$,

$$S_{30} = \frac{30}{2}[2(5) + (29)(4)]$$

$$= 15[10 + 116]$$

$$= 1890.$$

EXAMPLE 4

Evaluate $2 + 5 + 8 + 11 + \ldots + 98$.

SOLUTION

For the given series $a = 2$, $d = 3$.

Our first step is to find the number of terms in the series.

$$t_n = 2 + (n - 1)3$$

$$= 3n - 1$$

Now $3n - 1 = 98$

$$3n = 99$$

$$n = 33.$$

There are 33 terms in the series.

Using $S_n = n\left[\frac{t_1 + t_n}{2}\right]$,

$$S_{33} = \frac{33}{2}[2 + 98]$$

$$= \frac{33}{2}[100]$$

$$= 1650.$$

EXAMPLE 5

The sum of n terms of a series is given by $S_n = n^2 - 2n$. Determine t_{10}.

SOLUTION

$$t_{10} = S_{10} - S_9$$
$$= (100 - 20) - (81 - 18)$$
$$= 17$$

Exercise 5.4

PART A

1. Evaluate the indicated sum of each of the arithmetic series given below.

   a. $a = 6$, $d = 3$, find S_7.

   b. $a = -3$, $d = -2$, find S_{10}.

   c. $a = \frac{1}{2}$, $d = \frac{1}{2}$, find S_{20}.

   d. $t_1 = 11$, $t_{20} = 101$, find S_{20}.

   e. $t_1 = 20$, $t_{15} = -40$, find S_{15}.

   f. $t_1 = -50$, $t_{101} = 50$, find S_{101}.

2. Given $S_n = 2n^2$, determine the following:

   a. S_1, S_2, S_3, S_4, and S_5

   b. the first 5 terms of the corresponding sequence

   c. t_{20}

PART B

3. Determine S_{200} for each of the following arithmetic series.

   a. $3 + 6 + 9 + 12 + \ldots$

   b. $95 + 85 + 75 + 65 + \ldots$

   c. $1 + 1.1 + 1.2 + 1.3 + \ldots$

   d. $-10 - 2 + 6 + 14 + \ldots$

   e. $\frac{1}{4} + \frac{1}{2} + \frac{3}{4} + 1 + \ldots$

   f. $1 + 3 + 5 + 7 + \ldots$

4. Determine the sum of each of the following arithmetic series.

   a. $2 + 4 + 6 + 8 + \ldots + 1000$

   b. $-1 - 2 - 3 - 4 - \ldots - 500$

   c. $52 + 47 + 42 + 37 + \ldots - 48$

   d. $1.2 + 2.3 + 3.4 + 4.5 + \ldots + 21$

5. Given $S_n = 2n^2 + n$, find the first 4 terms and the general term for the corresponding arithmetic sequence.

6. A pile of logs is built up by laying 20 logs side by side and then piling succeeding rows on top so that each row has one fewer than the one below it. If this pile tapers to a single log in the top row, how many logs are there in the pile?

7. You are offered a job with an automatic raise factor. The first day you will earn $10.50, the second day $12, the third day $13.50, and so on. If the job lasts 21 days, how much will you earn?

8. In a theatre there are 14 seats in row 1, each row has 3 seats more than the one in front of it, and there are 14 rows of seats. How many seats are there in the theatre?

9. Determine the sum of all the two-digit multiples of 5.

10. Determine S_n for the arithmetic series $15 + 30 + 45 + 60 + \ldots$.

11. a. Show that the sum of the first n natural numbers is $\frac{n(n+1)}{2}$.

  b. How might you have used this to answer Question 10?

12. For the sequence given by $t_n = 3n - 1$, find S_n.

13. a. Evaluate $1 + 2 + 3 + 4 + \ldots + 50$.

  b. Evaluate $1 + 2 + 3 + 4 + \ldots + 100$.

  c. Use these results to evaluate $51 + 52 + 53 + \ldots + 100$.

14. a. Find a formula for the sum of the first n odd natural numbers.

  b. Find the formula for the sum of the first n even natural numbers.

  c. If you added these results, what would your result represent?

15. a. Given $S_n = 2n^2 - 3n$, determine t_n.

  b. Given $S_n = 3n^2 + 4n$, determine t_n.

PART C

16. For an arithmetic series, $t_1 = -2p + 3q$ and $t_5 = 6p + 7q$. Find an expression for S_n in terms of p, q, and n.

17. Given the sequence $1, -2, 3, -4, 5, -6, \ldots, (-1)^{n+1}n, \ldots$, find the average of the first 200 terms.

18. A sequence consists of 10 terms, all of which are positive integers. The first term is p and the second is q, with $q > p$. Each term thereafter is the sum of the 2 terms immediately preceding it, and the seventh term is 181. Given that there are exactly two such sequences, show that the 10 terms in each sequence have the same sum, and determine this sum. Show that the sum of n terms is $S_n = t_{n+2} - t_2$.

19. The first term in a sequence of numbers is $t_1 = 5$. Succeeding terms are defined by the statement $t_n - t_{n-1} = 2n + 3$ for $n \geq 2$. Find the value of t_{50}.

Carl Friedrich Gauss (1777–1855)

Carl Friedrich Gauss was a German physicist, mathematician, and astronomer, who made important contributions in all three fields.

His computational ability with numbers was recognized very early. It is thought that in his first arithmetic class, Gauss astonished his teacher by instantly solving what was intended to be a long "busy work" problem. Young Gauss was given the task of adding all the numbers from 1 to 100. He recognized a pattern and immediately gave the answer. Gauss' mathematical powers so overwhelmed his schoolmasters that by the time Gauss was ten years old, they admitted that there was nothing more they could teach the boy.

Gauss was the last of the great scholars whose interest spanned virtually every branch of mathematics. Since his time, the branches of mathematics have expanded so greatly that students generally specialize in a particular area.

Photo: Bettmann/CORBIS

Complex Numbers

Complex numbers were first used to solve equations by the Italian mathematician Girolamo Cardano (1501–1576). Cardano found the formula for solving cubic equations and required complex numbers to do this.

The German mathematician Carl Friedrich Gauss (1777–1855) published works of major importance in algebra, number theory, differential equations, non-Euclidean geometry, complex analysis, and theoretical mechanics. He paved the way for a general and systematic use of complex numbers.

Today complex numbers are used to solve many engineering problems.

The knowledge of complex numbers learned in algebra is usually sufficient for elementary problems involving electric circuits and mechanical vibrating systems.

The majority of non-elementary functions appearing in engineering mathematics are analytic functions. Advanced study on the theory of complex analytic functions is needed to solve some problems in the theory of heat, in fluid dynamics, and in electrostatics. The theory of complex analytic functions provides powerful and elegant methods for tackling these problems.

Photo: Kelly-Mooney Photography/CORBIS

5.5 Geometric Series

A **geometric series** is the indicated sum of a geometric sequence.
For the geometric sequence 3, 6, 12, 24, 48, the corresponding geometric series is
$3 + 6 + 12 + 24 + 48$.

EXAMPLE 1

Find the sum of 10 terms, S_{10}, of the geometric series
$3 + 3 \times 5 + 3 \times 5^2 + \ldots + 3 \times 5^9$.

SOLUTION

$3 + 3 \times 5 + 3 \times 5^2 + \ldots + 3 \times 5^9$

$a = 3, r = 5$

We can write an expression for the sum of 10 terms.

$$S_{10} = 3 + 3 \times 5 + 3 \times 5^2 + \ldots + 3 \times 5^9. \qquad ①$$

Multiply this equation by the value of r ($r = 5$).

$$5S_{10} = 3 \times 5 + 3 \times 5^2 + 3 \times 5^3 + \ldots + 3 \times 5^{10}$$

$$5S_{10} = \qquad 3 \times 5 + 3 \times 5^2 + \ldots + 3 \times 5^9 + 3 \times 5^{10} \qquad ②$$

$$S_{10} = 3 + 3 \times 5 + 3 \times 5^2 + \ldots + 3 \times 5^9 \qquad ①$$

Subtract ② − ① $4S_{10} = 3 \times 5^{10} - 3$

$$S_{10} = \frac{3 \times 5^{10} - 3}{4}$$

We can use the same method to find the sum of n terms of the general geometric series
with first term of a and common ratio of r.

$$S_n = a + ar + ar^2 + ar^3 + \ldots + ar^{n-2} + ar^{n-1} \qquad ①$$

Multiply by r $\qquad rS_n = \qquad ar + ar^2 + ar^3 + \ldots + ar^{n-2} + ar^{n-1} + ar^n \quad ②$

Subtract ② − ① $rS_n - S_n = ar^n - a$

$$S_n(r - 1) = a(r^n - 1)$$

$$S_n = \frac{a(r^n - 1)}{r - 1}, r \neq 1$$

For the **geometric series** with first term a and common ratio r,

the sum of n terms is given by $S_n = \frac{a(r^n - 1)}{r - 1}, r \neq 1$.

By subtracting ① − ② above, $S_n = \frac{a(1 - r^n)}{1 - r}, r \neq 1$. This form of the formula is used
when $-1 < r < 1$.

EXAMPLE 2

For the geometric series with $a = 7$ and $r = 2$, determine S_9.

SOLUTION

$$S_n = \frac{a(r^n - 1)}{r - 1}$$

For the given series, $a = 7$, $r = 2$, $n = 9$.

$$S_9 = \frac{7(2^9 - 1)}{2 - 1}$$

$$= \frac{7(512 - 1)}{1}$$

$$= 7(511)$$

$$S_9 = 3577$$

EXAMPLE 3

Determine the sum of the geometric series $15 - 45 + 135 - 405 + \ldots - 32\ 805$.

SOLUTION

For the given series, $a = 15$, $r = -3$.

First we find the number of terms in the series.

$$t_n = ar^{n-1}$$

$$= 15(-3)^{n-1}$$

Therefore, $15(-3)^{n-1} = -32\ 805$

$$(-3)^{n-1} = -\frac{32\ 805}{15}$$

$$(-3)^{n-1} = -2187.$$

Now $\quad\quad\quad\quad 2187 = 3^7.$

So $\quad\quad\quad\quad -2187 = (-3)^7$

$$(-3)^{n-1} = (-3)^7$$

$$n - 1 = 7$$

$$n = 8.$$

There are 8 terms in the series.

$$S_n = \frac{a(r^n - 1)}{r - 1}$$

For $a = 15$, $r = -3$, $n = 8$,

$$S_8 = \frac{15[(-3)^8 - 1]}{-3 - 1}$$

$$= \frac{15(6561 - 1)}{-4}$$

$$= \frac{15(6560)}{-4}$$

$$S_8 = -24\ 600.$$

EXAMPLE 4

For the geometric series with $a = 9$ and $r = \frac{1}{3}$, determine S_6.

SOLUTION

Since $-1 < r < 1$, use $S_n = \frac{a(1 - r^n)}{1 - r}$.

For the given series, $a = 9$, $r = \frac{1}{3}$, $n = 6$.

$$S_6 = \frac{9\left[1 - \left(\frac{1}{3}\right)^6\right]}{1 - \frac{1}{3}}$$

$$= \frac{9\left[1 - \frac{1}{729}\right]}{\frac{2}{3}}$$

$$= \frac{3}{2} \times 9 \times \left[\frac{728}{729}\right]$$

$$= \frac{364}{27}$$

Exercise 5.5

PART A

1. Determine the sum of each of the following geometric series.

    a. $a = 2, r = 3, n = 5$ **b.** $a = -4, r = -1, n = 20$

    c. $a = 20, r = 6, n = 5$ **d.** $a = 5, r = -2, n = 7$

2. Determine S_8 for each of the following geometric series.

    a. $5 + 10 + 20 + \ldots$ **b.** $1 + 3 + 9 + \ldots$

    c. $1 - 2 + 4 - \ldots$ **d.** $100 - 200 + 400 - \ldots$

PART B

3. Determine the sum of each of the following geometric series.

    a. $3 + 9 + 27 + \ldots + 59\,049$ **b.** $-4 + 12 - 36 + \ldots + 8748$

    c. $-4 - 12 - 36 - \ldots - 972$ **d.** $5 - 10 + 20 - \ldots + 1280$

4. The sum of n terms of the geometric series $-6 - 18 - 54 - \ldots$ is -726. What is the value of n?

5. Some of the series below are arithmetic and some are geometric. The first step in solving each problem is to determine which type of series you have.

   a. Determine S_8 for the series $\frac{1}{8} + \frac{1}{4} + \frac{1}{2} + \ldots$.

   b. Determine S_{10} for the series $\frac{1}{8} + \frac{1}{4} + \frac{3}{8} + \ldots$.

   c. Determine the sum of the series $4 + 8 + 12 + \ldots + 684$.

   d. Determine the sum of the series $4 + 8 + 16 + \ldots + 1024$.

6. You decide to have a party and invite 3 friends. You tell each friend to invite 3 more friends. If each of these friends invites 3 more friends, and they each invite 3 friends, how many people will be invited to the party, including yourself? Assume no person is invited by two different people.

7. A company's profits are increasing by 5% per year.

   a. If their profits were $500 000 in 1995, what were their profits in the year 2000?

   b. Find their total profit for the years 1995 to 2000.

8. A lottery is offering 8 prizes. The first prize is $50 and each succeeding prize is double the previous prize. What is the total amount of prize money?

9. A researcher is studying fruit flies. Under the laboratory conditions, each generation of fruit flies is 20% more numerous than the preceding generation.

   a. If the researcher started with 50 fruit flies, estimate the number in the sixth generation. (The original 50 fruit flies is the first generation.)

   b. Including the sixth generation, how many fruit flies in total could the researcher have studied?

10. You have an ear infection and are prescribed to take 250 mg of ampicillin four times a day. It is known that at the end of 6 h, about 4% of the drug is still in the body. What quantity of the drug is in the body after

    a. the third dose?

    b. the twentieth dose?

    c. the nth dose?

    Write your answer, correct to three decimals.

PART C

11. Show that the sum of $1 + \frac{1}{2} + \frac{1}{4} + \ldots + t_n$ is always less than 2, no matter what the value of n.

12. A ball is dropped from a height of 8 m. It rebounds $\frac{3}{4}$ of its height after each bounce.

    a. How far has the ball travelled when it touches the ground on the third bounce? *Hint:* Draw a diagram showing the three bounces. Find the sum of the downward distances and then find the sum of the upward distances. The total distance travelled by the ball is the sum of the downward and the upward distances.

    b. Find the total distance travelled by the ball when it touches the floor on the tenth bounce.

13. If $|r| < 1$ and n is a positive integer, then $|r^n|$ gets progressively smaller as n gets larger. Satisfy yourself that this is true. As n gets very large, $|r^n|$ approaches 0 in value. We write $|r^n| \to 0$ or $\lim_{n \to \infty} |r^n| = 0$. (This is read "The limit of $|r^n|$ as n becomes infinitely large is 0." You will learn more about this concept in a calculus class.)

    a. Use this result to show that if $|r| < 1$, the sum of an infinite number of terms of the geometric series $a + ar + ar^2 + \ldots$ is $\frac{a}{1-r}$.

    b. Determine the sum of an infinite number of terms for each of the following geometric series.

        i) $4 + 2 + 1 + \ldots$ **ii)** $4 - 2 + 1 + \ldots$

        iii) $5 + \frac{5}{3} + \frac{5}{9} + \ldots$ **iv)** $6 + 4 + \frac{8}{3} + \ldots$

14. a. Find all geometric sequences such that the sum of the first 2 terms is 2 and the sum of the first 3 terms is 3.

    b. For each of the sequences determined in part **a,** calculate the sum of all terms having value less than 1.

15. Consider the sequence $a_1, a_2, a_3, \ldots$ with general term a_n. Given that the sum of the first n terms of the sequence is $S_n = 2 + \frac{n}{3n+1}$,

    a. determine S_1, S_2, and S_3.

    b. determine a_1, a_2, and a_3.

    c. determine a_n.

    d. determine the sum to infinity of the sequence.

16. For n a positive integer, let $p = 1 + 3 + 3^2 + \ldots + 3^n$. Show that $1 + 9 + 9^2 + \ldots + 9^n$ is the sum of the integers from 1 to p.

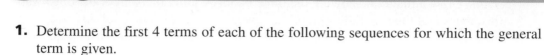

1. Determine the first 4 terms of each of the following sequences for which the general term is given.

   a. $t_n = 3n + 1$

   b. $t_n = n^2 - 3$

   c. $f(n) = 2^n + 1, n \in N$

   d. $t_n = \dfrac{2n}{n+1}$

   e. $f(n) = n^3 - 2$

   f. $f(n) = \dfrac{n^2 + n}{2}$

2. Determine t_3, t_4, and t_5 for the sequence given by $t_1 = 1$, $t_2 = 3$, $t_k = 2t_{k-1} - 3t_{k-2}$, $k \geq 3$.

3. Determine a general term for each of the following sequences.

   a. 2, 6, 10, 14, . . .

   b. 2, 6, 18, 54, . . .

   c. 13, 8, 3, −2, . . .

   d. $0, \dfrac{1}{4}, \dfrac{2}{9}, \dfrac{3}{16}, \ldots$

4. Each of the following sequences is either arithmetic or geometric. In each case, determine the type of sequence, the general term, and the indicated term.

   a. 5, 8, 11, 14, ... Determine t_{20}.

   b. 3, 9, 27, 81, ... Determine t_8.

   c. 54, 48, 42, 36, ... Determine t_{15}.

   d. $\dfrac{1}{3}, -\dfrac{1}{6}, \dfrac{1}{12}, -\dfrac{1}{24}, \ldots$. Determine t_8.

   e. −5, −8, −11, −14, ... Determine t_{18}.

5. Each of the following sequences is either arithmetic or geometric. In each case determine the type of sequence and the number of terms.

   a. 5, 12, 19, . . . , 222

   b. 8, 24, 72, . . . , 5832

   c. $81, 27, 9, \ldots, \dfrac{1}{243}$

   d. 8, 5, 2, . . . , −64

6. Determine t_{20} in the arithmetic sequence where $t_6 = 28$ and $t_{11} = 63$.

7. A superball is dropped from a height of 60 m. The ball bounces back $\frac{4}{5}$ of the distance it falls.

   a. To what height does the ball bounce after the first bounce?

   b. To what height, to one decimal place, does the ball bounce after the sixth bounce?

8. Each of the following series is either arithmetic or geometric. In each case, determine the type of series and the sum of each series.

   a. $-18 - 15 - 12 - \ldots$ Find S_{100}.

   b. $1 - 2 + 4 - \ldots$ Find S_{10}.

   c. $\frac{1}{3} + \frac{2}{9} + \frac{4}{27} + \ldots + \frac{64}{2187}$

   d. $5 + 9 + 13 + \ldots + 213$

9. If $(x + 1)$, $(-2x - 4)$, and $(x + 15)$ are 3 consecutive terms of an arithmetic sequence, determine x and the 3 terms.

10. The sum of n terms of the geometric series $1 - 3 + 9 - \ldots$ is $-132\,860$. What is the value of n?

11. The sum of the first n terms of a sequence is $n(n + 1)(n + 2)$. What is the value of the tenth term of the sequence?

12. How many terms in the arithmetic sequence 7, 14, 21, $\ldots$ are between 40 and 28 001?

13. If the terms 27, x, y, 8 form a geometric sequence, find x and y.

14. Determine the value of n for which $(2^1)(2^2)(2^3)(2^4) \ldots (2^n) = 2^{210}$.

1. You can do this problem by evaluating the given expressions. We challenge you to answer the question asked by manipulating the expressions and not by evaluating.

   a. Which is the larger, $2000 \times (1 + 2 + 3 + 4 + \ldots + 2001)$ or $2001 \times (1 + 2 + 3 + 4 + \ldots + 2000)$?

   b. Which is the larger, $100^2 - 99^2 + 98^2 - 97^2 + 96^2 - \ldots + 2^2 - 1^2$ or $100 + 99 + 98 + 97 + 96 + \ldots + 2 + 1$?

2. If a, b, and c form an arithmetic sequence, show that the equation $(b - c)x^2 + (c - a)x + (a - b) = 0$ has equal roots.

3. In a geometric sequence, any term is equal to the sum of the two terms following it. Determine the common ratio of the sequence.

4. In an arithmetic sequence having first term 3 and common difference not equal to 0, the second, seventh, and twenty-second terms are the first 3 terms in a geometric sequence. What term in the arithmetic sequence is the fourth term in the geometric sequence?

5. Consider the sequence $\frac{1}{1}, \frac{1}{2}, \frac{1}{3}, \frac{1}{4}, \ldots$, where all numerators are 1 and the denominators are the natural numbers.

   a. In this sequence, can you find an arithmetic sequence of 3 terms?

   b. Can you find one of 4 terms?

   c. Can you find one of 5 terms?

   d. Can you find one of 6 terms?

   e. Can you find one of length greater than 6?

6. All multiples of 2 and 5 are removed from the set of consecutive positive integers $1, 2, 3, 4, \ldots, 10n$.

   a. If $n = 10$, determine the sum of the remaining integers.

   b. Determine the sum of the remaining integers for any value of n.

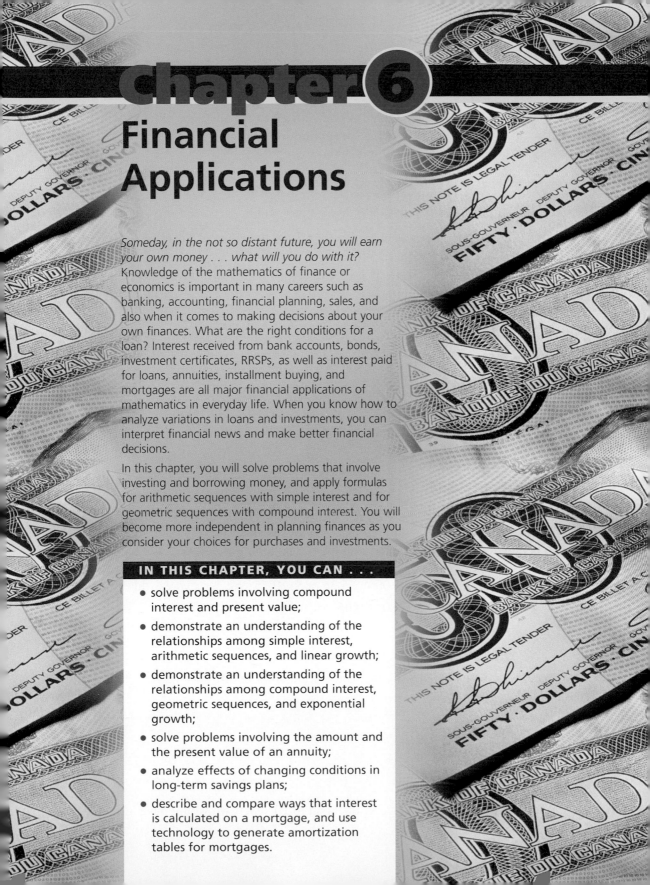

Chapter 6

Financial Applications

Someday, in the not so distant future, you will earn your own money . . . what will you do with it? Knowledge of the mathematics of finance or economics is important in many careers such as banking, accounting, financial planning, sales, and also when it comes to making decisions about your own finances. What are the right conditions for a loan? Interest received from bank accounts, bonds, investment certificates, RRSPs, as well as interest paid for loans, annuities, installment buying, and mortgages are all major financial applications of mathematics in everyday life. When you know how to analyze variations in loans and investments, you can interpret financial news and make better financial decisions.

In this chapter, you will solve problems that involve investing and borrowing money, and apply formulas for arithmetic sequences with simple interest and for geometric sequences with compound interest. You will become more independent in planning finances as you consider your choices for purchases and investments.

IN THIS CHAPTER, YOU CAN . . .

- solve problems involving compound interest and present value;
- demonstrate an understanding of the relationships among simple interest, arithmetic sequences, and linear growth;
- demonstrate an understanding of the relationships among compound interest, geometric sequences, and exponential growth;
- solve problems involving the amount and the present value of an annuity;
- analyze effects of changing conditions in long-term savings plans;
- describe and compare ways that interest is calculated on a mortgage, and use technology to generate amortization tables for mortgages.

We have seen that an arithmetic sequence can be used to calculate the amount of a loan or an investment, earning simple interest. **Compound interest** is interest paid on both interest previously earned and the original investment.

ACTIVITY

You invest $100 in a bond that pays 5% interest per annum.

a. Assume that the bond earns simple interest. Then the value of the bond at the end of each of the first four years is as follows:

Year	Calculation	Value of Bond
1	100.00 + 100.00 × 0.05	$105.00
2	105.00 + 100.00 × 0.05	$110.00
3	110.00 + 5.00	$115.00
4	115.00 + 5.00	$120.00

Copy this chart in your notebook and extend it to show the value of the bond at the end of 5, 6, 7, 10, 15, and 20 years.

Express these values in arithmetic sequence form and write an expression for the value of the bond after n years.

b. Assume that the bond earns 5% interest per annum compounded annually.

In one year, your $100 bond would earn $100.00 × 0.05 or $5.00 in interest. The value of the bond would then be $100.00 + $5.00 = $105.00.

What happens to the value of the bond in future years?
The bond continues to earn interest at 5% per annum, but now on $105.00, so the interest earned in year two is $105.00(0.05) or $5.25. By the end of the second year, the value of the bond is $105.00 + $5.25 or $110.25. We can express this by

$$\$105.00 + \$105.00 \times 0.05$$
$$= \$105.00 + \$5.25$$
$$= \$110.25.$$

We can set up a chart showing the value of the bond in different years.

Year	Calculation	Value of Bond
1	100.00 + 100.00 × 0.05	$105.00
2	105.00 + 105.00 × 0.05	$110.25
3	110.25 + 110.25 × 0.05	$115.76
4	115.76 + 115.76 × 0.05	$121.55

If we wanted to know the value of the bond in 20 years, we would have a lot of calculations to do.

Let's look for an easier way. Instead of calculating each number as we go, note that each of the expressions in the calculation column can be factored.

The value of the investment at the end of year one would be

$$100 + 100 \times 0.05 = 100(1 + 0.05)$$
$$= 100(1.05).$$

At the end of the second year, the value of the investment of $100(1.05)$ grows to

$$100(1.05) + 100(1.05)(0.05)$$
$$= 100(1.05)(1 + 0.05)$$
$$= 100(1.05)^2.$$

The value of the investment continues to grow at 5%. At the end of the third year, the value has grown to

$$[100(1.05)^2](1.05)$$
$$= 100(1.05)^3.$$

Now we see that we can restate the Calculation column in our chart, as follows:

Year	Calculation	Value of Bond
1	$100(1.05)$	$105.00
2	$100(1.05)^2$	$110.25
3	$100(1.05)^3$	$115.76
4	$100(1.05)^4$	$121.55

Copy this chart in your notebook and extend it to show the value of the bond at the end of 5, 6, 7, 10, 15, and 20 years.

Observe that the entries in the Calculation column form a geometric sequence. Use this to write an expression for the value of the bond after n years.

c. Using the expressions you obtained in parts **a** and **b,** draw graphs, on the same axes, to illustrate the difference between the values of the bond under the two methods of interest.

Noting that the value of a sum of money earning compound interest is given by a geometric sequence, we can write a general statement as below.

Compound Interest Formula

$$A = P(1 + i)^n$$

where P is the original principal, i is the rate of interest per period,
n is the number of periods the investment is held, and A is the total value
of the investment after n investment periods.

The value of a sum of money earning simple interest is given by an arithmetic sequence (or a linear function). The value of a sum of money earning compound interest is given by a geometric sequence (or an exponential function).

EXAMPLE 1

$10 000 is invested for five years at 6% per annum compounded semi-annually.

a. Determine the **amount of the investment** at the end of the five years.

b. Determine the interest earned in the five years.

SOLUTION

a. Using $A = P(1 + i)^n$,

P is the principal invested, so $P = 10\ 000$.

i is the interest rate per investment period. Since the interest rate is 6% per annum and the investment is compounded semi-annually, there are two investment periods per year, with

$$i = \frac{6}{2}\% = 3\% = 0.03 \text{ per investment period.}$$

n is the number of investment periods. The investment is compounded semi-annually for five years, so there are ten investment periods.

Then $A = 10\ 000(1 + 0.03)^{10}$

$= 10\ 000(1.03)^{10}$

$= 13\ 439.16.$

The amount of the investment is $13 439.16 after five years.

b. Interest earned in the five years is $13 439.16 − $10 000 = $3439.16.

EXAMPLE 2

Five years ago a sum of money was invested at 7% per annum compounded quarterly. If the amount of the investment is now $9000, how much money was deposited five years ago?

SOLUTION

Using $A = P(1 + i)^n$,

A is the value of the investment after n investment periods.

$A = 9000$

P is the principal invested.

i is the interest rate per investment period.

$$i = \frac{7}{4}\% = 1.75\% = 0.0175 \text{ per investment period.}$$

n is the number of investment periods.

$$n = 5 \times 4 = 20$$

Then $9000 = P(1 + 0.0175)^{20}$

$$9000 = P(1.0175)^{20}$$
$$P = \frac{9000}{(1.0175)^{20}}$$
$$P = 6361.42.$$

The amount invested five years ago was $6361.42.

EXAMPLE 3

Joe buys a new sofa priced at $800. He can pay $800 now or not make any payment now and pay $950 in one year. The salesperson tells Joe that in effect he will have a loan of $800 for one year, compounded monthly. What is the monthly interest rate that Joe would be paying?

SOLUTION

$A = P(1 + i)^n$

A is the value of the loan after n investment periods, $A = 950$.

P is the principal borrowed, $P = 800$.

i is the interest rate per investment period.

n is the number of investment periods, $n = 12$.

$$A = P(1 + i)^n$$
$$950 = 800(1 + i)^{12}$$
$$(1 + i)^{12} = \frac{950}{800}$$

Taking the 12th root of each side,

$$1 + i = \left(\frac{950}{800}\right)^{\frac{1}{12}}$$
$$1 + i \doteq 1.0144$$
$$i \doteq 0.0144.$$

Joe would be paying a monthly interest rate of 1.44% or 17.3% per annum compounded monthly.

EXAMPLE 4

Five thousand dollars is placed in an investment account. The investor checks later and his investment is worth $6092. If the interest rate is 5% per annum compounded semi-annually, how long has the money been invested?

SOLUTION

Using $A = P(1 + i)^n$, $A = 6092$, $P = 5000$, $i = \frac{0.05}{2} = 0.025$.

$$6092 = 5000(1 + 0.025)^n$$
$$\frac{6092}{5000} = (1.025)^n$$
$$1.2184 = (1.025)^n.$$

You can determine the value for n by using your graphing calculator or by using logarithms.

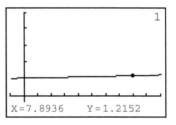

For a calculator solution, graph $y = 1.025^n$ and use the trace function to estimate the value of n giving $y = 1.2184$.

Depending on the setting on your calculator, answers will vary but the exponent is approximately 7.9. The money was invested for approximately 7.9 interest periods. Since the number of interest periods is a whole number, the money was invested for eight interest periods of six months each, or four years.

Using logarithms we obtain
$$n = \frac{\log 1.2184}{\log 1.0125}$$
$$\doteq 7.9999.$$

The money was invested for eight periods, or four years.

How Soon Does Money Double?

How long does it take, if money is compounded, for a sum of money to double in value? Clearly this depends on the rate of interest earned. In the activity below we consider this question.

ACTIVITY

a. If $100 is invested at 4% per annum compounded annually, determine, by using your calculator to estimate, the approximate time required for the money to double in value.

b. Repeat part **a** for interest rates of 5%, 6%, 8%, 10%, and 12%.

c. For each interest rate, consider the time required for the money to double. If the interest rate is $r\%$ and the doubling time is n, state a connection involving n and r.

In the investment business, the "Rule of 72" is used to approximate
the time required for money to double,
since $nr \doteq 72$, where r is the rate of interest and n is the time required.

Present Value

You may want to put money aside for your university or college education or for some
item you wish to purchase in the future. You would obviously like to know how much you
need to invest now to have a specific amount of money at some point in the future. The
amount of money that must be invested now to produce a specified amount of money at
some time in the future is called the **present value** of the amount.

EXAMPLE 5

Ryan is saving money for a car he wants to buy
in six years. He has a summer job and wants to
invest enough money now to have $5000 toward
his car in six years' time.

He can invest the money at $5\frac{1}{2}\%$ per annum
compounded annually. How much money
should he invest now, so that it will grow to
$5000 in six years? Another way to ask this is
"What is the present value of the $5000 investment?"

SOLUTION

Our formula for compound interest can be used to find the present value. In the formula
$A = P(1 + i)^n$, we have

> Amount $A = \$5000$.
>
> Interest rate per period $i = 5\frac{1}{2}\% = 0.055$.
>
> Number of compounding periods $n = 6$.
>
> P is the amount of money you should invest. This is the present value of the
> $5000.

Then $5000 = P(1 + 0.055)^6$

$$\frac{5000}{(1.055)^6} = P$$

$$3626.23 = P$$

Ryan should invest $3626.23 now to have approximately $5000 in six years.

We could also rearrange our formula to create an expression for present value.

> The present value P of some future amount A is $P = \dfrac{A}{(1 + i)^n}$ where i is the rate of
> interest per period and n is the number of periods the investment is held.

PART A

1. You invest $1000 for ten years. Determine the total interest earned if the investment earns 8% per annum

   a. compounded annually,

   b. compounded semi-annually,

   c. compounded monthly,

   d. compounded daily.

   Explain your answers.

2. Calculate approximately how long it will take for money to double if it is invested at the following interest rates per annum.

   a. 4%   b. 9%   c. 5.5%   d. 6.75%

3. Calculate the difference in return on an investment of $10 000 for ten years at 6% per annum compounded annually and compounded quarterly.

PART B

4. As a special to attract depositors, a bank offers 5% per annum compounded daily in comparison to its competing bank down the street, which offers $5\frac{1}{4}$% compounded annually. If you were investing $10 000 for six years, which bank should you choose?

5. If $1500 is invested now at 7% per annum compounded annually, when will it amount to $3150?

6. Becky borrows $5000 at 7% per annum compounded semi-annually. She is to repay the money in four years. How much money will Becky owe at the end of four years?

7. When William is born, William's grandfather places $1000 in trust for William's college tuition. The interest on the money is compounded annually. When William is 16 years old, he discovers that there is $2540.35 in the trust account. What is the annual interest rate being paid on the trust account?

8. A loan is to be repaid in five years. Interest is accumulating on the loan at the rate of 8% per annum compounded quarterly. The payment required to retire the loan in five years is $6686.76. Find the present value of the loan.

Thinking of the Future

The topics covered in this chapter will have a very important and direct influence on your life.

To build for your future and save for your retirement, you will likely be investing in investment certificates, term deposits, annuities, and RRSPs. These savings devices generally pay compound interest on the amount of money you invested.

Many items are purchased using credit cards. This really is a loan. The company or bank issuing the card charges you interest on the outstanding debt on the card. Expensive items like cars are often financed. You pay the cost of the car plus interest on the loan. A mortgage on a house is a very large loan on which you will pay interest. Many of you will leave college or university with outstanding student loans. And of course the biggest loan we all carry is the national debt on which we pay millions of dollars in interest every year.

How fast our investments grow and how much money we pay in interest charges has a direct bearing on what we can afford. As borrowers or investors, we need to look for ways to maximize the return on our money. This allows us to save for that exciting vacation, pay for our education, and generally improve the quality of our life.

Photo: COMSTOCK

---

9. Conrad buys a stereo. He pays $500 now and agrees to pay $2000 in three years.

   a. If interest on the money is compounded quarterly at 16% per annum, determine the cash value of the stereo now.

   b. How much money would Conrad save if he paid cash for the stereo when he bought it?

10. Crystal has an $8000 student loan. When Crystal leaves college, interest compounded quarterly starts accumulating on the loan. Crystal does not make any payments on the loan for three years. Her bank manager then calls her in and informs her that she now owes $10 146 on the student loan. What is the annual interest rate Crystal is being charged?

11. In 1626 the Dutch bought the Island of Manhattan from the natives for beads worth $24. If the natives had received $24 in cash, instead of the beads, and had invested the money at 2% per annum compounded annually, what would the investment be worth on the anniversary date of the investment in the year 2000?

12. A Roman centurion deposited a coin worth 10¢ in a bank at the beginning of the year A.D. 1. If the bank paid 3% per annum compounded semi-annually, what would the investment be worth at the beginning of the year 2000?

13. You inherit $5000 and want to save the money toward a car that you plan to buy in four years. You can invest the money for four years at $5\frac{1}{2}\%$ per annum compounded annually, or you can invest it for two years at 3% per annum compounded semi-annually, and then invest this new amount for the remaining two years at 8% per annum compounded annually. Which should you do?

PART C

14. Jerry and Jan want to start a house-painting business when they finish college in three years. They have calculated that they will need $10 000 to buy the equipment required for their business. Jan has $3000 that she can invest at 6% per annum compounded quarterly. Jerry can invest money at 7% per annum compounded annually. How much money must Jerry invest so that they will have the money required to start their business when they graduate?

15. New grandparents want to give their grandson a gift on his 18th birthday. They would like to give him the equivalent of what $5000 would buy on the day he was born.

a. If inflation is assumed to be 3% per year, how much money should he receive on his 18th birthday to be able to buy what $5000 would buy on the day he was born?

b. What is the amount of the single payment into a trust fund the grandparents should make on the day he was born, if the money is invested at 5% per annum compounded annually?

16. Money will accumulate faster if it is compounded at more frequent intervals. Two types of interest rates are sometimes quoted. The **nominal interest rate** is the rate usually quoted. The **effective interest rate** is the annual interest rate that would generate an equivalent amount of interest. If $1000 is invested at 12% per annum compounded monthly, the amount of the investment in one year is

$A = 1000(1 + 0.01)^{12}$

$\quad = 1126.83.$

Interest earned is $1126.83 - $1000 = $126.83.

The nominal interest rate is 12%.

Effective rate of interest $= \frac{126.83}{1000} \times 100 = 12.68\%.$

When the nominal rate is calculated more than once a year, the effective rate is higher. Find the effective rate of interest for the following nominal rates.

a. 6% per annum compounded semi-annually

b. 8% per annum compounded quarterly

Explain your answers.

Carrie wants to go on a trip to Europe. She plans to deposit $100 a month in a savings account at the end of each month for three years. A sequence of payments such as this is known as an **annuity.**

> An **annuity** is a series of equal payments or deposits
> made at regular intervals of time.

The **period of an annuity** is the time interval between two consecutive payments. The period for Carrie's annuity is one month. The **term of an annuity** is the total time involved in completing the annuity. The term of Carrie's annuity is three years or 36 months. Annuities are common in the business world. Car loan payments, mortgages, and monthly pension plan benefits are just three examples.

In this section, we discuss **ordinary annuities.** Ordinary annuities have payments made at the end of the payment period. The example at the beginning of this section is an ordinary annuity.

Let's calculate how much money Carrie would have for her trip if she deposited her $100 monthly payment in a savings account that paid interest at 6% per annum compounded monthly?

Solution

This annuity can be represented by a time diagram. Note $i = \dfrac{6\%}{12} = 0.5\% = 0.005$.

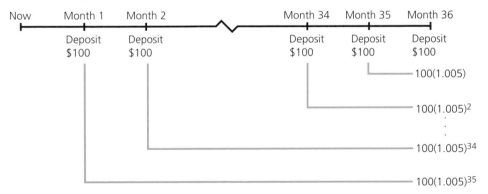

The initial $100 deposit will earn 6% per annum compounded monthly for 35 months. This $100 deposit will have a value of $100(1.005)^{35}$ dollars. The second monthly deposit is compounded monthly for 34 months. At the end of the annuity, this $100 deposit will have a value of $100(1.005)^{34}$ dollars.

At the end of the third year (36 months), the amount of money on deposit is given by

$$A = 100 + 100(1.005) + 100(1.005)^2 + \ldots + 100(1.005)^{34} + 100(1.005)^{35}.$$

This is a geometric series having $a = 100$, $r = 1.005$, $n = 36$.

$$S_n = \frac{a(r^n - 1)}{r - 1}$$

$$S_{36} = \frac{100(1.005^{36} - 1)}{1.005 - 1}$$

$$= 3933.61$$

Carrie will have saved $3933.61 for her trip to Europe.

EXAMPLE 1

On Gurinder's first birthday, his grandparents deposited $500 in a savings account that pays interest at 6.3% per annum compounded annually. They continue to make $500 deposits on each of his birthdays up to the age of 18 so that he can have some money for his university education. How much will be available on Gurinder's 18th birthday?

SOLUTION

The regular deposits can be represented on a time diagram.

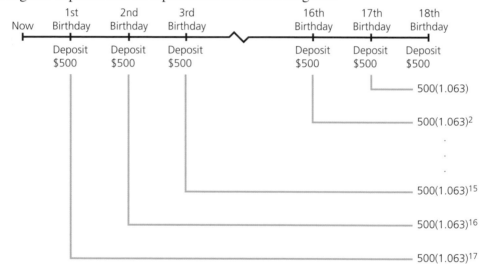

The amount of money on deposit on his 18th birthday is

$$A = 500 + 500(1.063) + 500(1.063)^2 + \ldots + 500(1.063)^{17}.$$

This is a geometric series with $a = 500$, $r = 1.063$, $n = 18$.

$$S_n = \frac{a(r^n - 1)}{r - 1}$$

$$S_{18} = \frac{500(1.063^{18} - 1)}{1.063 - 1}$$

$$\doteq 15\ 899.208\ 84$$

On his 18th birthday, Gurinder will have $15 899.21 in the account for his university education.

To encourage Canadians to save for their retirement, the government introduced the Registered Retirement Savings Plan (RRSP). Money deposited into an RRSP usually is invested in guaranteed investment certificates, mutual funds, or bonds. Any income earned is tax sheltered, meaning that no income tax is paid on this income for as long as it is in the RRSP. Example 2 illustrates the advantage of making deposits in an RRSP when you are young.

EXAMPLE 2

a. James deposits $1000 into his Registered Retirement Savings Plan (RRSP) on his birthday at ages 20, 21, 22, and 23. He makes no further deposits in his plan. If the plan pays 7% per annum compounded annually, determine the value of the plan when James celebrates his 65th birthday.

b. John, James' twin brother, made one deposit of $4000 into his RRSP on his 50th birthday. The investment earned 7% per annum compounded annually. Determine the value of John's plan when he is 65.

SOLUTION

a. A time diagram illustrating James' situation is shown.

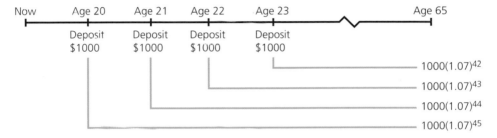

When James is 65 years old, the amount of money in his RRSP is given by
$$A = 1000(1.07)^{42} + 1000(1.07)^{43} + 1000(1.07)^{44} + 1000(1.07)^{45}.$$
This is a geometric series with $a = 1000(1.07)^{42}$, $r = 1.07$, $n = 4$.

$$S_n = \frac{a(r^n - 1)}{r - 1}$$

$$S_4 = \frac{1000(1.07)^{42}[1.07^4 - 1]}{1.07 - 1}$$

$$= 76\ 119.52$$

James will have $76 119.52 in his RRSP when he is 65 years old.

b. John made one deposit of $4000 when he was 50 years old. This investment earned 7% per annum compounded annually for 15 years.

$$\text{Amount at age } 65 = 4000(1.07)^{15}$$

$$= 11\ 036.13.$$

John has $11 036.13 in his RRSP when he is 65.

By depositing the $4000 early in his working career, James earned an additional $65 000 in interest.

EXAMPLE 3

Alison plans to buy a house in four years' time. To save for the $30 000 down payment, she decides to make equal monthly deposits at the end of each month in a savings account that pays 5.5% per annum compounded monthly. Determine the amount of the regular deposit if Alison wishes to be able to make the down payment immediately after her last deposit.

SOLUTION

Let $x represent the value of the monthly deposit. The time diagram shows the accumulated amounts of the deposits.

The period interest rate is $i = \frac{0.055}{12} \doteq 0.004\,583$. It is preferable, for accuracy in calculation, to avoid using rounded-off values such as this. A better procedure is to analyze the problem using $(1 + i)$ as a multiplier and to input the value $1 + i$ into your calculator's memory for use as needed.

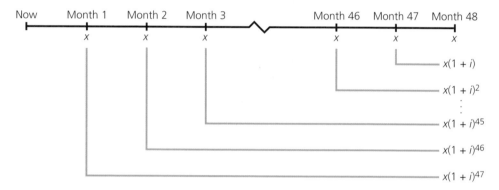

The accumulated **amount of the annuity** is
$$A = x + x(1 + i) + x(1 + i)^2 + \ldots + x(1 + i)^{46} + x(1 + i)^{47}.$$

This is a geometric series with
$$A = S_{48} = 30\,000,\ a = x,\ r = 1 + i,\ n = 48,\ 1 + i = \left(1 + \frac{0.055}{12}\right).$$

Input $\left(1 + \frac{0.055}{12}\right)$ in your calculator memory.

$$S_n = \frac{a(r^n - 1)}{r - 1}$$

$$30\,000 = S_{48} = \frac{x[(1 + i)^{48} - 1]}{(1 + i) - 1}$$

$$x = \frac{30\,000[(1 + i) - 1]}{(1 + i)^{48} - 1}$$

Using $(1 + i) = \left(1 + \frac{0.055}{12}\right)$ from your calculator memory,

$$x \doteq 560.194.$$

Alison should deposit $560.19 per month for the next four years in order to have $30 000 for the down payment.

Exercise 6.2

PART A

1. Henry deposits $150 into a savings account at the end of each month for the next six months.

Month 1	Month 2	Month 3	Month 4	Month 5	Month 6
Deposit $150	Deposit $150	Deposit $150	Deposit $150	Deposit $150	Deposit $150

Interest on the account is 6% per annum compounded monthly. How much money will be in the account in six months?

2. Draw a time diagram and calculate the amount of each of the following annuities.

   a. $2000 at the end of each year for 12 years at 5% per annum compounded annually

   b. 48 monthly payments of $300 each into an account that pays interest at 6% per annum compounded monthly

   c. $1000 at the end of every six months for seven years at 5% per annum compounded semi-annually

3. Determine the value of the regular payment x for each of the following ordinary annuities.

   a.
   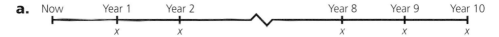

   4% per annum compounded annually for ten years to accumulate to $5000

   b.
   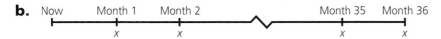

   4.8% per annum compounded monthly for three years to accumulate to $8500

4. William deposits $40 at the end of each month into a savings account. If the money earns 6% per annum compounded monthly, how much money is in the account at the end of 24 months? Draw a time diagram to help you solve the problem.

5. Craig deposits $20 every month in an account for his son. The account pays 6% per annum compounded monthly. How much money is in the account after 15 years? Draw a time diagram to help you solve the problem.

6. Describe how you can accumulate more money in a long-term savings plan. Consider the effect of changing the frequency of your deposits, the amount of the deposit, the interest rate, and the compounding period.

7. Cecilia deposits $1500 each year for three years into an RRSP. The plan earns 7% per annum compounded annually. If she makes deposits on her 18th, 19th, and 20th birthdays, what is the value of her RRSP when she is 65?

8. Stephen deposits $1500 each year for three years into an RRSP. The plan earns 7% per annum compounded annually. If he makes deposits on his 40th, 41st, and 42nd birthdays, what is the value of his RRSP when he is 65?

9. Mark's parents are saving money for his education. Each year, on Mark's birthday, they intend to deposit money into a savings account paying 8% per annum compounded semi-annually. If they plan to start the equal deposits on his fifth birthday, how much should they deposit each year to have accumulated $20 000 by his 18th birthday?

10. When Logan retires in four years, he plans to take a trip to Australia. He estimates it will cost him $8000 for the trip. How much money should he invest every six months if the investment will earn 6% per annum compounded semi-annually?

11. Joshua deposits $300 at the end of each month for three years. The account pays interest at 5.8% per annum compounded monthly.

   a. How much will be in the account after the last deposit?

   b. How much interest will Joshua earn on his deposit?

Toronto Stock Exchange
Photo: Dick Hemingway

6.3 Present Value of an Annuity

We have been investigating annuities where deposits have been made at specific time intervals. An annuity can also be purchased to provide a number of equal payments for a specified period of time. The amount of money required to purchase such an annuity is called the **present value of the annuity.**

We saw in Section 6.1 that we can calculate the amount of money P that must be invested now so that it will have value A in the future if an interest rate of i is applied for n periods. This is determined from

$$A = P(1 + i)^n$$

so $\quad P = \dfrac{A}{(1 + i)^n}.$

The present value of an annuity is the sum of the present values of each of the individual amounts specified in the annuity.

EXAMPLE 1

Mary wants to provide a $500 scholarship for an outstanding math student at her old high school. She wants one scholarship awarded every year for the next ten years, with the first scholarship given one year from now. The school can purchase an annuity, which provides $500 annual payments, at 6% per annum compounded semi-annually. How much money should Mary give to the school so that they can purchase the annuity?

SOLUTION

We draw a time diagram showing the schedule of payments and determine the present value of each amount. Note that payments are brought back to the present.

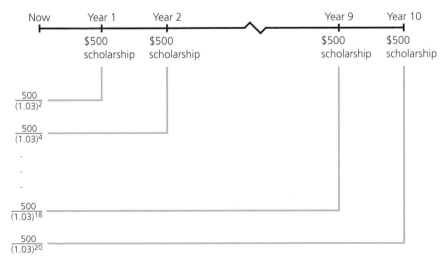

From the diagram, the present value of the last $500 is $\dfrac{500}{(1.03)^{20}}$.

The present value of all the scholarships or the annuity is

$$\frac{500}{(1.03)^2} + \frac{500}{(1.03)^4} + \frac{500}{(1.03)^6} + \ldots + \frac{500}{(1.03)^{18}} + \frac{500}{(1.03)^{20}}.$$

This is a geometric series with $a = \frac{500}{(1.03)^2}$, $r = \frac{1}{(1.03)^2}$, $n = 10$.

$$S_n = \frac{a(1 - r^n)}{1 - r}$$

$$S_{10} = \frac{\frac{500}{(1.03)^2}\left[1 - \left(\frac{1}{(1.03)^2}\right)^{10}\right]}{1 - \frac{1}{(1.03)^2}}$$

$$= 3664.40$$

If you wish to use an alternate calculation, you can add the present value amounts in reverse order, in which case $a = \frac{500}{(1.03)^{20}}$, $r = (1.03)^2$, $n = 10$, and $S_n = \frac{a(1 - r^n)}{1 - r}$.

Then $S_{10} = \frac{500}{(1.03)^{20}}\left[\frac{(1.03)^{20} - 1}{1.03^2 - 1}\right]$

$$= 3664.40$$

The present value of the annuity is \$3664.40, so Mary should give the school \$3664.40 to purchase the annuity.

EXAMPLE 2

Clara's Uncle Ron left her \$500 000. She buys a new car, some new clothes, and takes a trip. After quickly spending \$50 000, she decided to use the remaining \$450 000 to provide some income for the next 30 years. Clara buys an annuity at 6% per annum compounded annually. The annuity will provide equal annual payments to Clara. Determine the amount of each payment.

SOLUTION

Let the amount of each of the 30 annual payments be x dollars.

We can draw a time diagram to show the present value of each of the 30 payments Clara will receive.

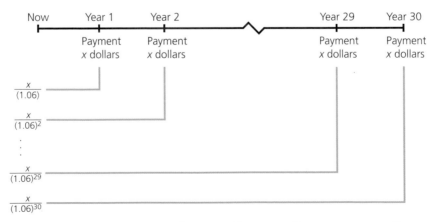

The present value of the annuity is $\frac{x}{(1.06)} + \frac{x}{(1.06)^2} + \frac{x}{(1.06)^3} + \ldots + \frac{x}{(1.06)^{29}} + \frac{x}{(1.06)^{30}}.$

This is a geometric series with $a = \dfrac{x}{(1.06)}$, $r = \dfrac{1}{(1.06)}$, $n = 30$.

$$S_n = \frac{a(1 - r^n)}{1 - r}$$

$$S_{30} = \frac{\dfrac{x}{(1.06)}\left[1 - \left(\dfrac{1}{1.06}\right)^{30}\right]}{1 - \dfrac{1}{1.06}}$$

$$\doteq 13.764\ 831\ 151\ 5x$$

The present value of the annuity is the amount of money Clara paid to purchase the annuity.

$$13.764\ 831\ 151\ 5x = 450\ 000$$

$$x \doteq 32\ 692.01$$

Clara will receive a payment of \$32 692.01 each year for the next 30 years.

The calculation in such examples is complicated because of the fractions. It is possible to solve the problem using the accumulated amount of an annuity if we also calculate interest on the original amount.

Consider the time diagram we used in the first solution. Instead of taking the present value of each payment, consider the accumulated value as indicated.

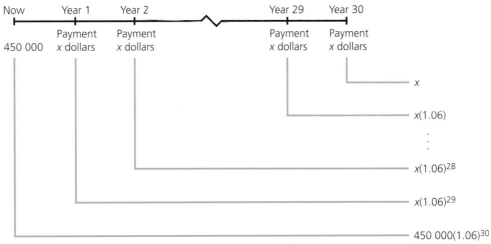

Note that we consider the accumulated value of the \$450 000 at the end of the 30-year period.

The accumulated value of the payment is

$$x + x(1.06) + x(1.06)^2 + x(1.06)^3 + \ldots + x(1.06)^{29}.$$

This is a geometric series with $a = x$, $r = 1.06$, $n = 30$.

$$S_n = \frac{a(r^n - 1)}{r - 1}$$

$$S_{30} = \frac{x(1.06^{30} - 1)}{1.06 - 1}$$

But this value is equal to the accumulated value of the $450 000.

$$\frac{x(1.06^{30} - 1)}{1.06 - 1} = 450\ 000(1.06)^{30}$$

$$\frac{x(1.06^{30} - 1)}{0.06} = 450\ 000(1.06)^{30}$$

$$x = \frac{450\ 000(1.06)^{30}(0.06)}{(1.06^{30} - 1)}$$

$$= 32\ 692.01 \text{ (the same value as in the first solution)}$$

In buying a car, most people will pay for it by making regular monthly payments. Paying for an article on a regular monthly basis is **installment buying.** We can calculate the regular monthly payment using the same method as in Example 2.

EXAMPLE 3

Brenda wants to buy a motorcycle for $5000. She has $1000 saved for the down payment and plans to borrow $4000 from the bank and repay the loan on a monthly basis over the next two years. If the bank charges 9% per annum compounded monthly, what is the value of the monthly payment?

SOLUTION

The loan is a two-year annuity with a present value of $4000.

Let the monthly payment be x dollars.

The interest rate per period is $i = \frac{0.09}{12}$ or 0.0075.

Draw a time payment diagram to represent these payments.

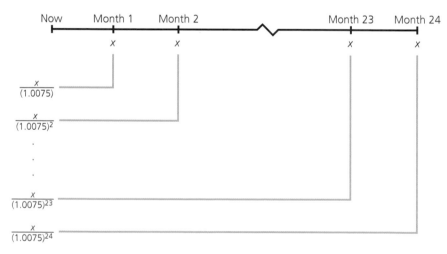

The present value of the annuity is $\frac{x}{1.0075} + \frac{x}{(1.0075)^2} + \frac{x}{(1.0075)^3} + \cdots + \frac{x}{(1.0075)^{24}}$.

This is a geometric series with $a = \frac{x}{1.0075}$, $r = \frac{1}{1.0075}$, $n = 24$.

$$S_n = \frac{a(1 - r^n)}{1 - r}$$

$$S_{24} = \frac{\frac{x}{1.0075}\left[1 - \frac{1}{(1.0075)^{24}}\right]}{1 - \frac{1}{1.0075}}$$

$$S_{24} \doteq x(21.889\ 146\ 14)$$

But this equals the value of the loan.

$$x(21.889\ 146\ 14) = 4000$$

$$x = \frac{4000}{21.889\ 146\ 14}$$

$$\doteq 182.738\ 969\ 1$$

The value of each payment is $182.74.

You can also solve this problem using the accumulated value. This leads to the equation

$$\frac{x(1.0075^{24} - 1)}{1.0075 - 1} = 4000(1.0075)^{24}.$$

Exercise 6.3

PART A

1. Draw a time diagram and calculate the present value of each of the following annuities.

a. 5 annual payments of $400 at 5% per annum compounded annually

b. 24 monthly payments of $350 at 6% per annum compounded monthly

c. $600 received at the end of every six months for $4\frac{1}{2}$ years from a fund earning 7% per annum compounded semi-annually

PART B

2. Alisha plans to attend college next year. Her grandfather wishes to deposit enough money today in an account paying 5.5% per annum compounded annually, so that she can make withdrawals of $5000 each year for four years beginning one year from now to pay for her tuition. How much should her grandfather invest today?

3. André purchases a stereo set and agrees to pay installments of $59.50 at the end of each month for the next three years. If he is charged interest at 12% per annum compounded monthly, determine the equivalent cash price of the stereo set and the total interest.

4. Majundar has won a Swell Oil Scholarship of $6000 per year for four years to attend university. How much money must Swell Oil invest now to provide the four payments starting one year from now if the money is invested at 6.6% per annum compounded monthly?

5. A giant screen television set is on sale for $3400. It can be purchased on the installment plan for three years, each payment to be made at the end of each month. If the interest is 10% per annum compounded monthly, determine the regular monthly payment.

6. Jay retires and takes a lump payment of $260 000 from the company pension plan. He purchases an annuity from which he will receive equal semi-annual payments for ten years. If money is worth 8% per annum compounded semi-annually, how much is each payment?

7. Dr. Morgan buys new equipment for his dental office. He will pay $500 per month for the next four years. What is the cash value of the dental equipment now, if interest charges are 18% per annum compounded monthly?

PART C

8. How much money must Anna deposit annually, starting on her 30th birthday, earning 5.5% per annum compounded annually to age 55, so that she can purchase an annuity at age 55, paying $1000 monthly (starting one month later) for her expected life span to age 85. When the annuity is purchased, assume money is worth 6% per annum compounded monthly.

9. a. Congratulations, you have just won $1 000 000 in a lottery. After spending $100 000, you decide to invest the remaining amount to provide you with a monthly income of $4500. If money can earn 6% per annum compounded monthly, for how many years can you withdraw this monthly income?

b. If you withdraw $10 000 per month, how long will it last?

10. Matko is 18 years old. He spends approximately $10 a week on cigarettes. He has decided to quit smoking and invest $50 a month into a savings account until he is 65.

a. Determine the accumulated amount Matko has saved by age 65, if money earns 8% per annum compounded monthly.

b. For the next ten years, after age 65, Matko wishes to withdraw an equal annual amount from his savings in part **a.** If interest is 7% per annum compounded annually, how much could Matko withdraw each year for the next ten years?

6.4 Mortgages

Most people cannot buy a home without borrowing money to do so. This loan, with the home as collateral, is called a **mortgage.** If the homeowner does not make the required mortgage payments, the bank or lending institution can foreclose on the mortgage and sell the home to recover their money.

A mortgage is an annuity with equal payments being made for the term of the mortgage, but in which the interest calculation is done differently. The equal payments consist of a principal and an interest component. Over the term of the mortgage, the principal portion increases and the interest portion decreases. However, the total monthly payment remains the same.

Photo: Dick Hemingway

The number of years it takes to repay a mortgage loan is called the **amortization period.** Mortgages usually have amortization periods between 5 and 25 years. If the amortization period is increased, there is a small reduction in the monthly payment required but a large increase in the total interest paid over the life of the mortgage.

Interest rates fluctuate from year to year. Because of these changes in the interest rates, mortgages are usually negotiated for terms from six months to five years. The length of the current mortgage agreement is called the **term of the mortgage.**

A mortgage may be amortized over a long period (such as 25 years) with a short term (such as two years). After the term expires, the balance of the principal owing can be repaid, or a new mortgage agreement can be entered into at the current interest rate.

Although a mortgage is similar to installment buying, the calculation of interest on the loan is different. For a mortgage, the interest can be compounded no more frequently than semi-annually. This means that we *cannot* apply ordinary annuity calculations to determine the monthly payments. Instead, we determine a monthly rate based on the published annual rate using the published rate on a semi-annual basis.

For example, if the published annual rate is 9% per annum compounded semi-annually, the mortgage monthly rate is determined from the equation

$$(1 + i)^{12} = (1.045)^2$$
$$1 + i = (1.045)^{\frac{2}{12}}$$
$$= (1.045)^{\frac{1}{6}}$$
$$\doteq 1.007\ 363\ 123.$$

The monthly interest rate is 0.007 363 123. Notice that this is slightly lower than the interest rate of 9% per annum compounded monthly, or 0.0075. Notice also that the difference is small, so it is important to carry as many digits as possible. For calculation purposes, it is preferable to input $(1 + i) = (1.045)^{\frac{1}{6}}$ in your calculator's memory since this avoids rounding-off problems.

You can learn more about mortgages by visiting some bank Web sites, such as the following:

www.tdbank.ca

www.cibc.com/mortgages

EXAMPLE 1

Patti and Paul purchase a home for $190 000. They have saved $40 000 for the down payment and arrange a $150 000 mortgage, amortized over 25 years. Paul thinks interest rates will be decreasing in a couple of years, so they choose a two-year term for their mortgage. The interest rate is 9% per annum compounded semi-annually. Determine their monthly mortgage payment.

SOLUTION

The mortgage is a 25-year annuity with a present value of $150 000. The interest is calculated at 9% per annum compounded semi-annually.

From above, use

$$(1 + i)^{12} = (1.045)^2$$

$$(1 + i) = (1.045)^{\frac{1}{6}}.$$

Let the monthly payment be x dollars.

In 25 years there will be 25×12 or 300 monthly payments.

We can represent each of the 300 monthly payments in a time diagram.

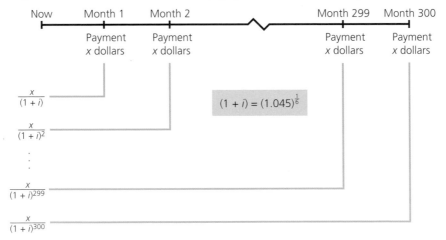

The present value of the annuity is

$$\frac{x}{(1 + i)} + \frac{x}{(1 + i)^2} + \frac{x}{(1 + i)^3} + \ldots + \frac{x}{(1 + i)^{299}} + \frac{x}{(1 + i)^{300}}.$$

This is a geometric series with $a = \dfrac{x}{(1 + i)}$, $r = \dfrac{1}{(1 + i)}$, $n = 300$.

$$S_n = \frac{a(1 - r^n)}{1 - r}$$

$$S_{300} = \frac{\dfrac{x}{(1 + i)}\left[1 - \left(\dfrac{1}{1 + i}\right)^{300}\right]}{1 - \dfrac{1}{1 + i}}$$

Now substitute $(1 + i) = (1.045)^{\frac{1}{6}}$.

$$S_{300} = \frac{\dfrac{x}{(1.045)^{\frac{1}{6}}}\left[1 - \dfrac{1}{\left[(1.045)^{\frac{1}{6}}\right]^{300}}\right]}{1 - \dfrac{1}{1.045^{\frac{1}{6}}}}$$

$$S_{300} \doteq 120.776\,244\,9x$$

The present value of the annuity is the amount of money Patti and Paul borrowed when they took out the mortgage.

$$120.776\,244\,9x = 150\,000$$

$$x \doteq 1241.97.$$

The monthly mortgage payment is $1241.97.

Let's compare this amount with the monthly payment using an interest rate compounded monthly.

Now the interest rate per investment period is $\dfrac{9}{12}\%$

$$= 0.75\%$$

$$= 0.0075.$$

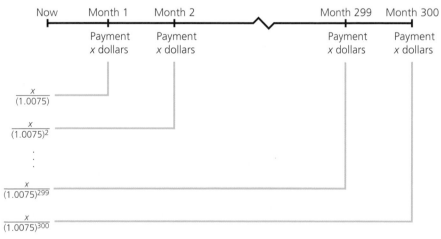

The present value of the annuity is

$$\frac{x}{(1.0075)} + \frac{x}{(1.0075)^2} + \frac{x}{(1.0075)^3} + \cdots + \frac{x}{(1.0075)^{299}} + \frac{x}{(1.0075)^{300}}.$$

This is a geometric series with $a = \dfrac{x}{(1.0075)}$, $r = \dfrac{1}{(1.0075)}$, $n = 300$.

$$S_n = \frac{a(1 - r^n)}{1 - r}$$

$$S_{300} = \frac{\dfrac{x}{(1.0075)}\left[1 - \left(\dfrac{1}{1.0075}\right)^{300}\right]}{1 - \dfrac{1}{1.0075}}$$

$$= 119.161\ 634x$$

The present value of the annuity is the amount of money Patti and Paul borrowed when they took out the mortgage.

$$119.161\ 634x = 150\ 000$$

$$x = 1258.79$$

The monthly mortgage payment is $1258.79.

Note that by calculating the monthly payment using a semi-annual compounded period, you can save $16.82 per month or $5046.00 if the interest rate does not change for the 25 years.

EXAMPLE 2

Patti and Paul's $150 000 mortgage is amortized over 25 years. The interest rate is 9% per annum compounded semi-annually. Their monthly payment is $1241.97. How much interest would they have paid by the end of the 25 years?

SOLUTION

Total of monthly payments = $1241.97 × 25 × 12

$$= \$372\ 591.00$$

Amount of interest paid = $372 591.00 − $150 000.00

$$= \$227\ 591.00$$

EXAMPLE 3

Patti and Paul's $150 000 mortgage is amortized over 25 years. The interest rate is 9% per annum compounded semi-annually. Their monthly payment is $1241.97. The term of the mortgage is two years.

a. Prepare a schedule of payments and determine how much is still owing at the end of one year.

b. How much is still owing at the end of the two-year term?

a. Part of each monthly payment goes to pay the interest owing, and the remainder of the monthly payment is used to reduce the principal.

Since $(1 + i) = (1.045)^{\frac{1}{6}}$ or $1.007\ 363\ 123$, the monthly interest rate is $0.007\ 363\ 123$.

For month 1:

$$\text{Interest} = \$150\ 000 \times 0.007\ 363\ 123$$
$$= \$1104.47$$

$$\text{Reduction of principal} = \$1241.97 - \$1104.47$$
$$= \$137.50$$

$$\text{Unpaid balance of principal} = \$150\ 000 - \$137.50$$
$$= \$149\ 862.50$$

For month 2:

$$\text{Interest} = \$149\ 862.50 \times 0.007\ 363\ 123$$
$$= \$1103.46$$

$$\text{Reduction of principal} = \$1241.97 - \$1103.46$$
$$= \$138.51$$

$$\text{Unpaid balance of principal} = \$149\ 862.50 - \$138.51$$
$$= \$149\ 723.99$$

We have to repeat this calculation for each of the 24 monthly payments. It would be much easier to set up a spreadsheet.

	A	B	C	D	E	F
2			Schedule of Mortgage Payments			
3	Payment	Principal	Monthly	Interest	Principal	Unpaid
4	Number		Payment			Balance
5	1	150000.00	1241.97	1104.47	137.50	149862.50
6	2	149862.50	1241.97	1103.46	138.51	149723.98
7	3	149723.98	1241.97	1102.44	139.53	149584.45
8	4	149584.45	1241.97	1101.41	140.56	149443.89
9	5	149443.89	1241.97	1100.37	141.60	149302.29
10	6	149302.29	1241.97	1099.33	142.64	149159.65
11	7	149159.65	1241.97	1098.28	143.69	149015.97
12	8	149015.97	1241.97	1097.22	144.75	148871.22
13	9	148871.22	1241.97	1096.16	145.81	148725.41
14	10	148725.41	1241.97	1095.08	146.89	148578.52
15	11	148578.52	1241.97	1094.00	147.97	148430.55
16	12	148430.55	1241.97	1092.91	149.06	148281.49
17	13	148281.49	1241.97	1091.81	150.16	148131.34
18	14	148131.34	1241.97	1090.71	151.26	147980.08
19	15	147980.08	1241.97	1089.60	152.37	147827.70
20	16	147827.70	1241.97	1088.47	153.50	147674.21
21	17	147674.21	1241.97	1087.34	154.63	147519.58
22	18	147519.58	1241.97	1086.20	155.77	147363.81
23	19	147363.81	1241.97	1085.06	156.91	147206.90
24	20	147206.90	1241.97	1083.90	158.07	147048.83
25	21	147048.83	1241.97	1082.74	159.23	146889.60
26	22	146889.60	1241.97	1081.57	160.40	146729.20
27	23	146729.20	1241.97	1080.39	161.58	146567.61
28	24	146567.61	1241.97	1079.20	162.77	146404.84

At the end of year 1, the amount owing is $148 281.49.

b. At the end of the two-year term, Patti and Paul still owe $146 404.84. Most of the money they have paid in monthly payments has gone to pay the interest charges. In the two years they have paid $26 212.12 in interest and $3595.16 on the principal.

A schedule of mortgage payments is called an **amortization schedule.**

Patti and Paul are now at the end of the two-year term in their mortgage agreement. They still owe $146 404.84. They must now **refinance** their mortgage. This means they must renegotiate their existing mortgage for this amount or, if they wish to pay off part of the principal, for a lesser amount.

EXAMPLE 4

Patti and Paul are refinancing their $146 404.84 mortgage. Paul's father recently died and left Paul $100 000. Paul decides to use most of this money to reduce his mortgage. He now wants a $50 000 mortgage amortized over five years and with a five-year term. The interest rate is now 6% per annum compounded semi-annually. Calculate the monthly payment on the mortgage.

SOLUTION

Convert 6% per annum compounded semi-annually to the equivalent rate compounded monthly.

6% compounded semi-annually is $\frac{0.06}{2} = 0.03$.

$$(1 + i)^{12} = (1.03)^2$$
$$1 + i = (1.03)^{\frac{1}{6}}$$
$$= 1.004\ 938\ 622$$

There will be 5×12 or 60 monthly payments.

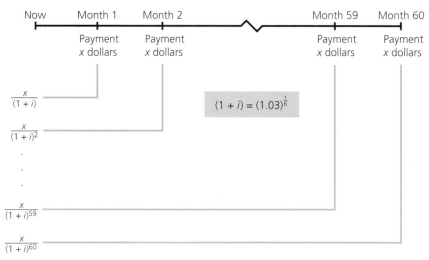

The present value of the annuity is

$$\frac{x}{(1 + i)} + \frac{x}{(1 + i)^2} + \frac{x}{(1 + i)^3} + \ldots + \frac{x}{(1 + i)^{59}} + \frac{x}{(1 + i)^{60}}.$$

This is a geometric series with $a = \dfrac{x}{(1+i)}$, $r = \dfrac{1}{(1+i)}$, $n = 60$.

$$S_n = \frac{a(1-r^n)}{1-r}$$

$$S_{60} = \frac{\dfrac{x}{(1+i)}\left[1 - \left(\dfrac{1}{1+i}\right)^{60}\right]}{1 - (1+i)^{\frac{1}{6}}}$$

Now substitute $(1+i) = (1.03)^{\frac{1}{6}}$.

$$S_{60} = \frac{\dfrac{x}{(1.03)^{\frac{1}{6}}}\left[1 - \left(\dfrac{1}{(1.03)^{\frac{1}{6}}}\right)^{60}\right]}{1 - \dfrac{1}{(1.03)^{\frac{1}{6}}}}$$

$$\doteq 51.817\ 305\ 21x$$

The present value of the annuity is the face value of the $50 000 mortgage.

$$51.817\ 305\ 21x = 50\ 000$$

$$x \doteq 964.928\ 604\ 4$$

The monthly mortgage payment is $964.93.

EXAMPLE 5

Use a spreadsheet to set up an amortization schedule for the $50 000 mortgage in Example 4.

SOLUTION

Only the beginning and the end of the amortization table are shown.

	A	B	C	D	E	F
1			Schedule of Mortgage Payments			
2	Payment	Principal	Monthly	Interest	Principal	Unpaid
3	Number		Payment			Balance
4	1	50000	964.93	246.93	718.00	49282.00
5	2	49282.00	964.93	243.45	721.48	48560.52
6	3	48560.52	964.93	239.89	725.04	47835.48
59	56	4756.39	964.93	23.50	941.43	3814.96
60	57	3814.96	964.93	18.85	946.08	2868.87
61	58	2868.87	964.93	14.17	950.76	1918.11
62	59	1918.11	964.93	9.48	955.45	962.66
63	60	962.66	964.93	4.76	960.17	2.49

Note that to pay off the mortgage, $2.49 will be added to the last payment.

PART A

1. A six-month term mortgage has monthly payments, including principal and interest, of $1252.40. Calculate the total paid in monthly payments during the six-month term.

2. A three-year term mortgage has monthly payments, including principal and interest, of $823.64. Calculate the total paid in monthly payments during the three-year term.

PART B

3. Ricardo purchased a condominium for $120 000. He paid $25 000 down and arranged a $95 000 mortgage amortized over 20 years. The mortgage has a three-year term at 6% per annum compounded semi-annually. Calculate the monthly payment.

4. Stephanie purchased a townhouse for $180 000. She paid $35 000 down and arranged a $145 000 mortgage amortized over 25 years. The mortgage has a five-year term at 9% per annum compounded semi-annually. Calculate the monthly payment.

5. a. Use a spreadsheet to create an amortization table for a $70 000 mortgage amortized over 15 years. The monthly payment, including principal and interest, is $587.92. Interest rate is 6% per annum compounded semi-annually.

    b. What is the unpaid balance of the principal at the end of five years?

6. Joe and Sheila have just arranged a $100 000 mortgage amortized over 25 years. The interest rate is 9% per annum compounded semi-annually. The monthly payment is $827.98.

    a. How much interest would Joe and Sheila pay over the 25 years?

    b. Use a spreadsheet to create an amortization schedule for the mortgage.

    c. How much would Joe and Sheila still owe at the end of year 10?

    d. Joe and Sheila decide to pay an extra $50 every month starting with their first mortgage payment. Adjust your spreadsheet to reflect the monthly payment of $877.98. How much would they owe at the end of year 10?

    e. When would the mortgage be paid off?

    f. How much in interest charges would Joe and Sheila save by paying an extra $50 each month?

7. a. Use a spreadsheet to create an amortization table for a $50 000 mortgage amortized over 20 years. Payments are to be made semi-annually. The semi-annual payment, including principal and interest, is $520.65. Interest rate is 8% per annum compounded semi-annually.

   b. What is the unpaid balance of the principal at the end of ten years?

8. Often mortgagors can choose the length of period of the mortgage.

   a. If the mortgage is $80 000 and the interest rate is 8% per annum compounded semi-annually, find both the monthly payment and the total cost of the mortgage for each of the following periods.

   i) 15 years

   ii) 25 years

   iii) 30 years

   b. What is the difference between the monthly payments for the mortgage over 15 years and the mortgage over 25 years? What is the saving in total costs for the mortgage over 15 years as compared to 25 years?

   c. What is the difference between the monthly payments for the mortgage over 25 years and the mortgage over 30 years? What is the saving in total costs for the mortgage over 25 years as compared to 30 years?

   d. What is the major benefit of a shorter period?

   e. Why might some mortgagors be unable to take advantage of a shorter period?

   f. Based on the results of this question, what advice could you offer a mortgagor wishing to reduce total costs?

9. The mortgage interest rate is one factor over which mortgagors have little control. Mortgage rates fluctuate from week to week, and from year to year.

   a. If the mortgage is $60 000 and the period is 25 years, find both the monthly payment and the total cost of the mortgage when the interest rate is compounded semi-annually at the following rates per annum.

   i) 6%

   ii) 8%

   iii) 10%

   b. Comment on how the prevailing interest rate for mortgages could influence which properties you could afford to buy.

   c. High rates of interest often cause sellers to lower their asking price. Why do you think this is true?

10. The size of the down payment you can make on a property may influence what property you are able to afford.

   a. The purchase price of a property is $150 000 with an interest rate of 8% per annum compounded semi-annually over a period of 20 years. Find both the monthly payment and the total cost of the property over the 20-year period, given each of the following down payments on this property.

   i) $50 000

   ii) $70 000

   iii) $100 000

   b. Using the results of your calculations, summarize how increasing the down payment influences the monthly payments and total cost of the property.

   c. What may be some disadvantage of saving for years, so as to have a large down payment, instead of purchasing a home with a small down payment?

11. Describe ways in which you could reduce the total amount of interest that you pay over the life of a mortgage. Consider the effect of changing the payment frequency, or the interest rate, or the length of time that it takes to pay off a mortgage.

PART C

12. a. Carol bought a cottage for $85 000. She paid $20 000 down and financed the rest with a $65 000 mortgage amortized over 15 years. The mortgage has a five-year term and payments are to be made every three months, with an interest rate of 8% per annum compounded semi-annually. Determine the quarterly payment.

   b. Use a spreadsheet to create an amortization table. Determine the principal at the end of the five-year term.

   c. At the end of the five-year term, Carol makes a lump sum payment on the principal of $20 000 and refinances the remainder of the mortgage. She arranges a ten-year amortization with yearly payments. The interest rate is 7% per annum compounded annually and the mortgage has a two-year term. Determine the payment Carol is to make annually.

1. You invest $5000 for five years. Determine the total interest earned if the investment earns 7% per annum

   a. compounded annually,

   b. compounded monthly.

2. Joel invests $7500 at 6% per annum compounded semi-annually, for five years.

   a. What is the amount of the investment at maturity?

   b. How much interest did Joel earn?

3. Calculate the present value of each of the following:

   a. $10 000 in five years at 8% per annum compounded annually

   b. $5000 in three years at 9% per annum compounded monthly

4. Janet bought a scooter. She paid $500 down and agreed to pay another $700 in two years. If money is worth 8% per annum compounded quarterly, what is the cash value of the scooter today?

5. Calculate the amount of each of the following annuities.

   a. $500 deposits at the end of each month for three years at 6% per annum compounded monthly

   b. $2000 deposits at the end of each year for nine years at 5.3% per annum compounded annually

6. Calculate the present value of an annuity, if $150 payments are to be made monthly for four years and the interest rate is 9% per annum compounded monthly.

7. Jatinder wishes to purchase a sound system. She can afford to make a down payment of $200 and monthly payments of $100 for 18 months. If the expected interest charges are 15% per annum compounded monthly, what is the cash value of the most expensive system she can afford to purchase today?

8. Brian makes $2500 deposits into his RRSP on his 20th, 25th, 30th, 35th, and 40th birthdays. The RRSP pays 7% per annum compounded annually. Determine the value of Brian's RRSP when he is 65 years old.

9. Wendy gives an endowment to her university to provide additional funds for research into a cure for cancer. Ten thousand dollars is to be given to the research team every year for ten years, starting one year from now. An annuity is purchased earning 8% per annum compounded semi-annually. How much money did Wendy give to the university to purchase the annuity?

10. Kimberly buys a ski chalet for $105 000. She pays $30 000 down and arranges a $75 000 mortgage amortized over 15 years. The interest rate is 9% per annum compounded semi-annually. Calculate the monthly payment.

11. a. Use a spreadsheet to create an amortization table for a $90 000 mortgage amortized over 20 years. Interest rate is 6% per annum compounded semi-annually. The monthly payment is $640.97.

b. How much is still owing on the mortgage

i) after five years, **ii)** after ten years.

12. Susan and Kyle bought a house for $180 000. They made a down payment of $45 000 and arranged a 35-year, 9% mortgage with their bank.

a. Calculate the monthly payment.

b. How much interest have they paid on their mortgage?

c. Calculate the monthly payment if they had arranged a 30-year mortgage.

d. How much could they have saved if they had arranged a 30-year mortgage instead of a 35-year mortgage?

e. What else could Susan and Kyle do to reduce the interest charges they will pay over the life of the mortgage?

f. Explain why each of these will reduce the interest cost.

1. a. Can you determine three consecutive positive integers whose sum is 150? 165? 170?

  b. Can you determine four consecutive positive integers whose sum is 125? 122? 160?

  c. Can you determine five consecutive positive integers whose sum is 150? 165? 170?

  d. What is the shortest string of consecutive positive integers that add up to 399? What is the longest string?

2. For a given arithmetic sequence, the ratio of the sum of the first m terms to the sum of the first n terms is $m^2:n^2$. Show that the ratio $t_m:t_n = (2m - 1):(2n - 1)$.

3. An acute-angled triangle has sides that are three consecutive integers. The altitude drawn to the second largest side divides that side into two parts. Show that these parts differ by 4 no matter what the consecutive integers are, as long as they are all at least 4.

4. Skiing Sam dearly wishes to take a year for non-stop skiing. To realize his dream, he estimates that he will need $25 000. Therefore, he plans to deposit $200 at the end of each month in an account paying 6.5% per annum compounded monthly. How long must Sam work before he can achieve his dream?

5. From $1 + x + x^2 + \ldots + x^{n-2} + x^{n-1} = \dfrac{x^n - 1}{x - 1}$, by multiplying both sides by $x - 1$ and reversing the order of the sum, we obtain
$$x^n - 1 = (x - 1)(x^{n-1} + x^{n-2} + \ldots + x^2 + x + 1).$$

  a. Use this fact to factor each of the following:

  i) $x^5 - 1$ **ii)** $x^5 - 32$

  iii) $a^3 - b^3$ **iv)** $x^{11} - y^{11}$

  v) $x^3 + 1$ **vi)** $x^5 + 32$

  b. Show that $(1 + x + x^2 + x^3 + x^4)(1 + x + x^2 + \ldots + x^9) = \dfrac{x^{15} - x^{10} - x^5 + 1}{(x - 1)^2}$.

1. Evaluate each of the following:

   a. $\left(-\dfrac{5}{3}\right)^{-3}$

   b. $(2^0 + 3^{-1})^2$

   c. $(-8)^{-\frac{2}{3}}$

   d. $\dfrac{32^4}{8^3 \times 16^2}$

   e. $3^{-1} + 4^0 - 8^{\frac{2}{3}}$

   f. $\dfrac{\sqrt{5} \times 5^{\frac{5}{2}} \times 125^{-1}}{25^{\frac{3}{2}}}$

2. Simplify each of the following:

   a. $\dfrac{(2m^2)^{-2}}{(3m^3)^{-3}}$

   b. $\dfrac{6(ab^2)^3}{(2a^2b)^2}$

   c. $\dfrac{9^2k^3}{27k^{-1}}$

   d. $\dfrac{2^{-1} - 2^{-2}}{2^{-2} + 2^{-1}}$

   e. $\dfrac{x^{2p-2g}}{x^{p-g}}$

   f. $\sqrt[3]{\dfrac{\sqrt{x} \cdot x^{\frac{2}{3}}}{x^{-1}}}$

3. Solve each of the following:

   a. $3^{x+4} - 4 = 23$

   b. $\left(\dfrac{1}{9}\right)^{x-2} = \left(\dfrac{1}{27}\right)^{x+1}$

   c. $(5^{x-1})^x = 125^2$

   d. $2(5^{2x}) = 5 - 3(5^x)$

   e. $2^{2x} - 12(2^x) + 32 = 0$

   f. $6^x = 321$

4. Determine the first 4 terms of each of the following sequences.

   a. $t_n = 5n - 3$

   b. $t_n = 2^n - 1$

   c. $f(n) = n^3 + 1, n \in N$

   d. $t_k = 2t_{k-1} + 3, t_1 = -2, k > 1$

5. Each of the following sequences is either arithmetic or geometric. In each case, determine the type of sequence, the general term, and the indicated term.

   a. $4, 7, 10, 13, \ldots$ Determine t_{12}.

   b. $3, -1, \dfrac{1}{3}, -\dfrac{1}{9}, \ldots$ Determine t_8.

   c. $14, 11, 8, 5, \ldots$ Determine t_{20}.

   d. $4, \dfrac{2}{5}, \dfrac{1}{25}, \dfrac{1}{250}, \ldots$ Determine t_9.

6. Each of the following series is either arithmetic or geometric. In each case, determine the type of series and the sum of the series.

   a. $16 - 8 + 4 - 2 + \ldots$ Find S_{12}.

   b. $16 + 19 + 22 + 25 + \ldots$ Find S_{52}.

   c. $16 + 11 + 6 + 1 + \ldots + (-294)$

   d. $16 + 32 + 64 + 128 + \ldots + 32\,768$

7. Determine t_{14} in the geometric series where $t_5 = 48$ and $t_9 = 768$.

8. The sum of n terms of the arithmetic series $4 + 1 - 2 - \ldots$ is -1850. Find n.

9. Determine the value of x such that $x + 3, x + 2$, and $x - 4$ are consecutive terms in each of the following:

   a. a geometric sequence

   b. an arithmetic sequence

10. Simon invests $6000 at 8% per annum compounded semi-annually for ten years.

   a. What is the amount of the investment at maturity?

   b. How much interest did Simon earn?

11. Calculate the present value of $9000 in four years at 6% per annum compounded monthly.

12. Stephen borrows $1000 from his uncle. He agrees to repay the loan in five years. Stephen's uncle tells him that he wants the interest compounded quarterly. When Stephen repays the loan in five years, he owes his uncle $1485.95. What annual interest rate did Stephen's uncle charge?

13. Tania is saving money for a trip to visit her grandmother in Germany. She deposits $100 at the end of each month. The account pays 4% per annum compounded monthly. How much money will be in the account after five years?

14. Joanna buys new furniture for her living room. She will pay installments of $99.88 at the end of each month for the next four years. If she is charged interest at 18% per annum compounded monthly, determine the equivalent cash price of the furniture and the total interest she will pay.

15. Christina wants to buy a condominium. The condominium costs $140 000. She has saved $35 000 for the down payment but wonders whether she can afford the monthly payments. A mortgage amortized over 25 years, with a five-year term, has monthly payments with an interest rate of 6% per annum compounded semi-annually.

   a. Determine Christina's monthly payment.

   b. Determine the total interest that would be paid over the life of the mortgage.

16. State the domain and range of each of the following:

   a. $\{(4, 0), (-3, 5), (6, 2), (4, -2)\}$ **b.** $f(x) = 2x^2 + 3$

   c. $f(x) = \dfrac{4}{x + 1} - 2$ **d.** $f(x) = \sqrt{x + 3} - 1$

17. Given $f(x) = 3 - 2x$, determine the equation of its inverse.

18. If $f(x) = -2(x - 3)^2$, determine $f(-2), f\left(\dfrac{1}{4}\right)$, and $f(3x)$.

19. Simplify each of the following:

   a. $3x^2(2x - 4x^3)$ **b.** $(3x^2 - 1)(x + 2)$

   c. $(2x - 1)(x^2 + 3x - 4)$ **d.** $2 - x(x + 1)^2$

20. Factor completely each of the following:

   a. $3x^2 - 75$ **b.** $2x^2 + x - 3$

   c. $x^4 + 8x^2 - 9$ **d.** $(x^2 - 3)^2 - 4(x^2 - 3)$

21. Simplify each of the following:

   a. $\dfrac{x^2 - 4}{2x - 4}$

   b. $\dfrac{4x^2 + 20x + 25}{4x^2 - 25} \div \dfrac{1}{2x^2 - 11x + 15}$

   c. $\dfrac{2x}{x^2 + 2x} - \dfrac{x + 2}{2x}$

22. Solve the following:

   a. $2x^2 - 10x - 11 = 0$ **b.** $x^2 - 3x - 10 \geq 0$ **c.** $4x^2 + 8x + 13 = 0$

23. The difference between three times a number and its square is 378. Find the number.

24. Find three consecutive even integers such that the sum of the square of the smallest integer and the product of the other two integers is 928.

25. A deck of uniform width is built around a rectangular pool 12 m by 20 m. If the area of the deck is 185 m², find the width of the deck.

26. Determine the nature of the roots of each of the following quadratic equations.

   a. $3x^2 + 2x - 4 = 0$ **b.** $x^2 - 5x + 8 = 0$

27. For each of the following, determine the coordinates of the vertex and indicate whether the function has a maximum or minimum value. Sketch the graph.

   a. $y = 3x^2 - 24x + 53$ **b.** $y = -2x^2 + 12x - 20$

28. The sum of two numbers is 60. Find the numbers if their product is a maximum.

29. A rectangular field is to be enclosed by a fence unit, then divided into two smaller plots by a fence parallel to one of the sides. If 1200 m of fencing is used, what are the dimensions of the field if its area is to be a maximum?

Chapter 7
Trigonometric Ratios

When would you need to calculate inaccessible heights and distances? Frequently, occasions arise in construction, surveying, and navigation where it is necessary to find lengths such as the distance across a river, the height of a cliff or mountain, the depth of water, the height of a building, the distance between aircraft, angle measures such as the angle of elevation to the sun, or the angle of depression in a gorge. Trigonometric ratios, combined with the Pythagorean Theorem, can be used to calculate these seemingly inaccessible heights and distances. Trigonometric ratios will give you a variety of approaches to solve such problems, then check solutions with variations in the steps. Trigonometry will provide tools to use measurements you know to determine measurements you do not know.

IN THIS CHAPTER, YOU CAN . . .

- apply sine, cosine, and tangent ratios to determine side lengths and angle measures in right triangles;
- determine angle measures for coterminal angles;
- determine the sine, cosine, and tangent of angles greater than 90°;
- solve problems in two dimensions and in three dimensions involving oblique triangles;
- apply the law of sines and the law of cosines.

Trigonometry as a subject grew from simple geometric concepts. The two geometric ideas that provide the foundations for trigonometry are the Theorem of Pythagoras and the Similar Triangles Theorem.

Pythagorean Theorem

If a triangle is right-angled, the square on the hypotenuse is equal to the sum of the squares on the other two sides; that is, $a^2 = b^2 + c^2$.

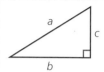

Similar Triangles Theorem

If two triangles ABC and PQR have corresponding angles equal,

$$\frac{AB}{PQ} = \frac{AC}{PR} = \frac{BC}{QR}$$

or, by rearranging any pair,

$$\frac{AB}{BC} = \frac{PQ}{QR}, \frac{AB}{AC} = \frac{PQ}{PR}, \frac{AC}{BC} = \frac{PR}{QR}.$$

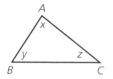

EXAMPLE 1

In the diagram, $\angle ABC = 90°$, AB is parallel to ED, $AB = 40$, $EC = 17$, and $DC = 15$. Determine the lengths of AC and BC.

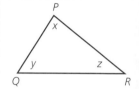

SOLUTION

Since ED is parallel to AB, $\angle EDC = 90°$.

Triangle ABC is similar to $\triangle EDC$ because $\angle ABC = \angle EDC$ and $\angle ECD$ is common to both triangles. Therefore $\frac{AB}{ED} = \frac{BC}{DC} = \frac{AC}{EC}$.

By Pythagoras, $\qquad ED^2 + DC^2 = EC^2$

$$ED^2 + 225 = 289$$

$$ED^2 = 64$$

$$ED = 8, \text{ since } ED > 0.$$

Then $\dfrac{AB}{ED} = \dfrac{BC}{DC}$

$\dfrac{40}{8} = \dfrac{BC}{15}$

$BC = \dfrac{15 \times 40}{8}$

$= 75.$

Also $\dfrac{AB}{ED} = \dfrac{AC}{EC}$

$\dfrac{40}{8} = \dfrac{AC}{17}$

$AC = \dfrac{17 \times 40}{8}$

$= 85.$

The required lengths are $AC = 85$, $BC = 75$.

In $\angle ABC$, at points P and R in arm BC, perpendiculars PQ and RS are drawn, as in the diagram. By the properties of similar triangles, $\dfrac{PQ}{BQ} = \dfrac{RS}{BS}$.

This is true no matter where the points P and R are chosen on BC.

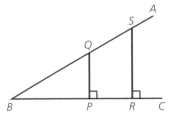

We conclude that for a given $\angle ABC$, the value of the ratio $\dfrac{PQ}{BQ}$ is unique for that angle. The same is true for the ratios $\dfrac{BP}{BQ}$ and $\dfrac{PQ}{BP}$.

These ratios define the primary trigonometric ratios.

The Primary Trigonometric Ratios

In any right triangle ABC, where $\angle C$ is the right angle, the primary trigonometric ratios are

$\sin B = \dfrac{\text{Opposite}}{\text{Hypotenuse}} = \dfrac{AC}{AB}$

$\cos B = \dfrac{\text{Adjacent}}{\text{Hypotenuse}} = \dfrac{BC}{AB}$

$\tan B = \dfrac{\text{Opposite}}{\text{Adjacent}} = \dfrac{AC}{BC}$

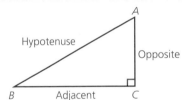

EXAMPLE 2

Triangle ABC has $AB = 7$, $BC = 4$, and $\angle ABC = 90°$.

a. Determine the exact value of AC.

b. Determine the three basic trigonometric ratios for $\angle BAC$ to three decimal places.

SOLUTION

a. By Pythagoras,

$AC^2 = 7^2 + 4^2$

$AC^2 = 65$

$AC = \sqrt{65}$, $AC > 0$.

b. $\sin \angle BAC = \dfrac{4}{\sqrt{65}} \doteq 0.496$

$\cos \angle BAC = \dfrac{7}{\sqrt{65}} \doteq 0.868$

$\tan \angle BAC = \dfrac{4}{7} \doteq 0.571$

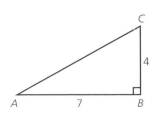

EXAMPLE 3

Radar tells an air-controller that an approaching plane is at an altitude of 800 m and that the angle of approach relative to the ground is 21.5°. Determine the distance PT from the plane to the control tower to the nearest metre.

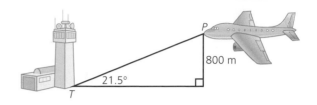

SOLUTION

In the diagram, P is the plane's position, T is the tower's position, and X is the foot of the perpendicular from P to the ground.

$$\sin T = \frac{PX}{TP}$$

Then $\sin 21.5° = \frac{800}{TP}$,

so $\quad\quad TP = \frac{800}{\sin 21.5°}$,

$\quad\quad\quad\quad \doteq 2182.8.$

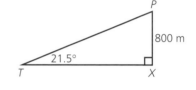

The plane is 2183 m from the tower.

EXAMPLE 4

A salvage company is using two ships to search for a wrecked ship on the ocean floor. The diagram shows the data collected when the two salvage ships are 2500 m apart and make a sonar contact with the wreck. Determine the depth of water above the wreck to the nearest metre.

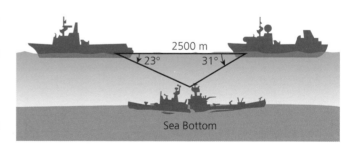

SOLUTION

In the diagram, A and B represent the positions of the ships, and C represents the top of the sunken ship.

Draw $CD \perp AB$.

Let CD be h metres and AD be x metres.

Then DB is $(2500 - x)$ metres.

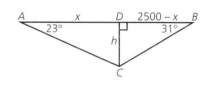

In $\triangle ACD$, $\tan 23° = \dfrac{h}{x}$ or $h = x \tan 23°$.

In $\triangle BCD$, $\tan 31° = \dfrac{h}{2500 - x}$ or $h = (2500 - x)\tan 31°$.

$$x \tan 23° = (2500 - x)\tan 31°$$
$$x \tan 23° = 2500 \tan 31° - x \tan 31°$$
$$x \tan 23° + x \tan 31° = 2500 \tan 31°$$
$$x(\tan 23° + \tan 31°) = 2500 \tan 31°$$
$$x = \dfrac{2500 \tan 31°}{\tan 23° + \tan 31°}$$
$$\doteq 1465.03$$
$$h = 1465.03 \tan 23°$$
$$\doteq 621.87$$

The water above the wreck is 622 m deep.

If the value of a trigonometric ratio is known, the measure of the angle can be determined by using your calculator.

If $\sin \theta = 0.438$, we determine that $\theta \doteq 26°$. We write this as $\sin^{-1} 0.438 \doteq 26°$ and we read it as "inverse sine of 0.438 is approximately 26°."

Recall that this same notation was used to indicate the inverse of a function $f(x)$. Is there a connection? Yes, there is. In fact, as you will see in the next chapter, $y = \sin \theta$ is itself a function.

EXAMPLE 5

In each of the following, determine the value of x to the nearest degree.

a.

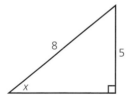

b.

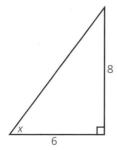

SOLUTION

a. $\sin x = \dfrac{5}{8}$ and $\sin^{-1}\left(\dfrac{5}{8}\right) \doteq 38.7°$.
Therefore x is $39°$ to the nearest degree.

b. $\tan x = \dfrac{8}{6}$ and $\tan^{-1}\left(\dfrac{8}{6}\right) \doteq 53.1°$.
Therefore x is $53°$ to the nearest degree.

Special Trigonometric Ratios

There are some angles that occur sufficiently frequently that their trigonometric ratios should be known exactly, rather than as decimal approximations.

The 45°, 45°, 90° Triangle

In a triangle with angles of 45°, 45°, 90°, let $AB = BC = 1$.

$AC^2 = 1 + 1$ and $AC = \sqrt{2}$.

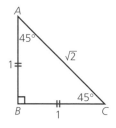

$$\sin 45° = \frac{1}{\sqrt{2}}, \cos 45° = \frac{1}{\sqrt{2}}, \tan 45° = 1$$

The 30°, 60°, 90° Triangle

In $\triangle ABC$ with angles 30°, 60°, 90°, extend BC to D such that $BC = CD$.

Since $\triangle ABC$ is congruent to $\triangle ACD$, $AB = AD$, $\angle CAD = 30°$, and $\angle ADC = 60°$. Therefore $\triangle ABD$ is equilateral.

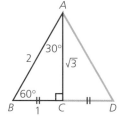

If we let $AB = 2$, then $BC = 1$.

$$AC^2 = AB^2 - BC^2$$
$$= 4 - 1$$
$$= 3$$
$$AC = \sqrt{3}$$

$\sin 30° = \frac{1}{2}$	$\sin 60° = \frac{\sqrt{3}}{2}$
$\cos 30° = \frac{\sqrt{3}}{2}$	$\cos 60° = \frac{1}{2}$
$\tan 30° = \frac{1}{\sqrt{3}}$	$\tan 60° = \sqrt{3}$

EXAMPLE 6

Evaluate $\sin^2 60° + \cos^2 60°$.

SOLUTION

Note that $\sin^2 60° = (\sin 60°)^2$.

$$\sin^2 60° + \cos^2 60° = \left(\frac{\sqrt{3}}{2}\right)^2 + \left(\frac{1}{2}\right)^2$$
$$= \frac{3}{4} + \frac{1}{4}$$
$$= 1$$

PART A

1. State the equal ratios in the following pairs of similar triangles.

a.    **b.** 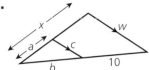   **c.**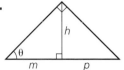

2. Solve for the variable x in each of the following:

a. 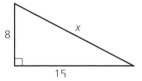   **b.** 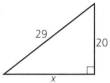   **c.**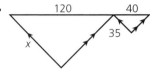

3. In each of the following, determine the indicated variable.

a. $\dfrac{p}{22} = \dfrac{26}{44} = \dfrac{39}{q}$ **b.** $\dfrac{h}{26} = \dfrac{52}{d} = \dfrac{140}{65}$

4. For each of the following, state the exact value for each of the primary trigonometric ratios of angle θ.

a. 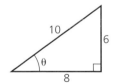   **b.** 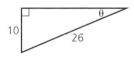   **c.**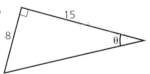

5. Use a calculator to determine each of the following, correct to four decimals.

a. cos 33°, sin 33°   **b.** sin 57.5°, tan 57.5°

c. sin 17.3°, cos 72.7°   **d.** sin 48.8°, cos 41.2°

6. For each of the following trigonometric equations, use a calculator to determine the value of the acute angle θ, correct to one decimal.

a. cos θ = 0.8390   **b.** sin θ = 0.8434

c. tan θ = 0.310 51   **d.** 2 sin θ = 1.317 982

7. Solve for the variable indicated in each of the diagrams given.

a.

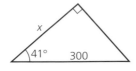

b.

c.

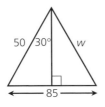

d.

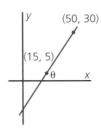

PART B

8. Solve for the lengths *a* and *b* in each of the following:

a.

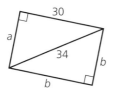

b.

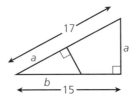

c.

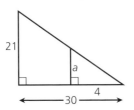

d.

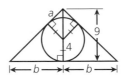

9. The distance from the base of a tall building to a point *A* is 150 m and ∠*BAC* = 53.1°. Determine the height of the building to the nearest metre.

10. A perpendicular *AW* is drawn to base *BC* in △*ABC*. If *AW* = 12, *BW* = 5, and *WC* = 35, determine if △*ABC* contains a right angle.

11. In each of the following, state the exact values for the three primary trigonometric ratios of angles β and α.

a.

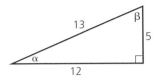

b.

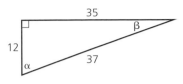

c.

d.

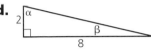

12. A group of students measured a distance of 50 m from the base of a pole *PB* to a point *A*. If ∠*BAP* = 16.8°, calculate the height of the pole, correct to one decimal.

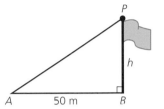

13. When the **angle of elevation** of the sun is 33°, a flagpole casts a shadow 20 m in length. Determine the length of its shadow when the elevation of the sun is 40°, correct to one decimal.

14. A telephone pole casts a shadow 6 m long when the angle of elevation of the sun is 62°. Determine the angle of elevation of the sun when the pole's shadow is 20 m long, correct to one decimal.

15. An architect is designing a house with a garden so the wall length *LB* is 30 m and a thick garden hedge *LA* is 55 m. From a window at *L*, what is the viewing angle if the back edge of the garden *AB* is perpendicular to the wall *LB*?

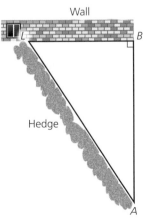

16. From the top of a building 60 m high, the **angle of depression** of a car on the road is 37°. (The angle of depression is the angle between the horizontal and the line of sight of an object.) Calculate the distance from the base of the building to the car.

17. From the top of a building 100 m high, the angle of depression to the foot of a second building is 24°. From the same point, the angle of elevation to the top of the second building is 7°. Calculate the height of the second building.

18. From a balloon that is 1000 m above the ground, an object is sighted on the ground at an angle of depression of 41°. An observer in a second balloon, positioned so that both balloons and the object on the ground are in the same vertical plane, is 600 m above the ground and views the object at an angle of depression of 28°. Determine all possible positions of the second balloon relative to the first balloon, and calculate the horizontal distance between them in each case.

19. Evaluate each of the following. Give exact answers.

    a. $\sin^2 30° + \cos^2 30°$ **b.** $\tan^2 60° + 2 \tan^2 45°$

    c. $(\sin 30°)(\cos 45°)(\tan 60°)$ **d.** $\dfrac{2}{\sin^2 45°} - \dfrac{3}{\cos^2 30°}$

    e. $2 \sin 45° + \dfrac{1}{2 \cos 45°}$ **f.** $(\sin 60° + \cos 30°)(\sin 60° - \cos 30°)$

    g. $\sin 30° + \cos 30°$ **h.** $(\tan^2 45°)(\sin 60°)(\tan 30°)(\tan^2 60°)$

20. a. Triangle ABC has a base $BC = 115$ cm, side $BA = 80$ cm, and $\angle B = 35°$. Determine the height h from the vertex A and the area of $\triangle ABC$.

    b. A result from early trigonometry states: "If in $\triangle ABC$ $CA = b$, $CB = a$, and $\angle C$ is known, then the area of $\triangle ABC = \frac{1}{2}ab \sin C$." Prove this theorem.

PART C

21. a. Two tangents from a point A are drawn to a circle with centre C and radius 12 cm. If the tangents make an angle of 43° with each other at A, then find the length of each tangent, correct to one decimal.

    b. Determine the area of quadrilateral $ABCD$ where B and D are the points of contact of the tangents with the circle, correct to one decimal.

22. Prove the radius of the inscribed circle in any triangle is equal to twice the area divided by the perimeter.

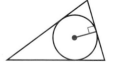

23. A sphere of radius 8 cm rests inside a conical funnel whose axis is vertical. The highest point of the sphere is 44 cm above the vertex of the cone. Determine the angle of the cone, correct to one decimal.

24. A bridge AB across a river valley is 140 m in length, where points A and B are at the same elevation. From point A the angle of depression to a point W on the riverbank below is 37°, and from point B the angle of depression to W is 64°. Determine the perpendicular distance from point W to line AB to the nearest metre.

Our work in trigonometry has so far been limited to angles between 0° and 90°. If an angle has its vertex at the origin in the *XY* plane and its initial arm along the positive *X*-axis, then the measure of an angle can be extended beyond 90°. We do this by rotating the terminal arm of the angle about the origin.

The *XY* plane is divided into four quadrants, numbered for convenience as shown.

The **angle of rotation** is the angle created when a line segment is rotated about its vertex to a terminal position.

An angle is in **standard position** if its vertex is at the origin in the *XY* plane, its initial arm is on the positive *X*-axis, and its terminal arm is a rotation of the initial arm about the origin.

If the rotation is **counterclockwise,** the angle has positive measure. If the rotation is **clockwise,** the angle has negative measure. The rotation is usually indicated by a directed arrow starting from the positive X-axis.

EXAMPLE 1

For each of the angles given, state the value of θ.

a. 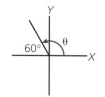   **b.** 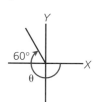   **c.** 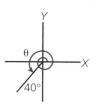   **d.**

SOLUTION

a. $\theta = 180° - 60°$

$ = 120°$

c. $\theta = 360° + 180° + 50°$

$ = 590°$

b. $\theta = -(180° + 60°)$

$ = -240°$

d. $\theta = -(360° + 70°)$

$ = -430°$

Two angles in standard position can have the same terminal arm. For example, an angle of 120° and an angle of 480° differ only in that the second one has a full rotation in addition to a rotation of 120°.

> Two angles in standard position are **coterminal** if they have the same terminal arm. Coterminal angles differ by a multiple of 360°.

For example, an angle coterminal with 80° is 360° + 80° or 440°. Other angles coterminal with 80° are 80° + (360° × 2) or 800° and 80° + (−360°) or −280°.

EXAMPLE 2

Determine two angles, one positive and one negative, coterminal with 115°.

SOLUTION

A positive angle is 115° + 360° = 475°.

A negative angle is 115° − 360° = −245°.

> For an angle whose measure is $p°$, any angle whose measure is $(p + 360k)°$, $k \in I$, is coterminal with the given angle.

EXAMPLE 3

For each of the following angles α and θ, determine whether or not the angles are coterminal.

  a. $\alpha = 57°, \theta = 1137°$

  b. $\alpha = 57°, \theta = -663°$

  c. $\alpha = 340°, \theta = -400°$

SOLUTION

a. $\theta - \alpha = 1137° - 57°$
$$= 1080°$$
$$= 3 \times 360°$$
The angles are coterminal.

b. $\alpha - \theta = 57° - (663°)$
$$= 720°$$
$$= 2 \times 360°$$
The angles are coterminal.

c. $\alpha - \theta = 340° - (-400°)$
$$= 740°$$
This is not a multiple of 360°,
so the angles are not coterminal.

Exercise 7.2

PART A

1. Find two angles, one positive and one negative, that are coterminal with θ, where θ is equal to the following:

  a. 30°   **b.** 320°   **c.** −150°   **d.** −225°

2. Determine which of the following pairs of angles are coterminal.

  a. 70° and −290°   **b.** 150° and −210°   **c.** 200° and 560°

  d. 270° and 630°   **e.** 4290° and 570°   **f.** −840° and −120°

3. Determine the size of coterminal angles α and θ, measured in degrees.

  a.   **b.**   **c.**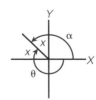

4. Each of the following angles is coterminal with some angle θ where θ is between 0° and 360°. Determine the value of θ.

a. 620° **b.** 795° **c.** 1999° **d.** 4321°

e. −117° **f.** −590° **g.** −650° **h.** −2001°

PART B

5. In the diagram, $\angle AOP = \theta$, $\angle OAP = \alpha$, and $OP = OA$. Determine each of the following:

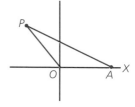

a. α if θ = 130°

b. θ if α = 15°

c. α and θ if θ = 6α

6. In the diagram, Q is the reflection of point P in the X-axis and the radius of the circle is 4. Determine the following:

a. the measure of ∠AOQ in a clockwise direction

b. an expression for the measure of all angles coterminal with ∠AOQ in a clockwise direction

c. the measure of ∠AOQ in a counterclockwise direction

d. an expression for the measure of all angles coterminal with ∠AOQ in a counterclockwise direction

e. the coordinates of P and Q

PART C

7. The triangle ABC is inscribed in a circle with centre E and radius R.

a. Prove that the length of any side of △ABC is equal to the product of the diameter and the sin θ, where θ is the angle subtended by the side selected.

b. Prove the area of △ABC is equal to $2R^2 \sin A \sin B \sin C$.

8. The diagonals of a quadrilateral are m and p units in length. If the diagonals intersect at an angle β, then prove the area of the quadrilateral is $A = \dfrac{mp}{2} \sin \beta$.

7.3 Primary Trigonometric Ratios for Any Angle

With the acceptance that angles can be of any measure, a reconsideration of the definitions of the trigonometric ratios that allows us to consider angles of any size is in order. We base the new definitions on angles in standard position.

If point $P(x, y)$ is chosen on the terminal arm of angle θ, then $OP = \sqrt{x^2 + y^2}$. Every point in the XY plane determines an angle.

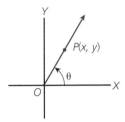

If $P(x, y)$ is a point on the terminal arm of an angle, and a circle with centre the origin is drawn through $P(x, y)$, then its radius $r = \sqrt{x^2 + y^2}$. We state new definitions for the trigonometric ratios.

The Primary Trigonometric Ratios

$$\sin \theta = \frac{y}{r}, \cos \theta = \frac{x}{r}, \tan \theta = \frac{y}{x}, x \neq 0$$

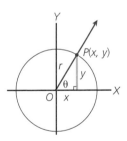

These definitions make it clear that there is a *unique* trigonometric ratio for every angle. Given an angle, we can use a calculator to determine its sine, cosine, or tangent. The value obtained is usually a decimal approximation. If given a point on an angle's terminal arm, we can determine its trigonometric ratios either as exact values or as decimal approximations. In practical problems an approximation is acceptable, but in theoretic problems an exact value is expected.

EXAMPLE 1

a. Determine the three exact values of the primary trigonometric ratios for angle θ in standard position determined by the point $P(6, -5)$.

b. Determine the measure of angle θ, where $0° \leq \theta \leq 360°$.

SOLUTION

a. $r = \sqrt{x^2 + y^2}$

$= \sqrt{(6)^2 + (-5)^2}$

$= \sqrt{61}$

$$\sin \theta = \frac{y}{r} \qquad \cos \theta = \frac{x}{r} \qquad \tan \theta = \frac{y}{x}$$

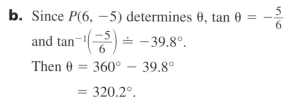

$$= -\frac{5}{\sqrt{61}} \qquad\qquad = \frac{6}{\sqrt{61}} \qquad\qquad = -\frac{5}{6}$$

b. Since $P(6, -5)$ determines θ, $\tan \theta = -\frac{5}{6}$

and $\tan^{-1}\left(\frac{-5}{6}\right) \doteq -39.8°$.

Then $\theta = 360° - 39.8°$

$= 320.2°$.

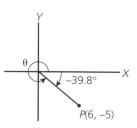

EXAMPLE 2

Point P is 15 cm from the origin and $\angle POX = 205°$. Determine the coordinates of P, giving your answer to one decimal.

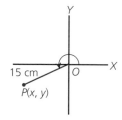

SOLUTION

Let P have coordinates (x, y).

$$\sin 205° = \frac{y}{r} = \frac{y}{15}$$

$$y = 15 \sin 205°$$

$$\doteq -6.339$$

Then y is -6.4.

$$\cos 205° = \frac{x}{r} = \frac{x}{15}$$

$$x = 15 \cos 205°$$

$$\doteq -13.595$$

Then x is -13.6.

The coordinates of P, to one decimal, are $(-13.6, -6.4)$.

What happens if we are given a particular trigonometric ratio? Suppose that we are given $\sin \theta = \frac{3}{5}$ and asked to determine the other ratios exactly. Our reasoning is as follows:

Since $\sin \theta = \frac{y}{r}$ for a particular point (x, y) and determines angle θ, let $y = 3$, $r = 5$, and $r^2 = x^2 + y^2$.

$$25 = x^2 + 9$$
$$x^2 = 16$$
$$x = \pm 4$$

There are two points possible: $P_1(4, 3)$ and $P_2(-4, 3)$.

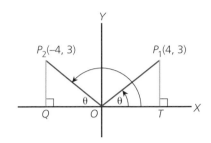

This means that there are two angles determined by points $P_1(4, 3)$ and $P_2(-4, 3)$, having $\sin \theta = \frac{3}{5}$. They are $\angle P_1OT$ and $\angle P_2OT$ in the diagram. How are these angles related?

Draw P_1T and P_2Q perpendicular to the X-axis.
$$OT = |OQ| = 4$$
$$TP_1 = QP_2 = 3$$
$$OP_1 = OP_2 = 5$$
$$\triangle OP_1T \cong \triangle OP_2Q$$

Therefore $\angle P_1OT = \angle P_2OQ = \theta$.

This means that $\angle P_2OT = 180° - \theta$.

Are there always two angles for a given trigonometric ratio in one rotation of $360°$?

Let $P(a, b)$ represent a point in quadrant I and $OP_1 = r$ be the terminal arm of angle θ as shown in the diagram.

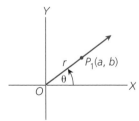

There are three other points having these coordinates and the appropriate sign as indicated.

From the diagram we note that

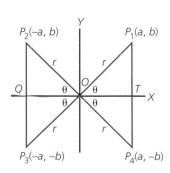

P_2 is the reflection of P_1 in the positive Y-axis,

P_3 is the reflection of P_2 in the negative X-axis,

and P_4 is the reflection of P_1 in the positive X-axis.

This means that $OT = OQ$, $OP_1 = OP_2 = OP_3 = OP_4 = r$, and $PT_1 = QP_2 = P_3Q = P_4T$.

The four triangles are congruent and $\angle P_1OT = \theta$, $\angle P_2OQ = \theta$, $\angle P_3OQ = \theta$, and $\angle P_4OT = \theta$.

Consider the four points separately and the primary trigonometric ratios of the angles formed by the terminal arm as shown in the diagrams.

Case 1

$$\sin \theta = \frac{b}{r}$$

$$\cos \theta = \frac{a}{r}$$

$$\tan \theta = \frac{b}{a}$$

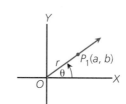

Case 2

$$\sin(180° - \theta) = \frac{b}{r} = \sin \theta$$

$$\cos(180° - \theta) = -\frac{a}{r} = -\cos \theta$$

$$\tan(180° - \theta) = -\frac{b}{a} = -\tan \theta$$

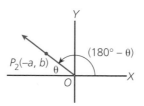

Case 3

$$\sin(180° + \theta) = -\frac{b}{r}$$
$$= -\sin \theta$$
$$\cos(180° + \theta) = -\frac{a}{r}$$
$$= -\cos \theta$$
$$\tan(180° + \theta) = \frac{b}{a}$$
$$= \tan \theta$$

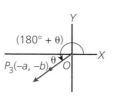

Case 4

$$\sin(360° - \theta) = -\frac{b}{r}$$
$$= -\sin \theta$$
$$\cos(360° - \theta) = \frac{a}{r}$$
$$= \cos \theta$$
$$\tan(360° - \theta) = -\frac{b}{a}$$
$$= -\tan \theta$$

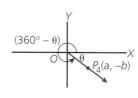

We can now make the following conclusions about the primary trigonometric ratios.

◆ If the terminal arm of the angle lies in quadrant I, all the ratios are positive.

◆ If the terminal arm of the angle lies in quadrant II, only the sine ratio is positive.

◆ If the terminal arm of the angle lies in quadrant III, only the tangent ratio is positive.

◆ If the terminal arm of the angle lies in quadrant IV, only the cosine ratio is positive.

A handy way of remembering this is from the mnemonic below.

Trig Ratio	Positive	Negative
	Terminal Arm in Quadrant	
sin θ	I, II	III, IV
cos θ	I, IV	II, III
tan θ	I, III	II, IV

S | A

T | C

EXAMPLE 3

Use a calculator to determine the two angles associated with cos θ = −0.627 between 0° and 360°.

SOLUTION

Since the given cosine value is negative, the terminal arms of the two angles must be located in quadrant II and quadrant III.

From the calculator, θ = cos⁻¹(−0.627) ≐ 129° to the nearest degree.

To determine the value of the angle with terminal arm in quadrant III, calculate the related acute angle associated with 129°, namely 180° − 129° = 51°.

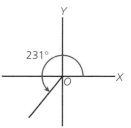

The angle with terminal arm in quadrant III is 180° + 51° = 231°.

If cos θ = −0.627, θ = 129° or 231°.

To determine the values of two angles between 0° and 360° associated with a given trigonometric ratio, use the following procedure.

❶ Determine the locations of the terminal arms.

❷ Determine the measure of the angle from the calculator.

❸ Determine the measure of the related acute angle.

❹ Determine the two measures of the two required angles.

EXAMPLE 4

Determine the measure of the two angles associated with each of the following, to the nearest degree between 0° and 360°.

a. $\sin \theta = -0.819$ **b.** $\tan \theta = -12.3$

SOLUTION

a. $\sin \theta = -0.819$

The terminal arms are in quadrants III and IV.

$$\theta = \sin^{-1}(-0.819)$$
$$= -55°$$

Related acute angle is 55°.

The measure of the two angles is $360° - 55° = 305°$ and $180° + 55° = 235°$.

Therefore $\theta = 235°$ or $305°$.

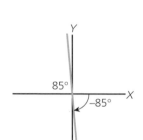

b. $\tan \theta = -12.3$

The terminal arms are in quadrants II and IV.

$$\theta = \tan^{-1}(-12.3)$$
$$= -85°.$$

Related acute angle is 85°.

The measures of the angles are $180° - 85° = 95°$ and $360° - 85° = 275°$.

Therefore $\theta = 95°$ or $275°$.

EXAMPLE 5

Give a graphical representation of the two angles between 0° and 360° associated with each of the following:

a. $\cos \alpha = \frac{3}{5}$ **b.** $\sin \theta = -\frac{2}{5}$

SOLUTION

a. Since $\cos \alpha = \frac{x}{r}$, let $x = 3$, $r = 5$.

$$y^2 = 25 - 9$$
$$= 16$$
$$y = \pm 4$$

The two points are $P(3, 4)$ and $Q(3, -4)$.

The two angles are $\angle XOP$ and $\angle XOQ$.

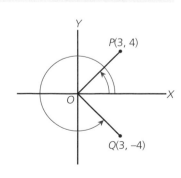

b. Since $\sin \theta = \dfrac{y}{r} = -\dfrac{2}{5}$, let $y = -2$, $r = 5$,

$$x^2 = 25 - 4$$
$$= 21$$
$$x = \pm\sqrt{21}$$

The two points are $P(\sqrt{21}, -2)$ and $Q(-\sqrt{21}, -2)$.
The two angles are $\angle XOQ$ and $\angle XOP$.

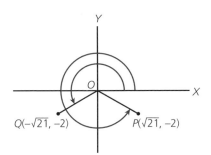

Exercise 7.3

PART A

1. θ is an angle in standard position. P is a point on the terminal arm of θ. Determine the primary trigonometric ratios of θ.

   a. $(9, 12)$ **b.** $(-8, 15)$ **c.** $(-5, -3)$ **d.** $(24, -10)$

2. Use a calculator to show that each of the following pairs of coterminal angles θ and β have the same values, correct to four decimals, for the sine ratio.

   a. $\theta = 100°$, $\beta = 460°$ **b.** $\theta = 245°$, $\beta = -115°$

   c. $\theta = -75°$, $\beta = -1155°$ **d.** $\theta = 2000°$, $\beta = 920°$

3. Repeat Question 2 for the tangent ratio.

4. For each of the following, state the exact values for $\sin \theta$, $\cos \theta$, and $\tan \theta$.

   a.

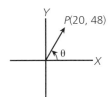

   b.

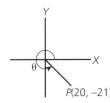

   c.

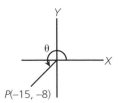

   d.

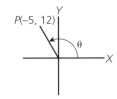

PART B

5. **a.** Given $\cos \theta = \frac{12}{13}$ with the terminal arm of θ in quadrant IV, determine $\tan \theta$.

  b. Given $\tan \alpha = \frac{15}{8}$ with the terminal arm of α in quadrant III, determine $\sin \alpha$.

  c. Given $\sin \theta = \frac{8}{10}$ and $0° \leq \theta \leq 360°$, determine $\tan \theta$.

  d. Given $\cos \beta = \frac{2}{5}$ and $0° \leq \beta \leq 360°$, determine $\sin \beta$.

6. **a.** In Question 5, part **a,** if $\theta \geq 360°$, then find θ and $\sin \theta$.

  b. In Question 5, part **b,** if α is a negative angle, then determine α and $\cos \alpha$.

7. **a.** If $\cos \theta = \frac{12}{13}$, $0° \leq \theta \leq 90°$, draw a sketch of θ and determine $\sin \theta$, $\tan \theta$, and $\frac{\sin \theta}{\cos \theta}$.

  b. If $\tan \beta = -\frac{8}{15}$, $90° \leq \beta \leq 180°$, draw a sketch of β and determine $\sin \beta$, $\cos \beta$, and $\frac{\sin \beta}{\cos \beta}$.

  c. Angle α is a third-quadrant angle such that $\cos \alpha = -\frac{8}{17}$. Sketch α and then determine $\sin \alpha$, $\tan \alpha$, and $\frac{\sin \alpha}{\cos \alpha}$.

  d. If $\sin \beta = -\frac{20}{29}$, $270° \leq \beta \leq 360°$, sketch β and determine $\cos \beta$, $\tan \beta$, and $\frac{\sin \beta}{\cos \beta}$.

  e. From the above four parts of Question 7, make a conclusion with respect to $\tan \theta$ and $\frac{\sin \theta}{\cos \theta}$. Prove your conjecture for any angle.

8. Determine the two angles, to the nearest degree, between $0°$ and $360°$ associated with each of the following trigonometric ratios.

  a. $\cos \alpha = 0.276$ **b.** $\sin \theta = 0.159$ **c.** $\cos \theta = -0.513$

  d. $\tan \theta = 5.23$ **e.** $\sin \alpha = -0.373$ **f.** $\tan \alpha = -0.618$

9. **a.** Find the exact values for the primary trigonometric ratios of $300°$.

  b. Find the exact values for the primary trigonometric ratios of $225°$.

10. In each of the following, determine the coordinates of P.

  a. $OP = 20$ and $\angle POX = 215°$ **b.** $OP = 125$ and $\angle POX = 310°$

  c. $OP = 8.5$ and $\angle POX = 165°$ **d.** $OP = 1$ and $\angle POX = \alpha$

11. The first-quadrant rotation $\angle POA = \beta$ has x-coordinate 8 in a circle with radius 17.

    a. Determine the primary trigonometric ratios of $\angle\beta$.

    b. Determine $\tan(180° - \beta)$ and $\sin(180° - \beta)$ exactly.

    c. Determine points P_3 and P_4, which define the angles in quadrants III and IV that are related to β.

    d. State the values of $\cos(180° + \beta)$ and $\sin(360° - \beta)$.

12. Write each of the following in terms of $\sin\theta$, $\cos\theta$, or $\tan\theta$.

    a. $\sin(180° - \theta)$ **b.** $\cos(180° - \theta)$ **c.** $\tan(180° - \theta)$

    d. $\sin(180° + \theta)$ **e.** $\cos(180° + \theta)$ **f.** $\tan(180° + \theta)$

    g. $\sin(360° - \theta)$ **h.** $\cos(360° - \theta)$ **i.** $\tan(360° - \theta)$

    j. $\sin(-\theta)$ **k.** $\cos(-\theta)$ **l.** $\tan(-\theta)$

PART C

13. $\triangle ABC$ is a right-angled triangle with $\angle C = 90°$. Prove the following:

    a) $\sin(90° - A) = \cos A$

    b) $\cos(90° - A) = \sin A$

    c) $\tan(90° - A) = \dfrac{1}{\tan A}$

14. Three congruent circles, each with radius 5, touch each other at points X and Y. Line XY is extended to meet the first circle at A and the third circle at M. If a tangent is drawn from A to meet the third circle at D, then determine the length of chord BC. (*Hint:* The right bisector of a chord passes through the centre of a circle.)

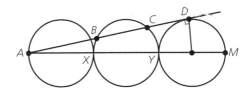

To this point we have used the trigonometric ratios in right-angled triangles. What can we do if no angles are right-angled? Is it possible to determine unknown parts of a triangle by using our knowledge of trigonometry? The theorem below tells us that this is frequently possible, depending on what information we are given.

The Law of Sines

If a triangle ABC has sides a, b, and c opposite angles A, B, and C, respectively, then $\dfrac{a}{\sin A} = \dfrac{b}{\sin B} = \dfrac{c}{\sin C}$.

Consider triangle ABC with all angles acute. From A draw AD perpendicular to side BC, and let the length of AD be h. Then $\dfrac{h}{c} = \sin B$ or $h = c \sin B$

and $\dfrac{h}{b} = \sin C$

or $h = b \sin C$.

$b \sin C = c \sin B$

or $\dfrac{b}{\sin B} = \dfrac{c}{\sin C}$.

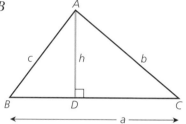

Drawing a second perpendicular from B to AC gives $a \sin C = c \sin A$ or $\dfrac{a}{\sin A} = \dfrac{c}{\sin C}$.

Therefore $\dfrac{a}{\sin A} = \dfrac{b}{\sin B} = \dfrac{c}{\sin C}$.

If one angle, say A, is obtuse, then as before

$b \sin C = c \sin B$

or $\dfrac{b}{\sin B} = \dfrac{c}{\sin C}$.

Now, the perpendicular from B meets CA extended at E. If BE has length l, then

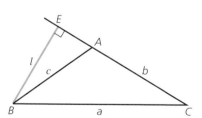

$l = a \sin C$

and $l = c \sin \angle BAE = c \sin(180° - A)$.

But $\sin(180° - A) = \sin A$,

so $l = c \sin A$.

Again, $a \sin C = c \sin A$,

$\dfrac{a}{\sin A} = \dfrac{c}{\sin C}$,

and $\dfrac{a}{\sin A} = \dfrac{b}{\sin B} = \dfrac{c}{\sin C}$.

If one angle, say B, is right-angled,

$$\frac{a}{b} = \sin A \text{ or } b = \frac{a}{\sin A}$$

and $\qquad \frac{c}{b} = \sin C \text{ or } b = \frac{c}{\sin C}.$

But $\qquad \sin 90° = 1$, so $b = \frac{b}{\sin B}.$

Again, $\dfrac{a}{\sin A} = \dfrac{b}{\sin B} = \dfrac{c}{\sin C}.$

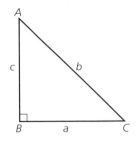

The relationship is true for all triangles.

EXAMPLE 1

Triangle ABC has $\angle A = 35°$, $\angle B = 47°$, and side $AB = 15$. Determine the other parts of the triangle, giving answers to one decimal. (Usually we simply say "Solve the triangle.")

SOLUTION

$$\angle C = 180° - \angle A - \angle B$$
$$= 180° - 35° - 47°$$
$$= 98°$$

By the Law of Sines $\dfrac{a}{\sin A} = \dfrac{b}{\sin B} = \dfrac{c}{\sin C}$,

and $\qquad \dfrac{a}{\sin 35°} = \dfrac{b}{\sin 47°} = \dfrac{15}{\sin 98°}.$

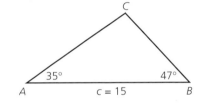

From the last two ratios $b = \dfrac{15 \sin 47°}{\sin 98°}$

$$b \doteq 11.08.$$

Also, $\qquad a = \dfrac{15 \sin 35°}{\sin 98°}$

$$a \doteq 8.69.$$

Then $BC = 8.7$, $AC = 11.1$, to one decimal, and $\angle ACB = 98°$.

Note that we must be given the correct combination of sides and angles if we wish to use the Law of Sines. If we are given the three sides of a triangle, we cannot determine its angles using the Sine Law. If given the three angles, we cannot determine the sides. If we are to use the Law of Sines, we must know both parts of one of the ratios and one part of each of the other ratios. Even then we can run into difficulties because, as we have seen, when given $\sin \theta$ there are two angles, θ and $(180° - \theta)$, each having the same Sine ratio. This means that there can be two acceptable triangles.

EXAMPLE 2

Solve $\triangle KLM$ with $KL = 20$, $LM = 11$, and $\angle K = 30°$. Write your answer, correct to one decimal.

Since $\dfrac{20}{\sin M} = \dfrac{11}{\sin 30°}$

$\qquad \sin M = \dfrac{20 \sin 30°}{11}$

$\qquad\qquad = \dfrac{20\left(\frac{1}{2}\right)}{11}$

$\qquad\qquad = \dfrac{10}{11}$

$\qquad \angle M = \sin^{-1}\left(\dfrac{10}{11}\right) \doteq 65.4°.$

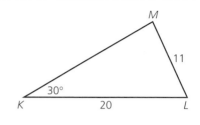

But $\angle M$ could also be $180° - 65.4° = 114.6°$.

We consider the two cases.

Case 1

$\angle M = 65.4°$

$\angle L = 180° - 30° - 65.4°$

$\qquad = 84.6°$

Then $\dfrac{l}{\sin 84.6°} = \dfrac{11}{\sin 30°}$

$\qquad\quad l = \dfrac{11 \sin 84.6°}{\sin 30°}$

$\qquad\qquad \doteq 21.902$

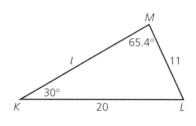

If $\angle M = 65.4°$, then $\angle L = 84.6°$, $KM = 21.9$.

Case 2

$\angle M = 114.6°$

$\angle L = 180° - 30° - 114.6°$

$\qquad = 35.4°$

Then $\dfrac{l}{\sin 35.4°} = \dfrac{11}{\sin 30°}$

$\qquad\quad l = \dfrac{11 \sin 35.4°}{\sin 30°}$

$\qquad\qquad \doteq \dfrac{11(0.5793)}{0.5}$

$\qquad\qquad = 12.7$

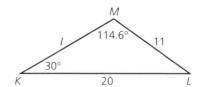

If $\angle M = 114.6°$, then $\angle L = 35.4°$, $KM = 12.7$.

Geometrically, this situation arises when the side opposite the given angle is less than the other given side, and results from the fact that an arc with L as centre cuts the side KM at two points. This is *not* a problem if the side opposite the given angle is greater than the other given side.

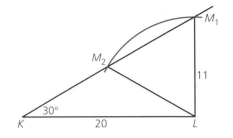

Surveyors and Civil Engineers

Surveyor's Transit

The Greek historian Herodotus described how land was divided among the citizens in Egypt (460–455 B.C.). Each citizen was given a quadrangle of equal size. Every year a tax was levied on the land. The Nile River flooded its banks annually and washed away some of the land. If a citizen lost some of his land, he would notify the authorities. An overseer was sent to the quadrangle of land to measure how much was left after the flood. The tax was then calculated on the remaining land. These overseers were some of the first surveyors.

Specialists whom the later Greeks called rope-stretchers performed the task of surveying. Their main tool was a rope with knots or marks at equal intervals. They would use this rope for measuring distances.

Today, for the construction of roads and bridges, distances must be calculated with great precision. Modern surveyors and civil engineers use an instrument called a transit to measure angles. This instrument uses a telescope for accurate sighting at long distances. Angles are then measured very accurately within the transit. Trigonometry is used to calculate inaccessible distances.

Governments, construction companies, exploration companies, and private survey companies employ surveyors and civil engineers.

Photo: Tim Wright/CORBIS

PART **A**

1. Calculate the measure of each indicated side or angle.

a.

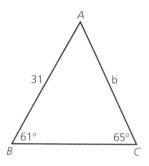

b.

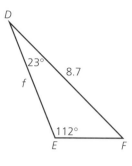

c.

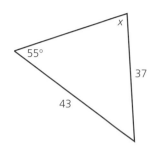

2. Solve each of the following for *a* or *b*, correct to one decimal.

a. $\dfrac{a}{\sin 43°} = \dfrac{7.5}{\sin 71°}$

b. $\dfrac{a}{\sin 55°} = \dfrac{230}{\sin 110°} = \dfrac{b}{\sin 15°}$

3. Determine acute angles θ and β in each of the following, correct to two decimals.

a. $\dfrac{\sin θ}{12} = \dfrac{\sin 120°}{18}$

b. $\dfrac{\sin θ}{100} = \dfrac{\sin β}{117} = \dfrac{\sin 73°}{130}$

4. On each of the following diagrams, three pieces of information are given. State whether it is possible to solve the triangle. If so, determine if more than one triangle is possible.

a.

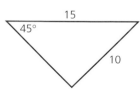

b.

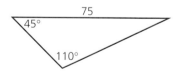

c.

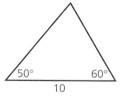

d.

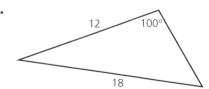

5. For each of the following acute-angled triangles, determine w, correct to one decimal.

a.

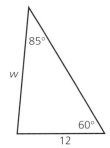

b.

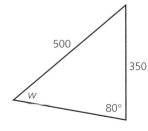

c.

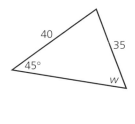

6. For each of the triangles in Question 5, determine the area, correct to one decimal.

PART B

7. A mountain is surveyed to determine whether the slope AT is long enough for a ski run. If the surveyor made the measurements shown, then determine the length TA to the nearest metre.

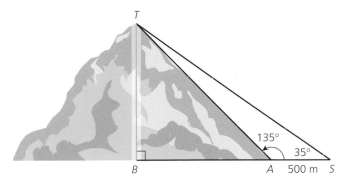

8. Triangle RST has $ST = 100$, $TR = 80$, and $\angle RST = 48°$. Determine all possible values for $\angle SRT$.

9. Solve each of the following triangles. Write your answers, correct to one decimal.

a.

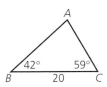

b.

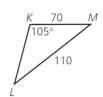

c.

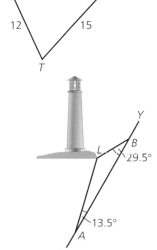

10. A ship is on course XY. At point A, the ship's captain measures the angle LAB from the ship to a lighthouse to be $13.5°$. Ninety minutes later, from position B, $\angle LBX = 29.5°$. If the ship is sailing at 18 km/h, find the distance LB to the nearest 100 m.

11. A roof truss has to span a distance $AB = 12$ m. The symmetrical design is shown in the diagram, where $\triangle TAB$ is isosceles and $\angle TAB = 35°$. Write your answers, correct to two decimals.

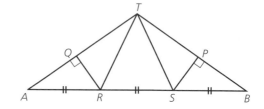

a. Determine the length of rafter AT.

b. Determine the length of brace RT.

c. Determine the length of the cross brace QR.

12. In $\triangle DEF$, $DE = 30$ cm, $\angle D = 80°$, and $\angle E = 55°$.

a. Determine the perimeter of $\triangle DEF$, correct to one decimal.

b. Determine the area of $\triangle DEF$, correct to one decimal.

13. Two ships A and B with underwater sonar have located the wreckage, W, of an old ship on the sea floor. When the distance AB between the ships is 8000 m, the angles of depression to the wreck are $\angle B = 18.5°$ and $\angle A = 56.5°$. If the sonar bearing from both ships is due west, determine the depth of water above the wreck to the nearest metre.

PART C

14. In any $\triangle ABC$, constants k, m, and p exist so that $k \sin A + m \sin B + p \sin C = 0$. Prove that $ka + mb + pc = 0$, where a, b, c are the lengths of the sides of the triangle.

15. Each of the five lines forming a regular star pentagram is 10 cm.

a. Find the radius of the circle that will circumscribe the pentagram.

b. Find the area of the pentagram.

16. In any $\triangle ABC$, prove that the perimeter is equal to $\frac{a}{\sin A}(\sin A + \sin B + \sin C)$.

17. In any $\triangle KPM$, prove that the following is always true.
$$k(\sin P - \sin M) + p(\sin M - \sin K) + m(\sin K - \sin P) = 0$$

The Law of Sines is not sufficient for solving all oblique triangles. For example, in the diagram at the right, you are asked to find the length of one side of a triangular plot of land with the given information. Label the triangle ABC.

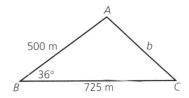

Using the Law of Sines, we would get

$$\frac{b}{\sin 36°} = \frac{500}{\sin C} = \frac{725}{\sin A}.$$

Unfortunately, each term in the proportion has a different unknown variable so the equation cannot be solved. We need to develop a new formula that will allow us to solve triangles in which we are given either two sides and the contained angle, or three sides.

Consider the two cases of $\triangle ABC$ where $\angle A$ is acute and $\angle A$ is obtuse.

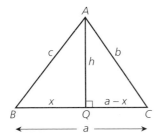
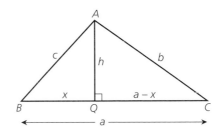

In both cases, construct $AQ \perp BC$ and let $AQ = h$, $BQ = x$, and $QC = a - x$.

In $\triangle ABQ$,

$$c^2 = h^2 + x^2$$

or $\qquad h^2 = c^2 - x^2.$

In $\triangle AQC$

$$b^2 = h^2 + (a - x)^2$$

or $\qquad h^2 = b^2 - (a - x)^2.$

Then $\qquad b^2 - (a - x)^2 = c^2 - x^2$

$$b^2 - a^2 + 2ax = c^2$$

$$b^2 = a^2 + c^2 - 2ax.$$

From $\triangle ABQ$, $\cos B = \frac{x}{c}$ or $x = c \cos B$.

Then $\qquad b^2 = a^2 + c^2 - 2a(c \cos B)$

or $\qquad b^2 = a^2 + c^2 - 2ac \cos B.$

By drawing perpendiculars from the vertices B and C, we can obtain alternate forms of this formula, namely

$$a^2 = b^2 + c^2 - 2bc \cos A$$

and $\qquad c^2 = a^2 + b^2 - 2ab \cos C.$

The Law of Cosines

For any $\triangle ABC$

$$a^2 = b^2 + c^2 - 2bc \cos A$$
$$b^2 = a^2 + c^2 - 2ac \cos B$$
$$c^2 = a^2 + b^2 - 2ab \cos C.$$

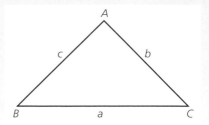

We now use the Law of Cosines to calculate the length of the third side in the triangular plot from the beginning of this section.

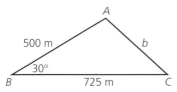

$$b^2 = a^2 + c^2 - 2bc \cos B$$
$$a = 725, c = 500, \angle B = 30°$$
$$b^2 = (725)^2 + (500)^2 - 2(725)(500)\cos 30°$$
$$\doteq 147\ 756.58$$
$$b \doteq 384.4$$

The length of the third side is approximately 384 m.

If we are given three sides of a triangle, we can use the Law of Cosines and the Law of Sines to determine the three angles.

EXAMPLE 1

In $\triangle RST$, $RS = 6$, $ST = 13$, and $TR = 9$. Solve the triangle. Write your answers, correct to one decimal.

SOLUTION

By the Law of Cosines

$$r^2 = s^2 + t^2 - 2st \cos R$$
$$r = 13, s = 9, t = 6$$
$$169 = 81 + 36 - 108 \cos R$$

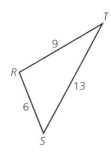

$$108 \cos R = -52$$
$$\cos R = -\frac{52}{108}$$
$$\angle R = \cos^{-1}\left(-\frac{52}{108}\right)$$
$$\doteq 118.8°$$

Use the Law of Sines to determine $\angle T$.

$$\frac{\sin T}{6} = \frac{\sin 118.8}{13}$$

$$\sin T = \frac{6 \sin 118.8}{13}$$

$$\doteq 0.4045$$

$$\angle T = \sin^{-1}(0.4045)$$

$$\doteq 23.9°$$

$$\angle S = 180° - (118.8° + 23.9°)$$

$$= 37.3°$$

The measures of the three angles are 118.8°, 23.9°, and 37.3°.

We can use the Law of Cosines to find the third side of a triangle if two sides and the contained angle are given.

EXAMPLE 2

An aircraft control tower at T was tracking the positions of two aircraft at points A and B. At the last check, A was 3.5 km from the tower and B was 7 km from the tower. If $\angle ATB = 16°$, find the distance between the aircraft, correct to one decimal.

SOLUTION

Let t represent the distance between the aircraft.

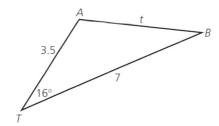

By the Law of Cosines

$$t^2 = a^2 + b^2 - 2ab \cos T$$

$$\doteq 49 + 12.25 - 14(3.5)(\cos 16°)$$

$$\doteq 14.148$$

and $t \doteq 3.761$.

The distance between the aircraft is 3.8 km.

PART A

1. Solve each of the following:

    a. $w^2 = 25 + 49 - 2(5)(7)\cos 50°$ **b.** $144 = 64 + c^2 - 2(8)c \cos 120°$

2. Determine the size of the angle, correct to two decimals.

    a. $\cos A = \dfrac{225 + 100 - 64}{300}$ **b.** $\cos \theta = \dfrac{12^2 + 11^2 - 18^2}{2(12)(11)}$

3. In each diagram, three pieces of information are given. State whether a unique answer is possible.

    a.
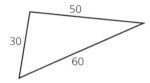

    b.
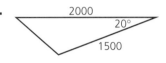

    c.
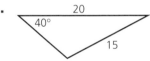

    d.

4. **a.** In Question 3, part **a,** find the angle opposite side 50.

    b. In Question 3, part **b,** find the side opposite the 20° angle.

    c. In Question 3, part **c,** find the values for the third side.

5. In each of the following, find the value of w or θ.

    a.
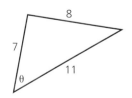
    b.
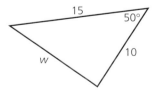
    c.

PART B

6. The triangle ABC has $AB = 50$, $BC = 40$, and $\angle BAC = 35°$. Determine all possible values for AC to one decimal. Sketch the different cases.

7. Solve each of the following triangles. Write your answers, correct to one decimal.

a.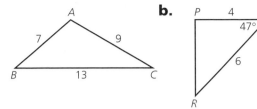

b.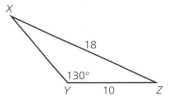

c.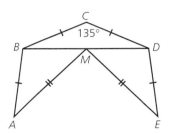

8. A roof truss is designed so $AB = 5$ m and rafter $AM = 7$ m, where M is the midpoint of cross brace BD.

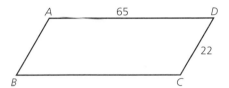

a. Determine the length for brace BD to the nearest centimetre.

b. Determine the angles at points A and E to the nearest degree.

9. Determine the area of parallelogram $ABCD$, given that diagonal $BD = 77$.

10. Three circles with centres A, B, and C are mutually tangent and no circle lies inside another circle. The circle with centre A has radius 3 cm and the circle with centre B has radius 5 cm.

a. If $\angle BAC = 60°$, then determine the radius of the third circle.

b. Find the area of $\triangle ABC$.

11. A trapezoid $ABDE$ has a point C on side BD so that $BC = 21$ and $CD = 18$. If $\angle ACB = 69°$ and $\angle DCE = 81°$, then determine AE, correct to one decimal.

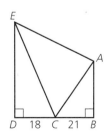

PART C

12. a. In any $\triangle ABC$, prove $a^2 - b^2 = c(a \cos B - b \cos A)$.

b. In the same $\triangle ABC$, prove $c = a \cos B + b \cos A$.

13. The angle between a space exploration satellite and the planet Jupiter is observed at the Victoria, British Columbia, observatory to be 34°. At the time of observation, the distance from Earth to Jupiter is ten times the distance from Earth to the space satellite. If you are an observer on planet Jupiter, then determine the angle between Earth and the space satellite, correct to three decimals.

14. Two concentric circles have centre C. Two points A and B are located on the smaller circle, with radius 10, so that $\angle ACB = 52°$. Point M is located on the larger circle so that $\angle CAM = 150°$ and $\angle CBM = 135°$. Find the radius of the larger circle, correct to one decimal.

15. The quadrilateral $ABCD$, with sides a, b, c, and d, is inscribed in a circle. Determine the cosine of angle C in terms of the lengths of the sides of the quadrilateral.

The Importance of Trigonometry

Air Traffic Control Tower

Over 3500 years ago, trigonometry was developed and used to solve problems in astronomy. Originally, trigonometry dealt only with the measurement of the sides and angles of triangles. It wasn't until the thirteenth century that trigonometry and astronomy were considered separate subjects.

Now trigonometry is used in many fields. In navigation, directions to and from a reference point are often given in terms of bearings. A bearing is an acute angle between a line of sight and the north-south line. Pilots and air traffic controllers must be very familiar with trigonometry. In physics, biology, and economics many quantities are periodic. These include such things as the oscillations of a pendulum or spring, alternating current, vibrations of the string in a musical instrument, periodic fluctuations in the population of a species, and periodic fluctuations in a business cycle. Many of these quantities can be described by harmonic functions. Harmonic functions are written using trigonometric functions.

Periodic functions that occur in engineering problems are often very complex. It is very useful to represent these functions in terms of simple periodic functions. Almost any periodic function can be represented by a trigonometric series.

As you can see from the many applications of trigonometry, you will likely be dealing with trigonometric functions in almost any area of future study.

Photo: Tim Wright/CORBIS

We now have available all the skills required to solve most problems in plane trigonometry. Problems from the world around us often require combining together the primary trigonometric definitions with either the Law of Sines or the Law of Cosines. Many problems require a knowledge of equation solving to be combined with geometric properties obtained from a diagram. The first example will illustrate how to use trigonometry on a three-dimensional problem.

EXAMPLE 1

To find the height of a cliff that is inaccessible, a surveyor measures a baseline AC of 400 m.
In the horizontal plane ABC, $\angle A = 27°$ and $\angle C = 35°$.
In the vertical plane BTC, $\angle BCT = 18°$. Determine the cliff height to the nearest metre.

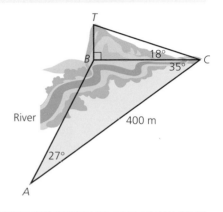

SOLUTION

In $\triangle ABC$

$$\frac{BC}{\sin 27°} = \frac{400}{\sin B}, \text{ where } \angle B = 180° - (27 + 35)° = 118°$$

$$BC = \frac{400 \sin 27°}{\sin 118°}$$

$$\doteq 205.7.$$

In vertical $\triangle TBC$, use the side BT opposite to $\angle C$ and BC is adjacent to $\angle C$.

$$\tan 18° = \frac{BT}{BC} = \frac{BT}{205.7}$$

$$BT = 205.7 \tan 18°$$

$$BT \doteq 66.8$$

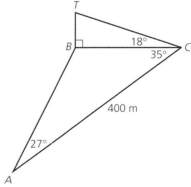

The height of the cliff is about 67 m.

Trigonometry is used in navigation to determine courses and distances. Compass directions are essential in navigation problems. The basic compass has a North-South axis (*NS*) perpendicular to the East-West axis (*EW*). The course taken by a ship or airplane is defined by describing an angle measured off the *NS* axis. On the diagram, *OA* describes the *N70°E* course. The distance *OB* is on a bearing of *S25°W*.

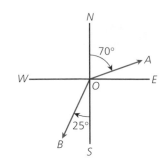

EXAMPLE 2

A smuggler leaves a private airfield at 06:00 and flies on a course of *N40°E* at 200 km/h. The plane is detected by radar at the police airport, which is located 150 km northwest of airfield *A*. At 06:30 the police airplane leaves its airport with the intention of intercepting the smuggler at 08:30. Determine the course and speed of the police airplane to the nearest unit.

SOLUTION

In the diagram, *AS* is the smuggler's flight path, *A* is the private airfield, *P* is the police airfield, and *I* is the projected point of interception.

The smuggler is flying at 200 km/h and in 2.5 h will fly 500 km.

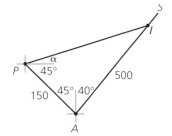

From the diagram, $\angle PAI = 85°$ and α represents the angle of flight of the police aircraft.

$$PI^2 = 150^2 + 500^2 - 2.150.500 \cos 85°$$
$$= 259\ 426.6386$$
$$PI \doteq 509.3$$

From the Law of Sines
$$\frac{\sin \angle IPA}{500} = \frac{\sin 85°}{509.3}$$
$$\sin \angle IPA \doteq 0.978\ 00$$
$$\angle IPA \doteq 77.96°$$
$$= 78°$$

Since $\alpha + 45° = 78°$, $\alpha = 33°$.

Since the police plane flies for 2 h, its speed must be 255 km/h and it must fly on a bearing of *N57°E*.

The development of trigonometry occurred along with the development of astronomy. The next example illustrates how useful trigonometry is for calculations in space.

EXAMPLE 3

Two observatories are located on the same longitude. Observatory A is at $42°$ North and observatory B is at $15°$ South. To measure the distance to the moon, each observatory arranges to measure the angle from the zenith to the moon at the same time. For A the angle is $TAM = 30.0°$ and for B the angle is $RBM = 27.95°$, where T and R are the zenith points.

If the radius of the earth is known to be 6435 km, find the distance between the earth and moon to the nearest 1000 km.

SOLUTION

The zenith points T and R are located directly above the observatories at A and B
so that $\angle TAC = 180° = \angle RBC$.

Thus $\angle CAM = 180 - 30 = 150°$

and $\angle CBM = 180 - 27.95 = 152.05°$.

Let line CE be the radius to the equator and $\angle ACE = 42°$ and $\angle ECB = 15°$.

Thus $\angle ACB = 57°$.

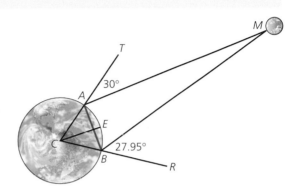

Join AB to create $\triangle ACB$ and use the Law of Cosines.

$$AB^2 = r^2 + r^2 - 2r^2 \cos 57°$$
$$= 2r^2 - 2r^2 \cos 57°$$
$$= 2r^2(1 - \cos 57°)$$

But $r = 6435$,

then $AB^2 = 2(6435)^2(1 - \cos 57°)$

$AB \doteq 6141$.

Now use the Law of Sines in $\triangle ABC$,

$$\frac{\sin CAB}{6435} = \frac{\sin 57°}{6141}$$

$$\sin CAB = \frac{6435(0.838\ 67)}{6141}$$

$$\doteq 0.878\ 82.$$

$$\angle CAB = \sin^{-1}(0.8788) \doteq 61.5°$$

Therefore
$$\angle BAM = 180° - (61.5° + 30°) = 88.5°$$
$$\angle CBA = 180° - (61.5° + 57°) = 61.5°$$
$$\angle ABM = 180° - (61.5° + 27.95°) = 90.55°$$
$$\angle AMB = 180° - (88.5° + 90.55°) = 0.95°.$$

In △ABM, the Law of Sines gives

$$\frac{AM}{\sin 90.55°} = \frac{BM}{\sin 88.5°} = \frac{6141}{\sin 0.95°}$$

$$AM = \frac{6141(\sin 90.55°)}{\sin(0.95°)}$$

$$\doteq 370\ 372$$

If we join C to M, where M is on the surface of the moon, the length $CM - CE$ gives the distance between the surface of the earth and the surface of the moon.

Use the Law of Cosines in △CAM to get

$$CM^2 = CA^2 + AM^2 - 2(CA)(CM)\cos 150°$$

$$= 6435^2 + 370\ 372^2 - 2(6435)(370\ 372)(\cos 150°)$$

$$= 1.413\ 449 \times 10^{11}.$$

$$CM \doteq 375\ 959$$

$$CM - CE \doteq 375\ 959 - 6435 = 369\ 524.$$

The distance between the surfaces is approximately 370 000 km.

Exercise 7.6

PART B

1. The buoys to mark the triangular course for a yacht race are located at points Y, T, C. If $YT = 3.5$ km, $\angle Y = 65°$, and $\angle T = 71°$, determine the length of the course, correct to one decimal.

2. A power line is to be installed under water from point A to point W. The surveyor measures $TA = 410$ m, $TB = 750$ m, $\angle T = 35°$, and $BW \doteq 26$ m. Determine the length of cable from A to W to the nearest metre.

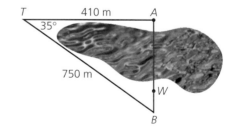

3. A grandfather clock has a pendulum that is 105 cm long. If the distance between the ends of the swing is $AB = 20$ cm, determine the value of θ to the nearest degree.

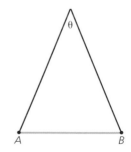

4. In quadrilateral *QUAD*, $\angle U = 65°$, $\angle UAQ = 35°$, $UA = 18$ cm, $\angle UAD = 85°$, and $AD = 24$ cm.

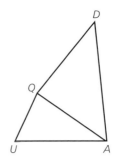

   a. Determine *UQ*.

   b. Determine *QD*.

   c. Determine the perimeter and area of *QUAD*, correct to one decimal.

5. A hexagonal tile has the angles at *A* and *D* equal to 50° and each of the four equal sides is 5 cm. If the diagonal *BE* = 12 cm, determine the area of the tile, correct to one decimal.

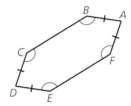

6. Two ships leave a harbour *H*, the first at 12:00 and the second at 12:30. The first ship sails on a course of *N30°E* at a speed of 20 km/h. The second ship sails on a course of *S20°E* at 15 km/h.

   a. Determine the distance between the two ships at 14:30 to the nearest kilometre.

   b. Determine the radar bearing from the first ship to the second ship.

7. The right-angled △*RST* has a point *Q* located inside the triangle so that $SQ = 20 = TQ$ and $\angle QST = 30°$. If $\angle R = 20°$, determine the length of *RQ*, correct to one decimal.

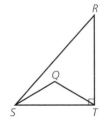

8. Two mountain peaks *R* and *T* can be observed from an alpine meadow. To estimate the straight-line distance between the peaks, members of a hiking party measure out a 500 m baseline *AB* in the meadow where *A*, *B*, *R*, and *T* are in the same plane. The angles from *A* and *B* to the peaks are then measured and $\angle BAR = 155°$, $\angle BAT = 33°$, $\angle ABT = 145°$, and $\angle ABR = 21°$. Determine the distance *RT* to the nearest 10 m.

9. Two fire towers are located 100 km apart on high hills *T* and *R*. The bearing from *T* to *R* is north east. A fire *F* is observed from tower *T* at *N10°E* and from tower *R* at *N75°W*. The town of Pretty Valley, at point *V*, is on a bearing of *N25°E* from *T* and *S70°W* from *R*. The observers report the wind is blowing the fire directly toward Pretty Valley at a rate of 8 km/h. How many hours do the officials have to evacuate the town?

10. A square tract of land $ABCD$, with side 10 km, has a hot spring located at a point P, where $\triangle PBC$ is an isosceles triangle with $\angle PBC = 20°$. The developer plans to build a resort beside the spring at P. Determine the length of the roads that must be constructed from A to P and B to P, to the nearest 10 m.

11. Two oil-drilling platforms are located in "Iceberg Alley" off the coast of Newfoundland. The CaP platform C is 150 km northwest of the Seashell platform S. Radar from the two platforms has detected a very large ice mass drifting south. From station C the left side L of the ice mass has a bearing $N50°E$ and the right side R is on a bearing $N53°E$. From station S, the left side L is on a bearing $N30°W$ and R is at $N28°W$. Determine the length LR of the ice mass showing on the radar screen, correct to the nearest 100 m.

PART C

12. A surveyor, trying to find the height of a steep canyon, observes that a prominent rock R at the base is almost vertically below a large tree T on the east ridge of the canyon. The west ridge is not flat, and the base line AB is established by measuring 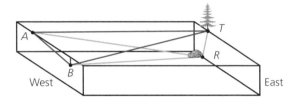 horizontally 50 m from A along with a vertical drop of 4 m. From A the angles measured are $\angle RAB = 77°$, $\angle RAT = 31°$, and $\angle TAB = 73°$. From point B, $\angle ABR = 81°$, $\angle ABT = 83°$. Determine the height of the canyon.

13. The port of Swan Harbour is located 200 km away from Merry Town Inlet on a bearing of $N50°E$ from Merry Town. A ship leaves Merry Town at 08:00 and sails $N15°W$ at 15 km/h. At the same time, a second ship leaves Swan Harbour on a course of $S80°W$ at a speed of 20 km/h. How close, to the nearest kilometre, are the two ships at 13:00?

14. One astronomical unit is defined to be 1 au = 149.5 × 10^6 km. A simple model of the solar system has Earth and Mars rotating about the sun in circular orbits, with the radius of Earth's orbit being 1 au. At point A in Earth's orbit, an astronomer observes Mars, M, to be located so that $\angle SAM = 116.5°$. It is known that Mars takes 687 Earth days to return to the same position in the Mars orbit. After 687 days, the astronomer observes that $\angle SBM = 140°$, where B is the position of Earth after 687 days. Use this data to find the distance from the Sun, S, to Mars (in astronomical units).

15. In any $\triangle ABC$, show that $\dfrac{\sin A - \sin B}{\sin A + \sin B} = \dfrac{a - b}{a + b}$ is true for all possible triangles.

1. Determine the value of *b* in each of the following:

a.

b.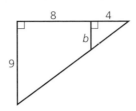

c.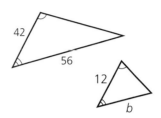

2. For each of the following, state the exact value for each of the primary trigonometric ratios of angle θ.

a.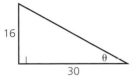

b.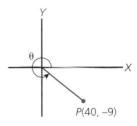

3. Find the value of each variable, correct to two decimals.

a.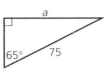

b.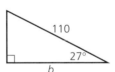

c.

d.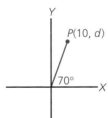

e.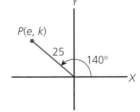

f.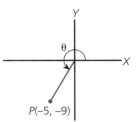

4. A kite string is 200 m long. If the kite string makes an angle of 72° with the ground, determine the height of the kite to the nearest metre.

5. Determine the two angles between 0° and 360° associated with each of the following trigonometric ratios.

    a. cos θ = 0.875   **b.** sin α = 0.259

    c. tan θ = 1.352   **d.** sin θ = −0.306

    e. cos α = −0.946   **f.** tan β = −0.4321

6. a. If $\sin \theta = -\dfrac{6}{10}$ with terminal arm in quadrant III, determine $\cos \theta$ exactly.

b. If $\cos \theta = -\dfrac{9}{41}$ with terminal arm in quadrant II, determine $\tan \beta$ exactly.

c. If $\tan \theta = -\dfrac{5}{7}$ with terminal arm in quadrant IV, determine $\sin \alpha$ exactly.

7. Solve the following for the variables shown, correct to one decimal.

a. $\dfrac{a}{\sin 57°} = \dfrac{11}{\sin 100°}$

b. $\dfrac{\sin \theta}{15} = \dfrac{\sin 81°}{20}$

c. $a^2 = 5^2 + 11^2 - 110 \cos 36.5°$

d. $\cos \alpha = \dfrac{5^2 + 9^2 - 13^2}{90}$

8. In each of the following, determine the values for the variables indicated, correct to one decimal.

a.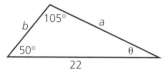

b.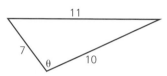

c.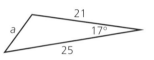

d.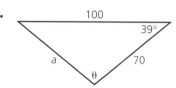

9. A guy wire is attached to a microwave transmitting tower at a point 40 m above the ground. The other end of the guy is anchored to the ground at a distance of 32 m from the base of the tower.

a. Determine the angle between the guy wire and the ground to the nearest degree.

b. Determine the length of the guy wire, correct to one decimal.

10. a. If angle α is a third-quadrant angle such that $\cos \alpha = -\dfrac{40}{41}$, sketch α and determine $\sin \alpha$ and $(\sin \alpha)^2 + (\cos \alpha)^2$ exactly.

b. If angle β is a second-quadrant angle such that $\sin \beta = \dfrac{4}{6}$, determine $\cos \beta$ and $(\sin \beta)^2 + (\cos \beta)^2$ exactly.

11. Solve each of the following triangles. State all answers correct to one decimal place.

a.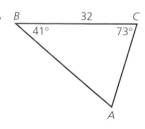

b.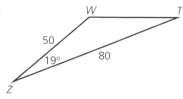

12. An observer in a search and rescue aircraft observes an object on the water at an angle of depression of 13°.

   a. If the aircraft is flying at a height of 600 m, determine the horizontal distance to the sighted object from the airplane to the nearest metre.

   b. If the airplane is flying at 240 km/h toward the object, determine how many seconds will pass before the aircraft will be directly above the sighted object to the nearest second.

13. The design for the seating section of a concert hall is based on the diagram where $\angle ECB = 160°$ and $EB = 26$ m. Determine the length of aisle AC to the nearest metre.

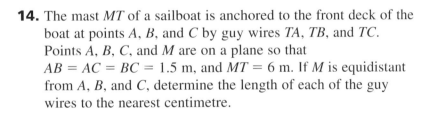

Photo: Superstock

14. The mast MT of a sailboat is anchored to the front deck of the boat at points A, B, and C by guy wires TA, TB, and TC. Points A, B, C, and M are on a plane so that $AB = AC = BC = 1.5$ m, and $MT = 6$ m. If M is equidistant from A, B, and C, determine the length of each of the guy wires to the nearest centimetre.

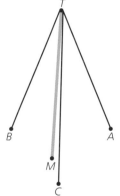

15. From a point A, three straight lines are drawn to points R, S, and T so that $\triangle ART$ is right-angled and points R, S, and T are collinear. If $AR = 12$, $AS = 13$, and $AT = 37$, determine the area of $\triangle AST$, correct to one decimal.

16. A surveyor has been hired by a resort developer to determine if a mountain is high enough to develop with a ski lift. The surveyor approaches the problem as follows:

In the valley below he sets up a transit at point K and sights an angle of elevation to a prominent rock R at the top to be 48°. He moves the transit to a second point M and measures the angle of elevation to R to be 45.3°.

   a. If distance KM is 100 m across level ground, and the line MK when extended meets a perpendicular from R, determine the height of the mountain.

   b. Find the length of the ski run from R to K.

1. Two identical squares are attached as shown, with the vertex of one attached to the centre of the other so that the top square can be rotated. How does the area of the overlapping region change as the top square rotates?

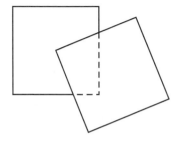

2. **a.** Can **33ab6** be a perfect square? (*a* and *b* are missing digits.)

   b. Can **301ab** be a perfect square?

   c. Is there a perfect square ending in 24? If so, how many are there?

3. What is the measure of the smaller angle between the hour hand and the minute hand of a 12-hour clock at 4:20?

4. In Problem 3, the lengths of the hour and minute hands are 4 cm and 6 cm, respectively. What is the distance between the tips of the hands at 4:20?

5. Given a circle with centre O, draw a chord AB in the circle. Choose any point C on the larger arc of the circle. Connect C to O and extend CO to a point P. Show that $\angle POA = 2\angle ACO$ and that $\angle POB = 2\angle BCO$, and hence that $\angle AOB = 2\angle ACB$. Is this result true if C is chosen on the smaller arc?

6. In a circle of radius 6, a triangle PQR is drawn having $QR = 8$ and $PQ = 10$. Determine the length of PR.

7. Show that $4x^3 + 8x^2y - xy^2 - 2y^3 = 0$ represents three straight lines. Taking these in pairs, determine the smaller angle between them.

Chapter 8
Trigonometric Functions

In science, medicine, and economics, trigonometric functions model changes that occur in repeating patterns. Diagnostic machines record medical data for heartbeats, breathing, and brain waves. Environmental data that follow repetitive increasing and decreasing cycles include orbits of planets, phases of the moon, tide levels, sunrise and sunset times, and a city's daily high temperatures. The motion of a chair on a Ferris wheel, the tip of a blade on a windmill, a pebble in a rotating wheel, and a swinging pendulum are modelled by what are called sinusoidal functions. Understanding how voltage for electric currents is represented by trigonometric functions is advantageous to electricians and engineers. Even cycles in business such as seasonal employment resemble these sinusoidal functions. In this chapter, you will learn how to analyze repeated changes represented by trigonometric functions and become aware of the role functions have in modelling cyclical changes.

IN THIS CHAPTER, YOU CAN . . .

- sketch graphs of $y = \sin x$, $y = \cos x$, and $y = \tan x$, and describe their properties;
- define the term *radian measure*, relate radian measure to degree measure, and use radian measure in solving equations and graphing;
- sketch the graph of a sinusoidal function;
- use technology to determine the effect of transformations on the graphs of $y = \sin x$ and $y = \cos x$;
- write the equation of a sinusoidal function, given its graph and its properties;
- solve problems related to models of sinusoidal functions in a variety of applications.

8.1 Periodic Functions

In earlier chapters, various functions have been discussed. One very important class of functions that is useful in both scientific and economic applications is the one describing changes that occur repeatedly. Events such as the phases of the moon, the height of the tide at an ocean port, and the strength of the electric current feeding an appliance are examples of such situations.

The diagram below shows several cycles of the electrical activity that takes place in a normal heart when measured on an electrocardiogram.

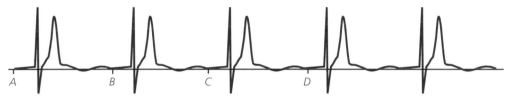

The cycle of increasing current followed by decreasing current is regular, corresponding to the contractions in the chambers of the heart. From the graph you can see that the portion of the graph from *A* to *B* repeats in the interval *B* to *C*, again in the interval *C* to *D*, and so on. We should note that these cycles show the patient at rest in the doctor's office. Under these conditions, the graph is periodic for a small number of beats.

EXAMPLE 1

Middle-distance runners are able to run at a steady pace for 800 m. At a constant rate of 5 m/s, Shanaz would run twice around a 400 m track in 160 s. The coach of the team is a mathematics teacher who measures the distance from the centre of the field to various points on the track.

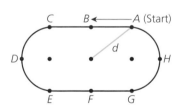

He records the time, when the runner passes these points, and distances in the table.

Photo: Dick Hemingway

Point	A	B	C	D	E	F	G	H	A
Time (*t*)	0	10	20	30	40	50	60	70	
Distance (*d*) From Centre	59.2	31.8	59.2	81.8	59.2	31.8	59.2	81.8	
Time (*t*)	80	90	100	110	120	130	140	150	160
Distance (*d*) From Centre	59.2	31.8	59.2	81.8	59.2	31.8	59.2	81.8	59.2

After entering this data in a graphing calculator, the graph at the right is obtained. The points shown are for the times recorded. If a graph showing the distance from the centre at any time is desired, we connect the points.

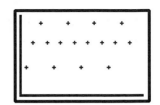

How should the points be connected? One possibility is to connect them with line segments. This implies that the distance from the centre of the track to a point on the track changes at a constant rate. However, it is obvious that the distance measured at the circular sections of the track will change at varying rates. To show this on a graph, the points should be connected with a curve as shown. The points A, B, C, D, E, F, G, and H on the graph correspond to the points A, B, C, ... , H indicated on the track, and a smooth curve has been drawn through the points.

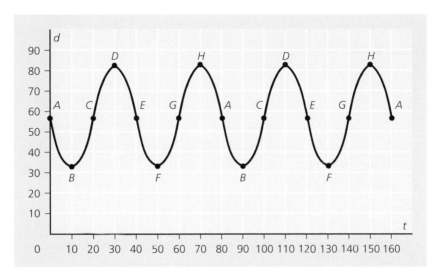

The function modelled is periodic with a period of 40 s.

From this graph we can read his distance from the centre at any time t between 0 and 160. For example, at $t = 65$ s, Shanaz is approximately 75 m from the centre.

The graph shows that the distance d is a function of time t, so $d = f(t)$. From the graph we observe that the distance from the centre at any time t will be the same 40 s later or at time $t + 40$. In function notation we write $f(t) = f(t + 40)$.

> A **periodic function** is a function whose values repeat at regular intervals, measured along the horizontal (or independent) axes.
> In function notation we write $f(x) = f(x + k)$.
> The period of the function is the value k.

Graphically, this means that if the graph of $y = f(x)$ is shifted horizontally k units, for some constant k, the new graph is identical to the original.

PART A

1. Which of the following functions are periodic? For these functions state the period.

a.

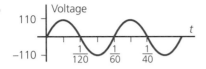

b.

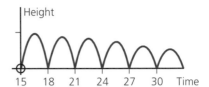

c.

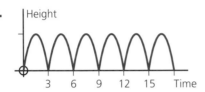

d.

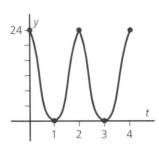

e.

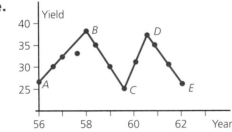

f.

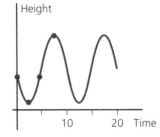

PART B

2. A part of southwestern Ontario is at 43° North Latitude. The following chart gives the number of hours of daylight on the 21st day of each month for 25 consecutive months.

a. Input the given data into a graphing calculator and plot the set of points. Make December your 0 entry, January entry 1, February entry 2, and so on, as shown in the table.

b. State whether the function appears to be periodic and what would be the length of one period.

c. Use the given chart to determine the difference in the number of daylight hours between the shortest day and the longest day in the year.

Month	Month Number	Number of Daylight Hours	Month	Month Number	Number of Daylight Hours
Dec.	0	9.0	Dec.	12	9.0
Jan.	1	9.5	Jan.	13	9.5
Feb.	2	10.8	Feb.	14	10.8
Mar.	3	12.2	Mar.	15	12.2
Apr.	4	13.7	Apr.	16	13.7
May	5	14.9	May	17	14.9
June	6	15.4	June	18	15.4
July	7	14.9	July	19	14.9
Aug.	8	13.7	Aug.	20	13.7
Sep.	9	12.2	Sep.	21	12.2
Oct.	10	10.8	Oct.	22	10.8
Nov.	11	9.5	Nov.	23	9.5
			Dec.	24	9.0

3. A point P travels counterclockwise along the circumference of a circle with radius 10 cm and centre at the origin. The point starts at $P_0(10, 0)$ and stops at intervals of $30°$ as it travels, creating $\angle P_1OP_0 = 30°$, $\angle P_2OP_0 = 60°$, $\angle P_3OP_0 = 90°$, $\angle P_4OP_0 = 120°$, and so on. If the point travels two complete circuits around the circle, the final point is P_{24}, which coincides with P_0.

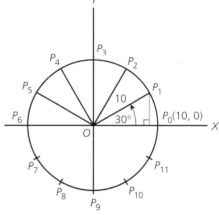

a. Using the trigonometric ratios you know, calculate the coordinates of points P_1 and P_2.

b. Using properties of reflection in the axes, determine the coordinates of points P_3, P_4, P_5, and so on to point P_{12} for the first circuit around the circle.

c. Using the Y-coordinates of the 13 points, copy and complete the following table.

Point	P_0	P_1	P_2	P_3	P_4		P_8	P_9	P_{10}	P_{11}	P_{12}
Angle	0°	30°	60°	90°	120°	...	240°				360°
Y-coordinate	0		8.6	10	8.6		−8.6				0

d. Extend and complete the table in part **c** to determine the Y-coordinates of the other 12 points P_{13} to P_{24} obtained on the second circuit of point P around the circle.

e. Using your graphing calculator, plot the 25 points and observe the graph. When inputting the first coordinate of the ordered pairs into your calculator, use the values 0, 1, 2, 3, . . . , 24 to correspond to the angle measure 0°, 30°, 60°, ...

f. Is the graph of this function periodic? If so, what is the length of the period?

g. If the radius of the circle is changed to 30 cm, explain what changes will occur in the graph.

h. Repeat this activity for the X-coordinate of points P_0, P_1, P_2, and so on to P_{24}.

4. Sunrise at Niagara
The chart below gives the sunrise times on the 21st of each month at Niagara Falls for two years (all times are Eastern Standard).

Month	Month Number	Sunrise Time	Month	Month Number	Sunrise Time
Dec.	0	07:44	Dec.	12	07:44
Jan.	1	07:42	Jan.	13	07:42
Feb.	2	07:06	Feb.	14	07:06
Mar.	3	06:18	Mar.	15	06:18
Apr.	4	05:25	Apr.	16	05:25
May	5	04:47	May	17	04:47
June	6	04:37	June	18	04:37
July	7	04:55	July	19	04:55
Aug.	8	05:28	Aug.	20	05:28
Sep.	9	06:02	Sep.	21	06:02
Oct.	10	06:37	Oct.	22	06:37
Nov.	11	07:16	Nov.	23	07:16
			Dec.	24	07:44

a. Using a graphing calculator, input the data and graph the function. What is an efficient way of inputting the given times?

b. Is this function periodic?

c. Determine the time difference between the earliest sunrise of the year and the latest sunrise.

5. The graph below shows the height of the tide at Prince Rupert, British Columbia, for a day in September 1999. The heights are in metres and the times are in hours.

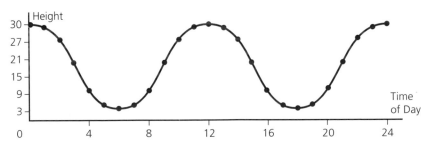

a. Determine if the function is periodic. State your reasons.

b. Calculate the increase in the height of water between low tide and high tide.

c. How many hours are required for one complete cycle of the tide?

6. The graph describes your height h, in metres, above the ground, where t is the time elapsed in minutes, as you ride a Ferris wheel.

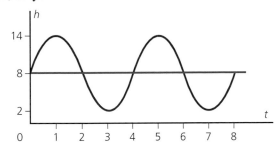

a. Does this graph represent a periodic function? If yes, what is the period?

b. What is the maximum height of the Ferris wheel above the ground?

c. How high is the boarding platform above the ground if it is located at the lowest point of the Ferris wheel?

d. What is your position on the Ferris wheel at time $t = 0$?

e. Draw the graph illustrating this situation given that at time $t = 0$, you are at the bottom of the Ferris wheel.

f. Draw the graph illustrating this situation given that at time $t = 0$, you are at the top of the Ferris wheel.

PART C

7. A piecewise function is defined as $f(x) = \begin{cases} x - 5, \text{ for } 0 \leq x < 10 \\ x - 15, \text{ for } 10 \leq x < 20 \\ x - 25, \text{ for } 20 \leq x < 30. \end{cases}$

a. Make a graph of the function.

b. Determine if the function is periodic.

In Section 8.1 we discovered that periodic functions repeat themselves on a regular basis. In Question 3 in Exercise 8.1, we noted the relationship between the coordinates of a point on a circle and the angle at the centre of the circle.

If the radius of the circle is r, the point $P(x, y)$ determines angle θ. We saw that an angle θ in standard position in a circle of radius r allows us to determine the coordinates of a point $P(x, y)$ on the terminal arm by using the definition $\dfrac{y}{r} = \sin \theta$ and $\dfrac{x}{r} = \cos \theta$.

If we let $r = 1$, then $y = \sin \theta$ and $x = \cos \theta$.

In fact, using a circle with radius 1, the coordinates of a point determined by angle θ are $(\cos \theta, \sin \theta)$.

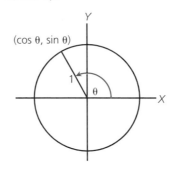

Since there is exactly one value of $\sin \theta$ for any given θ, as we have already seen, when we create ordered pairs $(\theta, \sin \theta)$ we make two observations:

1. As $\sin \theta$ is a distinct value for given θ, the definition of a function is satisfied.

2. As $\sin \theta = \sin(\theta + 360°) = \sin(\theta + 360k°)$ for k any integer, $y = \sin \theta$ is a periodic function.

Investigating the Sine Function $y = \sin \theta$

1. Use your graphing calculator to graph $y = \sin \theta$. Set your calculator in degree mode and use domain $0 \le \theta \le 720$.

2. Is the function $y = \sin \theta$ periodic? If so, what is the period?

3. To understand the shape of the graph, create a table of values using the ordered pairs $(\theta, \sin \theta)$ for values of $\theta = 0°, 30°, 60°, \ldots, 360°$, and plot these points.

4. It is useful to develop the ability to sketch the graph of the function. A quick sketch can be made by noting the basic shape and using the points given in the table of values below.

θ	0°	30°	90°	150°	180°	210°	270°	330°	360°
sin θ	0	0.5	1	0.5	0	−0.5	−1	−0.5	0

Successive periods of the graph follow by repetition.

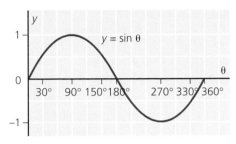

Investigating the Cosine Function y = cos θ

1. Use your graphing calculator to graph $y = \cos \theta$. Set your calculator in degree mode and use domain $0 \leq \theta \leq 720$.

2. Is the function $y = \cos \theta$ periodic? If so, what is its period?

3. To examine its graph, repeat the steps taken in connection with the graph of $y = \sin \theta$, first preparing a table of ordered pairs $(\theta, \cos \theta)$ for $\theta = 0°, 30°, 60°, \ldots, 360°$.

4. A working sketch can be obtained by using the table of values below.

θ	0°	60°	90°	120°	180°	240°	270°	300°	360°
cos θ	1	0.5	0	−0.5	−1	−0.5	0	0.5	1

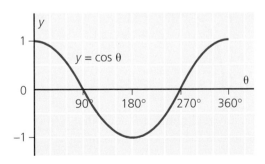

Notice that the shape of the graphs $y = \sin \theta$ and $y = \cos \theta$ are exactly the same if we draw more than one period, but that critical points (where they achieve maximum and minimum values and intercepts) are different. We refer to these as **sinusoidal functions.** Both functions have a maximum value of 1, a minimum value of −1, and a period of 360°.

Investigating the Tangent Function $y = \tan \theta$

1. Use your graphing calculator to graph $y = \tan \theta$. Set your calculator in degree mode and domain $-90 \le \theta \le 720$.

2. Is the function $y = \tan \theta$ periodic? If so, what is its period?

3. As we have seen, there is a unique value for $\sin \theta$ and $\cos \theta$ for a given angle θ.

$$\tan \theta = \frac{\sin \theta}{\cos \theta}, \text{ because } \tan \theta = \frac{y}{x} = \frac{\frac{r}{x}}{\frac{y}{r}} = \frac{\sin \theta}{\cos \theta}.$$

Since $\tan \theta = \frac{\sin \theta}{\cos \theta}$, this means that there is a distinct value for $\tan \theta$ for a given angle θ, except when $\cos \theta = 0$, which means that $\tan \theta$ is undefined (a calculator likely displays E or Error). This problem arises when $\theta = 90°$, $270°$, and in general when $\theta = (90 + 180k)°$ for k an integer. To examine its graph, proceed as before, using a table of ordered pairs $(\theta, \tan \theta)$ for $\theta = 0°, 30°, 60°, \ldots, 360°$ and plot these points.

4. For simple sketches, note that if θ is less than but near to $90°$, the value of $\tan \theta$ is very large and positive, while if θ is near to but larger than $90°$, the value of $\tan \theta$ is very large and negative. We note, then, that the vertical line at $\theta = 90°$ is an asymptote for the curve. This repeats at $\theta = 270°$ and for every $\theta = (90 + 180k)°$, k an integer. With this knowledge and the table of values below, a quick sketch can be drawn.

θ	$-90°$	$-45°$	$0°$	$45°$	$90°$	$135°$	$180°$	$225°$	$270°$
$\tan \theta$	und.	-1	0	$+1$	und.	-1	0	1	und.

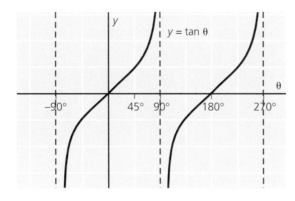

The domain of the tangent function is any real number θ such that $\theta \ne (90 + 180k)°$, k an integer. The range of the tangent function is the set of all real numbers. The period is $180°$.

Sketching these three graphs is straightforward. It is frequently desired that we locate points on a graph with a specific property, and this requires some care.

EXAMPLE

Draw a sketch of $y = \sin \theta$, locate four points having coordinates $\left(\theta, \frac{\sqrt{3}}{2}\right)$, and write a general statement for all points having y-coordinate $\frac{\sqrt{3}}{2}$ on the graph.

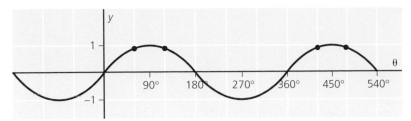

We know that $\sin 60° = \dfrac{\sqrt{3}}{2}$.

Then $\sin(180 - 60)° = \sin 120° = \dfrac{\sqrt{3}}{2}$.

Since angles of θ and $\theta + 360°$ are coterminal, $\sin 420° = \sin 480° = \dfrac{\sqrt{3}}{2}$.

The points $(60°, \sin 60°)$, $(120°, \sin 120°)$, $(420°, \sin 420°)$, $(480°, \sin 480°)$ all have $\sin \theta = \dfrac{\sqrt{3}}{2}$. A general statement for all points having $\sin \theta = \dfrac{\sqrt{3}}{2}$ is $\theta = (60 + 360k)°$ or $\theta = (120 + 360k)°$, $k \in I$.

Civil Engineering

Civil engineers use trigonometric functions to represent the vibrations that occur in structures.

In 1940, a two-lane suspension bridge 1500 m long was built across the Tacoma Narrows in the state of Washington. Vertical oscillations of the roadbed occurred in light breezes. The vibrations increased and local residents quickly named the bridge "Galloping Gertie" because of its spectacular movements. The up-down motion of the bridge soon reached ranges of one metre. Drivers that crossed the bridge could see cars disappear and reappear as the wave passed through. Fascinated by Galloping Gertie, thousands of people flocked to the bridge to experience the sensation of crossing the rolling centre span.

Four months after the bridge was completed, the vertical motion had increased substantially. Then the bridge started a twisting motion. As the structure twisted, the roadway rocked back and forth with the high side rising 8.5 m above the low side. The bridge soon broke apart and fell into the Narrows.

The Tacoma Narrows Bridge is a remarkable example of how things can go wrong if vibrations are not carefully analyzed.

The final moments of the bridge were captured on film and can be downloaded from the Internet.

Photo: Bettmann/CORBIS

PART B

Note: θ is measured in degrees.

1. Draw a sketch of $y = \cos\theta$, locate three points having the coordinates $\left(\theta, \frac{1}{2}\right)$, and write a general statement for the θ-values of all points having these coordinates.

2. Draw a sketch of $y = \sin\theta$. For each of the following, locate three points on the graph having the coordinates given, and write general statements for θ-values of all points having the given y-coordinate.

   a. $(\theta, 0)$ **b.** $\left(\theta, -\frac{1}{2}\right)$

   c. $(\theta, -1)$ **d.** $\left(\theta, -\frac{\sqrt{3}}{2}\right)$

3. Draw a sketch of $y = \cos\theta$. For each of the following, locate three specific points on the graph having the coordinates given, and write general statements for θ-values of all points having the given coordinates.

   a. $(\theta, 0)$ **b.** $(\theta, 1)$

   c. $\left(\theta, -\frac{\sqrt{3}}{2}\right)$ **d.** $\left(\theta, \frac{\sqrt{3}}{2}\right)$

4. Draw a sketch of $y = \tan\theta$. For each of the following, locate three specific points on the graph having the coordinates given, and write a general statement for θ-values of all points having the given coordinates.

   a. $(\theta, 0)$ **b.** $(\theta, 1)$

   c. $(\theta, \sqrt{3})$ **d.** $\left(\theta, -\frac{1}{\sqrt{3}}\right)$

A pendulum with a long arm *TA* requires 6.2 s to complete one complete cycle from *A* to *C* and back to *A*. As the pendulum swings, the perpendicular distance *p* from the head of the pendulum to the vertical line *BT* is a function of time. The graph for two cycles of this distance versus time function is shown.

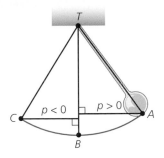

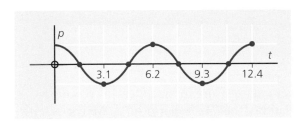

The graph appears to have the same shape as a cosine function, but the horizontal scale is totally different.

This example illustrates the fact that in many problems, the independent variable is *time* rather than an angle. For most problems in applied trigonometry, angles measured in degrees are acceptable, but in analytical trigonometry, it is more convenient to introduce a different unit of angle measurement.

If we think of a point moving along the circumference of a circle, there is a clear connection between the length of an arc traced out and the time taken. There is also a clear connection between the length of the arc and the angle subtending that arc. We link these by defining a new unit of angle measurement in terms of the length of the arc subtending the angle.

This unit of measurement is the **radian.**

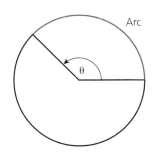

> A **radian** is the measure of an angle subtended at the centre of a circle
> by an arc equal in length to the radius of the circle.

Now the connection between the length of the arc and the angle subtending the arc is $a = r\theta$, where θ is measured in radians.

In the diagram, the arc AB has length equal to the radius. The measure of $\angle AOB$ is 1 radian.

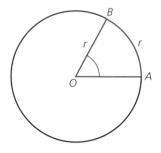

$$\frac{\angle AOB}{360°} = \frac{\text{arc } AB}{\text{circumference}}$$

$$\frac{1 \text{ radian}}{360°} = \frac{r}{2\pi r} = \frac{1}{2\pi}$$

Therefore 1 radian $= \dfrac{180°}{\pi}$.

The magnitude of a radian is independent of r and is the same for all circles. This also establishes the relation between radians and degrees.

$$\pi \text{ radians} = 180°$$

$$1 \text{ radian} = \frac{180°}{\pi}$$

$$1 \text{ degree} = \frac{\pi}{180} \text{ radians}$$

Using these relations, it is easy to convert from one measure to the other. In recording angle measures in radians, we use the abbreviations "rad" or superscript r. We can refer to an angle of 0.4 radians, or 0.4 rad, or 0.4^r.

EXAMPLE 1

Convert each of the following from degree measure to radian measure.

a. 30° **b.** 45° **c.** 135° **d.** 240°

SOLUTION

Since $1° = \dfrac{\pi}{180}$ rad,

a. $30° = \dfrac{30\pi}{180}$ rad $= \dfrac{\pi}{6}$ rad

b. $45° = \dfrac{45\pi}{180}$ rad $= \dfrac{\pi}{4}$ rad

c. $135° = \dfrac{135\pi}{180}$ rad $= \dfrac{3\pi}{4}$ rad

d. $240° = \dfrac{240\pi}{180}$ rad $= \dfrac{4\pi}{3}$ rad

Note that $\dfrac{\pi}{6}$ rad $\doteq 0.523\ 599^r$ and that $\dfrac{3\pi}{4}$ rad $\doteq 2.356\ 195^r$, and in general we can always obtain approximate decimal values for radian measures. This is usually not advantageous.

EXAMPLE 2

Convert each of the following to degree measure, correct to one decimal.

a. 4^r **b.** $\frac{5\pi}{6}$ rad **c.** 2.53^r

Since $1^r = \frac{180°}{\pi}$,

a. $4^r = \frac{4 \times 180°}{\pi} \doteq 229.2°$ **b.** $\frac{5\pi}{6}$ rad $= \frac{5\pi}{6} \times \frac{180°}{\pi} = 150°$

c. $2.53^r = 2.53 \times \frac{180°}{\pi} \doteq 145.0°$

An immediate result of angle measurement in radians is in the graphs of the trigonometric functions. When using degree measure, the maximum and minimum values of $y = \sin x$ are $+1$ and -1, respectively, but it takes 360 units of measurement for one period. The result is that one never sees a graph of the function where one unit vertically is equal to one unit horizontally. This problem disappears with radian measure, for a complete period requires 2π units, approximately 6.28 units.

It is practical now to draw a sketch with equal units horizontally and vertically. We divide the horizontal axis in units of $\frac{\pi}{2}$ and note that key points for sketching one period of $y = \sin x$ are $A(0, 0)$, $B\left(\frac{\pi}{2}, 1\right)$, $C(\pi, 0)$, $D\left(\frac{3\pi}{2}, -1\right)$, and $E(2\pi, 0)$. When we write angles in terms of π, we assume that the units are radians and omit the radian symbol.

The graph is

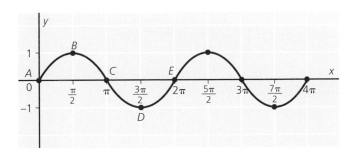

A major advantage in using radian measure is the existence of a correspondence between angle and linear measurement. Because the radian ties the arc length to the angle, the variable x in $y = \sin x$ can be viewed either as an angle or as a linear measurement. Replacing x with t means that we can describe the motion of the pendulum in the earlier example by the equation $p = \cos t$.

This means that when we consider a variable x, we can consider its values to be either an angle, measured in radians, or a linear measurement, the arc length subtended by the angle. The variable becomes simply a real number. When we graph a trigonometric function, the horizontal axis now represents real numbers in exactly the same way as it does in other functions we have considered.

EXAMPLE 3

What is the exact value of $\cos \frac{\pi}{3}$?

SOLUTION

$$\frac{\pi^r}{3} = \frac{\pi}{3} \times \frac{180°}{\pi}$$
$$= 60°$$

$$\cos \frac{\pi}{3} = \frac{1}{2}$$

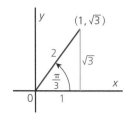

EXAMPLE 4

Evaluate $\sin^2 \frac{\pi}{3} + \cos^2 \frac{3\pi}{4}$.

SOLUTION

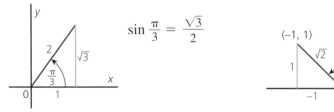

$$\sin \frac{\pi}{3} = \frac{\sqrt{3}}{2}$$

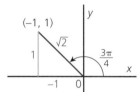

$$\cos \frac{3\pi}{4} = -\frac{1}{\sqrt{2}}$$

$$\sin^2 \frac{\pi}{3} + \cos^2 \frac{3\pi}{4} = \left(\frac{\sqrt{3}}{2}\right)^2 + \left(-\frac{1}{\sqrt{2}}\right)^2$$
$$= \frac{3}{4} + \frac{1}{2}$$
$$= \frac{5}{4}$$

PART A

1. Convert the following angles to radian measure. Leave your answer in simplest rational form.

   a. $60°$ **b.** $150°$ **c.** $225°$ **d.** $350°$ **e.** $300°$

   f. $495°$ **g.** $-120°$ **h.** $-855°$ **i.** $90°$ **j.** $270°$

2. Convert the following angles from radian measure to degree measure.

   a. $\frac{\pi}{3}$ **b.** $\frac{\pi}{4}$ **c.** $\frac{\pi}{5}$

   d. $\frac{7\pi}{6}$ **e.** 2.9^r **f.** 4.3^r

3. Use a calculator to evaluate the following. Make sure your calculator is in the correct mode.

   a. $\cos \frac{\pi}{5}$ **b.** $\tan \frac{\pi}{10}$ **c.** $\sin \frac{7\pi}{3}$

   d. $\cos 731°$ **e.** $\sin 147°$ **f.** $\tan 830°$

   g. $\cos 5.2^r$ **h.** $\tan 11.5^r$ **i.** $\sin 62.75^r$

PART B

4. **a.** Use a graphing calculator to graph the cosine function in radian mode for $-2\pi \le x \le 2\pi$.

   b. Sketch the cosine function $y = \cos x$, $0 \le x \le 2\pi$. (Use 3 squares for π units horizontally.)

5. Evaluate each of the following. Give exact answers.

   a. $\sin \frac{\pi}{6}$ **b.** $\cos \frac{\pi}{4}$

   c. $\tan \frac{\pi}{3}$ **d.** $\cos \frac{2\pi}{3}$

   e. $\sin \frac{5\pi}{6}$ **f.** $\cos \frac{4\pi}{3}$

   g. $\tan \frac{7\pi}{4}$ **h.** $\sin \frac{3\pi}{4}$

   i. $\cos \frac{\pi}{3} + \sin \frac{\pi}{6}$ **j.** $\sin^2 \frac{\pi}{4} + \cos^2 \frac{\pi}{6}$

   k. $\sin \frac{7\pi}{6} + \tan \frac{5\pi}{4}$ **l.** $\cos^2 \frac{\pi}{3} + \sin^2 \frac{\pi}{3}$

6. Use a graphing calculator to graph $y = \tan x$, $-\pi \le x \le 2\pi$.

7. Verify that $\sin^2 t + \cos^2 t = 1$ for the following values of t.

    a. $\dfrac{\pi}{3}$ **b.** $\dfrac{3\pi}{4}$ **c.** 5π **d.** 1.625^r

8. For each of the following, determine θ, correct to 3 decimals, $0 \le \theta \le 2\pi$.

    a. $\cos \theta = -0.575$ and θ is in quadrant II

    b. $\sin \theta = 0.892$ and θ is in quadrant II

    c. $\sin \theta = -0.123$ and θ is in quadrant III

    d. $\tan \theta = 2.882$ and θ is in quadrant III

    e. $\cos \theta = 0.158$ and θ is in quadrant IV

    f. $\tan \theta = -0.483$ and θ is in quadrant IV

9. In a circle with radius 5 cm, the arc length is $\dfrac{3\pi}{4}$ cm. Determine the measure of the angle at the centre of the circle.

10. In a circle, the arc length is 30 cm and the angle at the centre is $\dfrac{\pi}{3}$. What is the radius of the circle?

PART C

11. For what values of θ is $\sin \theta \le \cos \theta$, $0 \le \theta \le 4\pi$?
Hint: Use the graphs of $y = \sin \theta$ and $y = \cos \theta$.

12. A circle with centre $(0, 2)$, radius 2, rolls along the positive x-axis. The centre moves to the point $(5, 2)$. Point P, initially at $P(0, 0)$, rolls along the circumference to the point $Q(x, y)$. What are the coordinates of Q? Write your answer, correct to two decimal places.

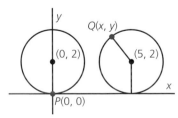

 **Horizontal and Vertical Translations**

In our study of transformations in Chapter 1, we developed techniques for relating the graph of $y = f(x) + k$ to the graph of $y = f(x)$. We observed that the graph of $y = f(x) + k$ is shifted k units vertically relative to the graph of $y = f(x)$. The direction of the shift depends on whether k is positive or negative. In the following activity, you will apply this technique to the graphing of sinusoidal functions.

ACTIVITY 1

1. a. Graph each of the following on the same set of axes. Set your domain at $0 \le x \le 2\pi$.

   i) $y = \sin x$ **ii)** $y = \sin x + 2$

   iii) $y = \sin x - 3$

   b. How is the graph of $y = \sin x + 2$ related to the graph of $y = \sin x$?

   c. How is the graph of $y = \sin x - 3$ related to the graph of $y = \sin x$?

   d. How is the graph of $y = \sin x + k$ related to the graph of $y = \sin x$? Consider the cases where $k > 0$ and where $k < 0$.

2. a. Graph each of the following on the same set of axes. Set your domain at $0 \le x \le 2\pi$.

   i) $y = \cos x$ **ii)** $y = \cos x + 3$

   iii) $y = \cos x - 2$

   b. How is the graph of $y = \cos x + 3$ related to the graph of $y = \cos x$?

   c. How is the graph of $y = \cos x - 2$ related to the graph of $y = \cos x$?

   d. How is the graph of $y = \cos x + k$ related to the graph of $y = \cos x$? Consider the cases where $k > 0$ and where $k < 0$.

ACTIVITY 2

From our earlier study of transformations, we also observed that the graph of $y = f(x + p)$ is translated horizontally p units to the right or left relative to the graph of $y = f(x)$.

1. Graph each of the following pairs of functions on the same set of axes. Set your domain at $-\pi \le x \le 3\pi$.

   a. $y = \sin x$ and $y = \sin\left(x - \frac{\pi}{2}\right)$.

   How is the graph of $y = \sin\left(x - \frac{\pi}{2}\right)$ related to the graph of $y = \sin x$?

b. $y = \sin x$ and $y = \sin\left(x + \frac{\pi}{6}\right)$.

How is the graph of $y = \sin\left(x + \frac{\pi}{6}\right)$ related to the graph of $y = \sin x$?

c. $y = \sin x$ and $y = \sin\left(x - \frac{\pi}{3}\right)$.

How is the graph of $y = \sin\left(x - \frac{\pi}{3}\right)$ related to the graph of $y = \sin x$?

d. How is the graph of $y = \sin(x + p)$ related to the graph of $y = \sin x$? Consider the cases where $p > 0$ and where $p < 0$.

In trigonometry a horizontal translation is called a **phase shift**. The phase shift of the function $y = \sin\left(x - \frac{\pi}{2}\right)$ is $\frac{\pi}{2}$.

2. What is the phase shift for each function graphed in Question 1, parts **b** and **c**?

3. Graph each of the following pairs of functions on the same set of axes. Set your domain at $-\pi \le x \le 3\pi$.

a. $y = \cos x$ and $y = \cos\left(x + \frac{\pi}{2}\right)$.

What is the phase shift? How is the graph of $y = \cos\left(x + \frac{\pi}{2}\right)$ related to the graph of $y = \cos x$?

b. $y = \cos x$ and $y = \cos\left(x - \frac{\pi}{3}\right)$.

What is the phase shift? How is the graph of $y = \cos\left(x - \frac{\pi}{3}\right)$ related to the graph of $y = \cos x$?

c. How is the graph of $y = \cos(x + p)$ related to the graph of $y = \cos x$? Consider the cases where $p > 0$ and $p < 0$.

EXAMPLE 1

Graph $y = \cos x + 2$, $0 \le x \le 2\pi$.

SOLUTION

The graph of $y = \cos x$ is translated vertically 2 units upward to obtain the graph of $y = \cos x + 2$.

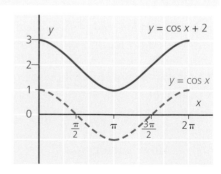

EXAMPLE 2

For the function $y = \sin\left(x + \frac{2\pi}{3}\right)$.

a. What is the phase shift?

b. Graph the function in the domain $-\pi \leq x \leq 2\pi$.

SOLUTION

a. The phase shift is $\frac{-2\pi}{3}$. This means that the graph of $y = \sin x$ is translated horizontally $\frac{2\pi}{3}$ units to the left.

b. First graph $y = \sin x$ in the given domain, and then translate it horizontally $\frac{-2\pi}{3}$ units.

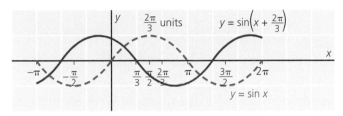

In the next example, we will combine horizontal and vertical translations.

EXAMPLE 3

For the function $y = \cos\left(x - \frac{\pi}{3}\right) - 2$:

a. What is the phase shift?

b. Graph the function in the domain $-\pi \leq x \leq 3\pi$.

SOLUTION

a. The phase shift is $\frac{\pi}{3}$.

b. First graph $y = \cos x$, and then translate it horizontally $\frac{\pi}{3}$ units and then vertically 2 units downward.

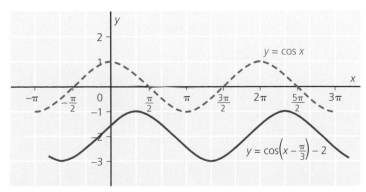

PART A

1. Describe the transformation(s) required to transform the curve $y = \sin x$ to each of the given curves.

a. $y = \sin x + 3$

b. $y = \sin\left(x - \frac{3\pi}{4}\right)$

c. $y = \sin\left(x - \frac{\pi}{3}\right) - 3$

d. $y = \sin\left(x + \frac{\pi}{2}\right) + 2$

2. Describe the transformation(s) required to transform the curve $y = \cos x$ to each of the given curves.

a. $y = \cos x - 5$

b. $y = \cos\left(x - \frac{\pi}{6}\right)$

c. $y = \cos\left(x + \frac{3\pi}{4}\right) - 3$

d. $y = \cos\left(x - \frac{2\pi}{3}\right) + 2$

3. Consider each of the functions drawn below as a sine function. Write an equation that represents each function.

a.

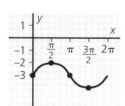

b.

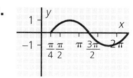

c.

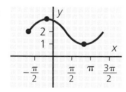

d.

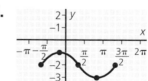

4. Consider each of the functions drawn in Question 3, parts **b** and **c**, as a cosine function. Write an equation that represents each function.

5. Graph one period of each of the following:

a. $y = \sin x - 1$

b. $y = \cos x + 2$

c. $y = \sin\left(x + \frac{\pi}{3}\right)$

d. $y = \cos\left(x - \frac{2\pi}{3}\right)$

e. $y = \sin\left(x + \frac{4\pi}{3}\right) - 2$

f. $y = \cos\left(x - \frac{5\pi}{6}\right) + 3$

PART B

6. For each of the following functions, determine the maximum and minimum values for h.

   a. $h = \sin t + 7$

   b. $h = \cos t - 3$

7. Determine the translation that transformed $y = \cos \theta$ to the following:

   a. $y = \cos\left(\theta - \frac{\pi}{2}\right)$

   b. $y = 5 + \cos\left(\theta + \frac{4\pi}{3}\right)$

8. A sinusoidal curve that has been translated vertically has an equation of the form $y = \sin \theta + p$. If the image curve passes through the point given, then find the equation.

   a. $(\pi, 11)$

   b. $\left(-\frac{3\pi}{2}, -5\right)$

   c. $\left(\frac{7\pi}{6}, 4\right)$

9. Graph the curves defined by each of the following equations for a domain from 0 to 3π.

   a. $y = \sin\left(\theta - \frac{\pi}{6}\right)$

   b. $y = \cos\left(x - \frac{\pi}{4}\right) + 3$

   c. $y = \sin\left(x - \frac{5\pi}{6}\right) + 2$

   d. $y = \cos\left(x + \frac{7\pi}{6}\right) - 1$

10. **a.** A sine function was translated horizontally to produce the curve shown. State two different translations that could have been used to define this curve.

    b. A cosine function was translated horizontally to produce the curve shown. State two different translations that might be used to define this curve.

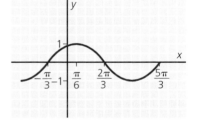

11. If $Q\left(b, -\frac{1}{\sqrt{2}}\right)$ is on the graph of $h = \cos\left(t - \frac{7\pi}{2}\right)$, then find two possible values for b.

12. A cosine function on the interval $0 \le x \le 3\pi$ has a vertical shift of -10 and the first minimum point is $P\left(\frac{7\pi}{6}, -11\right)$. Determine the equation of the function.

13. A sine function on the interval $0 \le x \le 3\pi$ has a vertical shift of 5 and the first maximum point is $P\left(\frac{5\pi}{6}, 6\right)$. Determine the equation of the function.

PART C

14. Determine the values of t on the domain $-\pi \le t \le 2\pi$ so that the function defined by $h = \frac{5}{2} + \cos\left(t - \frac{\pi}{3}\right)$ will have $h \le 3$.

8.5 Horizontal and Vertical Dilatations

The amount *I* of electric current supplied to a house is dependent on the length of time the current is being used and is represented by a function whose graph is sinusoidal. Alternating current has a frequency of 60 cycles per second, so the period of this function is $\frac{1}{60}$ of a second.

The graph illustrates two complete periods of this situation that attain a maximum current strength of 15 amperes.

In this section, we will examine how to graph sinusoidal functions of the form
$y = a \sin kx$ and
$y = a \cos kx$.

The following activities will use the transformations from Chapter 1 that result in a vertical or a horizontal dilatation.

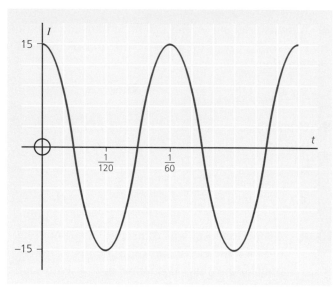

A graphing calculator or computer software may be useful in completing the following activities.

ACTIVITY 1

1. a. Graph each of the following on the same set of axes. Set your domain at $-2\pi \le x \le 2\pi$.

    i) $y = \sin x$

    ii) $y = 2 \sin x$

    iii) $y = 3 \sin x$

    iv) $y = \frac{1}{2} \sin x$

  b. How is the graph of $y = 2 \sin x$ related to the graph of $y = \sin x$?

  c. How is the graph of $y = 3 \sin x$ related to the graph of $y = \sin x$?

  d. How is the graph of $y = \frac{1}{2} \sin x$ related to the graph $y = \sin x$?

  e. How is the graph of $y = a \sin x$ related to the graph of $y = \sin x$? Consider the cases where $a > 1$ and $0 < a < 1$.

2. a. Graph each of the following on the same set of axes. Set your domain at $-2\pi \le x \le 2\pi$.

  i) $y = \cos x$

  ii) $y = 2 \cos x$

  iii) $y = \frac{1}{2} \cos x$

  b. How is the graph of $y = 2 \cos x$ related to the graph of $y = \cos x$?

  c. How is the graph of $y = \frac{1}{2} \cos x$ related to the graph of $y = \cos x$?

  d. How is the graph of $y = a \cos \theta$ related to the graph of $y = \cos x$? Consider the cases where $a > 1$ and $0 < a < 1$.

In general, multiplying $\sin x$ or $\cos x$ by a constant a results in a vertical stretch or contraction of the graph of $y = \sin x$ or $y = \cos x$. The amount of stretch or contraction is called the **amplitude.** For the function $y = 3 \sin \theta$, the amplitude is 3.

3. State the amplitude of each function listed in Questions 1 and 2 in Activity 1.

ACTIVITY 2

Graph each of the following pairs of sinusoidal functions on the same set of axes and set the domain as $-2\pi \le x \le 2\pi$.

1. $y = \sin x$ and $y = \sin 2x$

  a. How is the graph of $y = \sin 2x$ related to the graph of $y = \sin x$?

  b. What is the period of the function $y = \sin 2x$?

2. $y = \sin x$ and $y = \sin \frac{1}{2}x$

  a. How is the graph of $y = \sin \frac{1}{2}x$ related to the graph of $y = \sin x$?

  b. What is the period of $y = \sin \frac{1}{2}x$?

3. $y = \cos x$ and $y = \cos 3x$

  a. How is the graph of $y = \cos 3x$ related to the graph of $y = \cos x$?

  b. What is the period of $y = \cos 3x$?

In general, multiplying the variable x in the function $y = \sin x$ or $y = \cos x$ by a constant causes a horizontal contraction of the graph so that the period of the image is $\frac{1}{k}$ that of the original, where $k > 1$. If $0 < k < 1$, then $\frac{1}{k} > 1$ and the graph of the image is stretched horizontally with a period $\frac{1}{k}$ that of the original.

4. Complete the following table for different values of k in the functions $y = \sin kx$ and $y = \cos kx$.

Function	Value of k	Period
$y = \sin x$	1	2π
$y = \sin 2x$		
$y = \sin \frac{1}{2} x$		4π
$y = \cos x$		
$y = \cos 3x$		
$y = \sin kx$		
$y = \cos kx$		

The **period** for the graph of $y = \sin kx$ or the graph of $y = \cos kx$, $k > 0$, is

$$\frac{2\pi}{k} \text{ or } \frac{360°}{k}.$$

EXAMPLE 1

State the amplitude and period of the function $y = 3 \cos \frac{2}{5}x$.

SOLUTION

The amplitude is 3.

Since $k = \frac{2}{5}$, the period is $\frac{2\pi}{\frac{2}{5}}$ or $2\pi \times \frac{5}{2} = 5\pi$.

EXAMPLE 2

a. Graph $y = 2 \sin 3x$ over two cycles.

b. State the coordinates of the maximum points and minimum points.

SOLUTION

a. $y = 2 \sin 3x$ represents a sine function graph that has an amplitude of 2 and a period of $\frac{2\pi}{3}$.

To obtain the graph, $y = \sin x$ is stretched vertically by a factor of 2 and contracted horizontally by a factor of 3, with a period of $\frac{2\pi}{3}$.

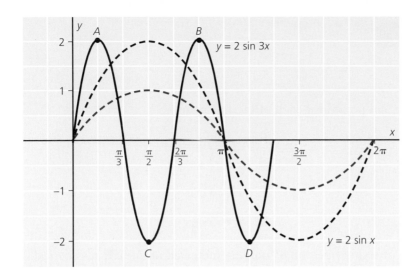

When we sketch a graph of a function such as $y = 2 \sin 3x$, it is not necessary to also graph $y = \sin x$ and $y = 2 \sin x$.

b. Label the maximum points A and B. The coordinates are $A\left(\frac{\pi}{6}, 2\right)$ and $B\left(\frac{5\pi}{6}, 2\right)$. Label the minimum points C and D. The coordinates are $C\left(\frac{\pi}{2}, -2\right)$ and $D\left(\frac{7\pi}{6}, -2\right)$.

EXAMPLE 3

Sketch one complete cycle of the function $y = 10 \cos \frac{1}{2}x$, $-2\pi \leq x \leq 2\pi$.

SOLUTION

The amplitude of the function $y = 10 \cos \frac{1}{2}x$ is 10 and the period is $\dfrac{2\pi}{\frac{1}{2}} = 4\pi$.

Set a vertical and horizontal scale as shown in the diagram.

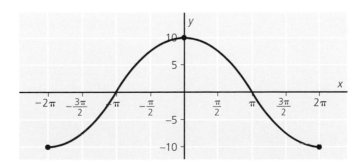

PART A

1. State the amplitude and period for each of the following:

    a. $y = 3 \sin x$ **b.** $y = \cos \frac{2x}{3}$

    c. $y = -5 \sin 6\theta$ **d.** $y = \frac{5}{3} \cos \frac{x}{3}$

    e. $y = -7 \sin 5x$ **f.** $y = 6 \cos \frac{3x}{5}$

    g. $y = 4 \sin 2\pi\theta$ **h.** $y = 5 \cos\left(\frac{2\pi}{3}x\right)$

2. Write the equation of a sine function having the indicated amplitude and period.

    a. amplitude 2, period 2π **b.** amplitude 1, period π

    c. amplitude $\frac{1}{3}$, period $\frac{2\pi}{3}$ **d.** amplitude 7, period 4π

    e. amplitude 3, period 3π **f.** amplitude $\frac{3}{5}$, period 3

3. How is the graph of $y = \cos kx$ related to the graph of $y = \cos x$ as k varies? Consider the cases where $k > 1$ and where $0 < k < 1$.

4. Each of the graphs below represents a sinusoidal function. Write an equation that represents each function.

    a.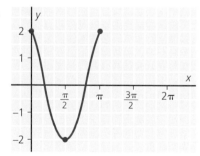

    b.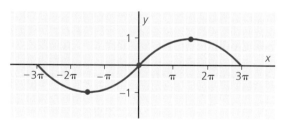

    c.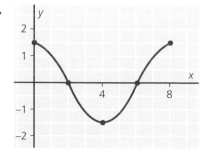

5. a. Graph each of the following functions for one complete cycle.

  b. For each function, state the coordinates of the maximum and minimum points.

  i) $y = 2 \cos x$ **ii)** $y = \frac{3}{2} \sin x$

  iii) $y = 3 \cos 2x$ **iv)** $y = 2 \sin \frac{1}{2}x$

  v) $y = 3 \sin 4x$ **vi)** $y = \cos 3x$

  vii) $y = -\sin \frac{3x}{2}$ **viii)** $y = \cos \frac{2x}{3}$

6. The point $\left(\frac{\pi}{4}, 1\right)$ lies on the function in the form $y = \sin kx$. Determine k.

7. The point $(2\pi, -1)$ lies on the function in the form $y = \cos kx$. Determine k.

8. a. Sketch one period of the function $f(t) = 50 \sin\left(\frac{t}{5}\right)$.

  b. On the same axes, sketch the function $g(t) = 25 \cos\left(\frac{t}{3}\right)$ for $0 \le t \le 10\pi$.

  c. Determine the number of intersection points of the functions f and g for $0 \le t \le 10\pi$.

  d. Check your work on a graphing calculator.

PART C

9. a. Graph $y = \sin x + \cos x$, $-2\pi \le x \le 2\pi$.

  b. The graph obtained in part **a** is sinusoidal and can be expressed as $y = a \sin(x - p)$ or $y = a \cos(x - p)$. Determine the values of a and p in each case.

10. a. Graph $y = 2 \sin x + 3 \cos x$, $-2\pi \le x \le 2\pi$.

  b. The graph obtained in part **a** is sinusoidal and can be expressed as $y = a \sin(x - p)$ or $y = a \cos(x - p)$. Determine the values of a and p in each case.

  c. By examining the value for a here and in the previous question, can you see any possible relationship between the coefficients 2 and 3 and the value of a?

  d. If $y = r \sin x + s \cos x$, and we can express this function as $y = a \sin(x - p)$ or $y = a \cos(x - p)$, determine a possible relationship between a, r, and s.

11. For any parallelogram $ABCD$, prove the sum of the squares on the diagonals is twice the sum of the squares on two adjacent sides.

8.6 Setting the Scales on the Axes System

From previous work, we know that a sinusoidal function having an equation in the form $y = a \sin k\theta$ or $y = a \cos k\theta$ will have an amplitude defined by parameter a and a period defined by $\frac{2\pi}{k}$. For example, in $y = \sin(30\pi t)$ the period is $\frac{2\pi}{30\pi} = \frac{1}{15}$, and we must, even if using a calculator, take this into consideration when setting scales on the axes. The following examples illustrate some techniques that are useful when setting scales on axes.

EXAMPLE 1

Graph two periods of $y = \sin\left(\frac{\pi}{2}x\right)$ on a graphing calculator.

SOLUTION

The period of the function is $\dfrac{2\pi}{\left(\frac{\pi}{2}\right)} = 4$.

For two periods we require $0 \le x \le 8$.

Note that the scale along the x-axis is $0, 1, 2, 3, \ldots, 8$.

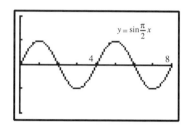

EXAMPLE 2

Graph two periods of $y = 2 \cos 3\pi x$ on a graphing calculator.

SOLUTION

The period of $y = 2 \cos 3\pi x$ is $\frac{2\pi}{3\pi} = \frac{2}{3}$.

In graphing this function, care must be taken in selecting the scale in order to obtain a reasonable graph.

For example, if we select the default trigonometric scale on the graphing calculator, we obtain the graph shown.

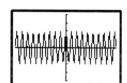

Since the period is $\frac{2}{3}$ and we would like two cycles, we select the domain as $0 \le x \le 1.3$ since $\frac{2}{3} \times 2 = \frac{4}{3}$ or approximately 1.3.

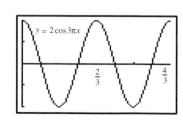

EXAMPLE 3

Graph one cycle of the function $y = 3 \sin\left(\frac{x}{3}\right)$.

SOLUTION

The period is $\frac{2\pi}{\frac{1}{3}} = 6\pi$.

A reasonable scale is to let $\pi = 2$ units along the x-axis.

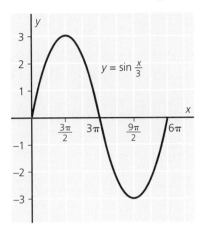

Exercise 8.6

PART A

1. For each of the following, state the period and amplitude.

   a. $y = 7 \sin(10x)$

   b. $P = 220 \cos(100\pi t)$

   c. $h = 3 \sin\left(\frac{\pi}{6}t\right)$

   d. $y = 10 \cos\left(\frac{5x}{8}\right)$

2. For each part of Question 1, determine sensible units to set the scales for the period and the amplitude.

3. Use a graphing calculator to graph each of the functions in Question 1.

PART B

4. For each of the following, determine the period, amplitude, and a suitable number of units for the horizontal and vertical scales. Graph each function for one period.

   a. $y = 25 \sin(5\pi x)$

   b. $I = 0.25 \cos(20t)$

   c. $h = 150 \sin\left(\frac{\pi}{10}t\right)$

   d. $y = \frac{3}{5} \cos\left(\frac{4}{7}x\right)$

5. The electrical current that passes through a 20-amp breaker switch is defined by $I = 20 \sin(120\pi t)$. Sketch this function for $0 \leq t \leq 0.05$, where t is time in seconds.

6. Sketch the function defined by $I = 100 \cos(120\pi t)$ for $0 \leq t \leq \frac{1}{30}$.

7. Sketch the function $v = 120 \sin(120\pi t)$, $0 \leq t \leq 0.017$.

PART C

8. In any $\triangle ABC$, prove the perimeter is equal to
   $$(b + c)\cos A + (a + c)\cos B + (a + b)\cos C.$$

9. From the centre O of two concentric circles, radii OAM and OBP are drawn, where A and B are on the inner circle and M and P are on the outer circle and $\angle AOB = \theta$ is acute. Prove the ratio of the areas of the segments subtended by θ is equal to the square of the ratio of the radii.

Multiple Transformations

In this section, we examine the effect of combining transformations studied in the previous sections.

EXAMPLE 1

For the function $y = 2\cos\left(x - \frac{\pi}{3}\right) - 1$, do the following:

a. State the amplitude, period, and phase shift of the given function.

b. Graph the function in the domain $-2\pi \le x \le 2\pi$.

c. State the coordinates of the maximum and minimum points.

d. What is the range of this function?

SOLUTION

a. The amplitude is 2, the period is 2π, and the phase shift is $\frac{\pi}{3}$.

b. *Step 1:* Translate the graph of $y = \cos x$ to the right $\frac{\pi}{3}$ units.

Step 2: Stretch the graph of $y = \cos\left(x - \frac{\pi}{3}\right)$ vertically by a factor of 2.

Step 3: Translate the graph of $y = 2\cos\left(x - \frac{\pi}{3}\right)$ vertically 1 unit down.

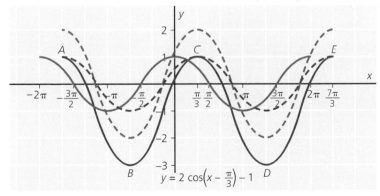

c. The maximum points are at A, C, and E. A is $\left(-\frac{5\pi}{3}, 1\right)$, C is $\left(\frac{\pi}{3}, 1\right)$, and E is $\left(\frac{7\pi}{3}, 1\right)$. The minimum points are B and D. B is $\left(-\frac{2\pi}{3}, -3\right)$ and D is $\left(\frac{4\pi}{3}, -3\right)$.

d. The range is $-3 \le y \le 1$, $y \in R$.
The **sinusoidal axis** of the graph of a sinusoidal function is halfway between the maximum and minimum values. In the above example, the maximum of the function is 1 and the minimum value is -3. The sinusoidal axis is located at $y = \frac{1-3}{2}$ or $y = -1$.

EXAMPLE 2

For the function $y = \sin 3\left(x + \frac{\pi}{2}\right) + 2$, $-2\pi \leq x \leq 2\pi$, $x \in R$, do the following:

a. State the amplitude, period, and phase shift of the function.

b. Graph the function.

c. Determine the equation of the sinusoidal axis.

d. State the coordinates of two maximum and two minimum points.

e. State the domain and range.

SOLUTION

a. The amplitude is 1. The period is $\frac{2\pi}{3}$.
The phase shift is $-\frac{\pi}{2}$.

b. *Step 1:* Since the period is $\frac{2\pi}{3}$, contract the graph of the function $y = \sin x$ horizontally by a factor of $\frac{1}{3}$.

Step 2: Translate the graph of the function $y = \sin 3x$ to the left $\frac{\pi}{2}$ units.

Step 3: Translate the graph of the function $y = \sin 3\left(x + \frac{\pi}{2}\right)$ vertically 2 units upward.

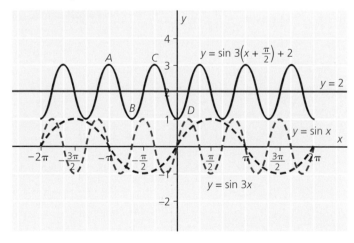

c. Since the maximum value of the function is 3 and the minimum value is 1, the sinusoidal axis occurs at $y = \frac{1 + 3}{2}$ or $y = 2$.

d. Two possible maximum points occur at A and C. A is $(-\pi, 3)$ and C is $(-\frac{\pi}{3}, 3)$.
Two possible minimum points occur at B and D. B is $(-\frac{5\pi}{6}, 1)$ and D is $(0, 1)$.

e. The domain is $-2\pi \leq x \leq 2\pi$, $x \in R$, and the range is $1 \leq y \leq 3$, $y \in R$.

For the sinusoidal functions $y = a \sin k(x - p) + d$ and $y = a \cos k(x - p) + d$, $|a|$ is the amplitude. Note that the value of a can be positive or negative, but the amplitude itself is always positive.

♦ The period is $\frac{2\pi}{k}$.

♦ The phase shift is p.

♦ d represents the vertical translation of the curves $y = a \sin k(x - p)$ and $y = a \cos k(x - p)$.

EXAMPLE 3

For the function $y = -2 \cos\left(3x + \frac{\pi}{2}\right) - 1$, $-\pi \le x \le \pi$, $x \in R$, do the following:

a. State the amplitude, period, and phase.

b. Graph the function.

SOLUTION

Rewrite the equation of the function in the form $y = a \cos k(x - p) + d$.

$$y = -2 \cos 3\left(x + \frac{\pi}{6}\right) - 1$$

a. amplitude $= |-2|$ period $= \frac{2\pi}{3}$ phase shift $= -\frac{\pi}{6}$
    $= 2$

b. Since the curve of $y = -2 \cos 3x$ is translated $\frac{\pi}{6}$ units to the left and 1 unit down, we can begin with the graph of $y = -2 \cos 3x$.

(Note that the graph of $y = -2 \cos 3x$ is the reflection of $y = 2 \cos 3x$ in the x-axis).

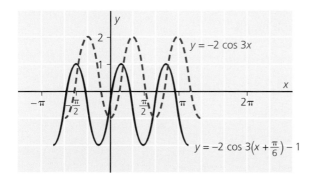

PART A

1. The general equation of any sine function is $y = a \sin k(x - p) + d$. In your own words, explain how a, k, p, and d affect the graph of this function.

2. For each of the following, state the amplitude, period, phase shift, and vertical translation.

   a. $y = 20 \sin 3x + 10$

   b. $y = \frac{1}{5} \cos 4\left(x + \frac{\pi}{2}\right) - 3$

   c. $y = \sin\left(2x + \frac{\pi}{8}\right)$

   d. $y = 5 \cos 2(x - \pi) - 3$

   e. $y = \frac{1}{3} \sin\left(3x + \frac{\pi}{4}\right) + 10$

3. Each of the following represents the graph of a sine function. State the following:

   i) the amplitude

   ii) the period

   iii) a possible phase shift

   iv) the vertical translation

   v) the equation of the sinusoidal axis

   vi) a minimum value of the function and the value of x for which it occurs

   vii) a maximum value of the function and the value of x for which it occurs

   viii) an equation that represents each graph

   a.

   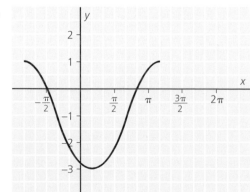

b.

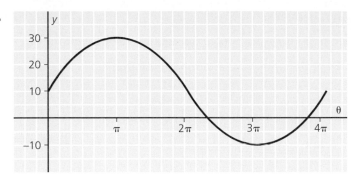

c.

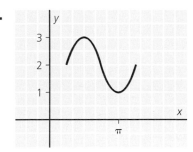

d.

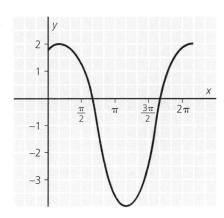

4. For each graph in Question 3, write an equation that represents it as a cosine function.

PART B

5. For each of the following functions,

    i) determine the amplitude, the phase shift, the period, and the vertical translation;

    ii) graph one complete period;

    iii) state the domain and range.

    a. $y = 5 \sin 2x + 3$

    b. $y = 2 \cos 3\left(x - \frac{\pi}{3}\right)$

    c. $y = 3 \sin 2\left(x - \frac{2\pi}{3}\right) - 5$

    d. $y = \cos \frac{1}{2}\left(x - \frac{\pi}{4}\right) + 2$

6. For each of the following functions,

    i) graph two complete periods of the curves defined by each function;

    ii) state the coordinates of the minimum and maximum points.

  a. $y = 5 \sin 3\left(x - \frac{\pi}{6}\right) + 8$

  b. $y = 10 \cos 4x - 6$

  c. $y = 3 \cos \frac{1}{2}\left(x - \frac{\pi}{3}\right) + 4$

  d. $y = 30 \sin\left(2x + \frac{\pi}{3}\right) - 3$

  e. $y = -2 \sin\left(2x + \frac{\pi}{2}\right) - 2$

7. Given the function $y = \sin 2\pi x$,

  a. determine the period;

  b. graph the function for one complete period.

8. Graph each of the following functions for one complete period.

  a. $y = \cos 2\pi x$ **b.** $y = \cos \frac{\pi}{3}x$

  c. $y = \sin \frac{\pi}{2}x$ **d.** $y = \cos 4\pi x$

9. a. Graph $y = (\sin x)^2$, $-2\pi \leq x \leq 2\pi$.

  b. What is the period of $y = (\sin x)^2$?

PART C

10. a. Graph $y = \sin 2x + \cos 3x$, $0 \leq x \leq 4\pi$.

  b. What is the period of this function? Explain your answer.

11. a. Graph $y = \sin 4x + \sin 6x$, $0 \leq x \leq 2\pi$.

  b. What is the period of this function? Explain your answer.

12. What is the period of $y = \cos 3x + \cos 5x$?

13. What is the period of $y = \sin 6x - \cos 15x$?

14. Explain how you would determine the period of $y = \sin Ax + \cos Bx$.

The basic graphs along with their transformations allow us to model a wide array of real life applications. In practical applications, the sizes of periods and amplitudes vary from very small to very large numbers.

EXAMPLE 1

A small windmill has its centre 6 m above the ground and blades 2 m in length. In a steady wind, a point P at the tip of one blade makes a complete rotation in 12 s.

a. If the rotation begins at the highest possible point, determine a function that gives the height of point P above the ground at a time t.

b. What is the height of point P at 5 s and 40 s?

c. At what time is P exactly 7 m above the ground?

SOLUTION

a. If we define points (t, h) for the position of P, where t is time and h is the height of P above the ground, then for the first rotations five positions of P are $(0, 8)$, $(3, 6)$, $(6, 4)$, $(9, 6)$, and $(12, 8)$. The points are shown on the graph extended to $t = 24$ s.

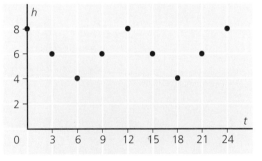

Since P is a point on a rotating circle, the graph represents a sinusoidal function as illustrated by the dotted curve.

This function can be written in the form $h = a \cos k(t - p) + d$, where $a = 2$.

Since the period is 12, then $\frac{2\pi}{k} = 12$ and $k = \frac{\pi}{6}$.

There is no phase shift; therefore, $p = 0$ and the vertical translation is 6 so $d = 6$.

The equation is $h = 2 \cos \frac{\pi}{6}t + 6$.

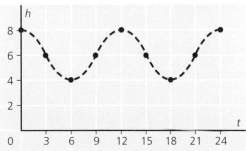

b. When $t = 5$, $h = 2 \cos \frac{\pi}{6}(5) + 6$ or approximately 4.3 m.

When $t = 40$, $h = 2 \cos \frac{\pi}{6}(40) + 6$ or 5 m.

c. When $h = 7$, $7 = 2 \cos \frac{\pi}{6}t + 6$

$$2 \cos \frac{\pi}{6}t = 1$$

$$\cos \frac{\pi}{6}t = \frac{1}{2}$$

$$\frac{\pi}{6}t = \cos^{-1}\left(\frac{1}{2}\right)$$

$$\frac{\pi}{6}t \doteq 1.047\ 198$$

$$t = \frac{1.047\ 198 \times 6}{\pi}$$

$$= 2$$

Therefore $h = 7$ at $t = 2$ s and $t = 10$ s for the first rotation.
But if the windmill continues to rotate, the height is 7 m at $t = (2 + 12k)$ s and
$t = (10 + 12k)$ s, where $k \in I$ and represents the number of rotations.

If we use our Trigonometric Regression or Sin Reg function on our calculator, the
equation is $y = 2 \sin(0.524x + 1.571) + 6$, a more difficult equation to work with.
To write this relationship in terms of a sine function, we assigned a phase shift.

EXAMPLE 2

In the Bay of Fundy, docks have been built high
above the ocean floor because of the extreme tides.
On an average day, the depth of the water around a
harbour dock changes from 1.5 m at low tide at
03:00 hours to 15.5 m at high tide at 09:00. The data
recorded in the table show the depth of water in a
24-hour period. The measurements begin at 24:00,
which we will record as time 0, and the remaining
depths are recorded in 3-hour intervals up to 24.

Photo: COMSTOCK/W. Griebeling

Time in Hours	0	3	6	9	12	15	18	21	24
Depth in Metres	8.5	1.5	8.5	15.5	8.5	1.5	8.5	15.5	8.5

a. Input the data into a graphing calculator and plot the set of points.

b. If the tidal cycle is sinusoidal, use your Sin Reg function on your calculator to
determine an equation for the set of the points in the form $y = a \sin k(t - c) + d$.

c. Graph the function obtained in part **b.**

d. What is the period and phase shift of the function?

e. Will it be safe for a ship to enter the harbour between 15:00 and 16:00 if the ship
requires at least 3.5 m of water?

a.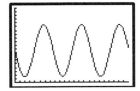

b. The sinusoidal function obtained is $y = 7 \sin(0.52x - 3.14) + 8.5$.

c.

d. Rewrite $y = 7 \sin(0.52x - 3.14) + 8.5$ in the form $y = 7 \sin 0.52(x - 6.04) + 8.5$.

The period is $\frac{2\pi}{0.52}$ or $\frac{2 \times 3.14}{0.52} \doteq 12$.

The phase shift is 6.04.

e. To determine if it is safe to enter the harbour, we can find the height of the water at $t = 15$ and $t = 16$.

At $t = 15$ $\qquad y = 7 \sin[0.52(15) - 3.14] + 8.5$

$\qquad\qquad\qquad \doteq 1.51$.

Since the height is less than 3.5 m, it is not safe to enter the harbour at 15:00 hours.

At $t = 16$ $\qquad y = 7 \sin[0.52(16) - 3.14] + 8.5$

$\qquad\qquad\qquad = 2.25$.

At 16:00 hours it is not safe to enter the harbour since the height of the tide is less than 3.5 m. Therefore, it is not safe to enter the harbour at any time between 15:00 and 16:00.

PART B

1. For a certain day, the depth of water h, in metres, at a seaport in British Columbia at time t, in hours, is given by the formula $h = 3.5 \sin \frac{\pi}{6.2}(t - 3) + 7.8$, where $0 \le t \le 24$.

    a. Graph this function on a graphical calculator using the domain 0 to 24.

    b. Determine the period and amplitude of its graph and explain the practical significance of these quantities.

    c. Calculate the depth of the water at 09:30 and 12:00 on this particular day.

    d. Determine the minimum depth of the water.

    e. At what times during the day does low tide occur?

    f. Graph $y = 6$ on the same set of axes.

    g. Use the trace function to determine the times when there will be at least 6 m of water in the harbour.

2. Write the sinusoidal function in Example 2 in terms of the cosine function.

3. The graph below represents the height above the ground of a pebble caught in the tread of a tire rotating at a constant rate.

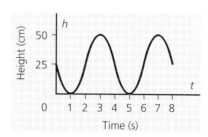

    a. What is the amplitude and period of the function represented by the graph?

    b. What do these quantities represent?

    c. Write an equation for the height above the ground at time t of this pebble.

    d. What is the height of the pebble at 3.5 s?

4. A Ferris wheel at an amusement park has a diameter of 18 m and travels at a rate of 5 revolutions every 2 min. At the bottom of the ride, the passenger is 2 m above the ground.

    a. Determine a function that represents the height h above the ground at time t, if $h = 20$ at $t = 0$.

    b. Determine the passenger's height above the ground at time $t = 78$ s.

5. A *CBR* was used to experimentally measure the distance from the *CBR* head to a swinging pendulum. The graph obtained for three periods of the pendulum is shown.

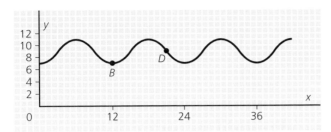

a. Determine the amplitude and period of the function.

b. What is the farthest distance the pendulum gets from the head of the *CBR*?

c. State the equation of the sinusoidal axis.

d. In what position is the pendulum for points *B* and *D* on the sinusoidal axis?

e. Find an equation to represent this motion.

f. When the pendulum is in the "rest position," what is the distance to the *CBR* head?

g. If a *CBR* is available, conduct an experiment to duplicate the above results.

6. If a rocket were launched from Cape Canaveral, *C*, to orbit the earth in an equatorial orbit in 90 min, then the path would appear approximately as shown (if we assume there is *no* rotation of the earth). The distance from Canaveral to the equator is about 2400 km. If the vertical axis were placed through *C*, determine an equation to represent the changing position of the rocket relative to the earth's surface and the equator. Time should be in hours.

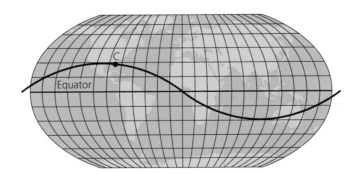

7. At Whitehorse in the Yukon, the following facts are available from the Internet.

◆ On June 21 the sun rises at 03:20 (Standard Time).

◆ On December 21 the sun rises at 10:14 (Standard Time).

The function describing sunrise times is sinusoidal with a period of 365 days.

a. Determine an equation for this function.

b. Determine the time of sunrise for July 1.

c. Which days of the year will have sunrise at 05:30?

d. How many days of the year have the sunrise after 09:00?

e. Is this function continuous? Explain.

8. A crank of length 30 cm, starting from an initial horizontal position P, rotates in a positive sense at a uniform rate of one revolution per second.

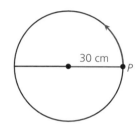

a. Determine a function that gives the height of the end of the crank from the initial position line at any time t.

b. How far is the end of the crank from the initial line in 0.7 s?

c. At what time(s) between 0 and 1 is the crank 12 cm from the initial position line?

9. a. A high-voltage electric current with a maximum voltage of 1200 volts is transmitted at the rate of 300 cycles per second. Find a sinusoidal function that represents current in terms of time t.

b. A transformer station reduces the voltage by half and the frequency by one-fifth. Find the equation for the new current.

10. A turbine in a hydroelectric power station is made to rotate by water coming down the penstock and pushing on the turbine blades. If the radius of the turbine from the centre of the axle to the tip of the blade is 5 m and the turbine is spinning at 500 rpm, determine a function that gives the horizontal displacement of a point P on the tip of one of the blades relative to the vertical line through the axle of the turbine at time t.

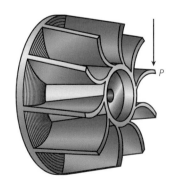

PART C

11. For the function $y = \sin\left(t - \frac{\pi}{2}\right) - 5$, where $0 \leq t \leq 3\pi$ radians, determine the values of t for which $y \leq -4.5$.

12. By changing the units on the axes, the graph of $y = \sin x$, $0 \leq x \leq 2\pi$, can become the graph of $y = 5 \sin 2x$. How should the units be altered to bring this about?

13. A crank 20 cm long, starting at $60°$ from the horizontal, sweeps out a positive angle at the uniform rate of 10 rad/s.

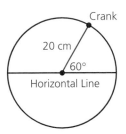

a. Determine a function that gives the height of the end of the crank from the horizontal line at any time t.

b. How far is the end of the crank from the horizontal line after 0.1 s?

c. At what times is the crank 12 cm from the horizontal line?

14. A tuning fork emits a sound wave that is described by the function $P_1 = 0.15 \sin(60\pi t)$. A second tuning fork emits a sound wave described by $P_2 = 0.15 \sin(120\pi t)$. If the two tuning forks are struck simultaneously, then the sound wave created will be a superposition of the two sounds. The diaphragm in our ear will hear the sound defined by $f(t) = P_1 + P_2$. Graph function $f(t)$.

15. An architect is designing a south-facing house to have a roof overhang of 75 cm. The wall dimensions are shown.

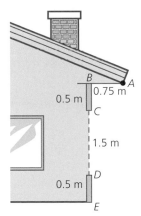

a. Find the measure of the angle of elevation, θ, of the sun so no direct sunrays will shine through the window. Write your answer, correct to one decimal.

b. There are no trees or other buildings in front of this window. Sunlight enters the room as soon as the sun is $5°$ above the horizon and continues throughout the day, provided there are no clouds. Determine the length of the interior floor that will receive direct sunlight when the angle of elevation of the sun is $45°$.

c. Build an algebraic model to describe the different lengths of interior floor that will receive direct sunlight throughout the part of the day that has angles of elevation from $5°$ up to $65°$.

d. Explain why your algebraic model will not work if the angle of elevation is $0°$.

PART A

1. Convert the following to radian measure.

   a. 45°
   b. 120°
   c. 210°

   d. 840°
   e. −150°
   f. −700°

2. Convert the following to degree measure.

   a. $\frac{\pi}{6}$ rad
   b. $\frac{9\pi}{8}$ rad
   c. $-\frac{7\pi}{4}$ rad

   d. $-\frac{13\pi}{3}$ rad
   e. 3 rad
   f. −5.2 rad

3. State the period and amplitude for each of the following:

   a. $y = 3\sin(5x)$
   b. $y = \frac{2}{3}\cos(12t)$
   c. $h = -4\sin(4\pi t)$

   d. $y = \frac{2}{7}\cos(12t)$
   e. $h = 8\cos(\frac{3\pi}{5}t)$
   f. $y = 7\sin(60\pi t)$

4. State the vertical translation and phase shift for each of the following:

   a. $y = \cos\left(x - \frac{\pi}{6}\right) + 3$
   b. $y = \sin 2(x - \pi) - 4$

   c. $y = \sin\left(3x + \frac{\pi}{2}\right) + 1$
   d. $y = \cos\left(t - \frac{\pi}{4}\right) + 2$

   e. $y = \cos 3\left(x + \frac{\pi}{6}\right) - 5$
   f. $y = \sin(5x - \pi) + 4$

5. Each of the diagrams below is the graph of a sinusoidal function.

   i) For each graph, state the period, amplitude, vertical translation, and sinusoidal axis.

   ii) Express each as a sine function.
   iii) Express each as a cosine function.

   a.
   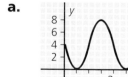

   b.
   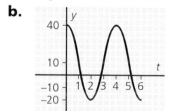

   c.
   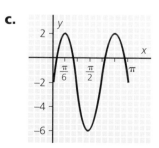

   d.
   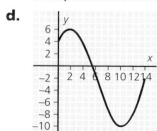

PART B

6. Sketch two complete periods for each of the following:

   a. $y = 3 \sin 2x$

   b. $y = 8 \cos 3x$

   c. $y = 2 \cos\left(\frac{\pi}{3}x\right) + 1$

   d. $y = 3 \cos\left(x + \frac{\pi}{3}\right) - 2$

   e. $y = 2 \sin\left(3x - \frac{\pi}{2}\right) + 4$

   f. $y = \frac{1}{2}\cos(2\pi t) + 3$

7. Determine the value for q such that the curve $y = \cos \alpha + q$ passes through the following points.

   a. $(2\pi, 3)$

   b. $\left(\frac{7\pi}{2}, -4\right)$

8. If $A\left(a, \frac{\sqrt{3}}{2}\right)$ is on the graph of $y = \sin\left(x + \frac{\pi}{4}\right)$, find two possible values for a.

9. What is the period of the function $y = a \sin\frac{2\pi}{T}t$, where t represents time in seconds and T is a positive constant? Determine a value for T such that the function has an amplitude of 5 and a period of 10 s. Now determine the value of y to one decimal place when $t = 4$ s.

10. The function $y = \sin(x - p) + d$ has a vertical translation of -3 and passes through the point $\left(\frac{\pi}{6}, -2\right)$. Determine the values of p and d.

11. A Ferris wheel has a radius of 10 m and rotates at the rate of one revolution every 48 s. At the bottom of the ride, the passenger is 2 m above the ground. You start your ride from the bottom of the wheel.

    a. Determine a function that represents your height h above the ground at any time t.

    b. Determine your height above the ground in 2.5 min.

    c. At what times is your height 9 m above the ground? Write your answer correct to the nearest second.

12. The pendulum in a grandfather clock makes 20 complete swings A–B–A per minute. The distance between the points A and B, which are at the maximum end of the swing, is 40 cm.

    a. Make a sketch to show two complete cycles of the pendulum, starting from point A.

    b. Determine an equation that will predict the perpendicular distance p from the vertical line as a function of time.

    c. What is the position of the pendulum after 40 s?

1. The set of integers 1, 2, 3 has the property that each of the numbers is a divisor of their sum.

   a. Find a set of five integers with this property (they're not consecutive).

   b. Find a set of seven integers with this property.

2. Given that x, y, and z are positive integers, determine their values if

$$x + \cfrac{1}{y + \cfrac{1}{z}} = \frac{13}{9}.$$

3. Recall that in the last Problems Page, you showed that a chord in a circle subtends an angle at the circle circumference that is half the angle the chord subtends at the centre. Use this to show that if the four vertices of a quadrilateral are on the circumference of a circle, then the opposite angles of the quadrilateral sum to 180°.

4. For any triangle ABC inscribed in a circle, with radius r, show that

$$\frac{a}{\sin A} = \frac{b}{\sin B} = \frac{c}{\sin C} = 2r.$$

5. A quadrilateral $ABCD$ has its vertices on a circle. $AB = 2$, $BC = 3$, $CD = 4$, and $DA = 6$. What is the length of AC?

6. An airplane leaves an aircraft carrier and flies due south at 400 km/h. The carrier proceeds 60° west of north at 32 km/h. If the plane has enough fuel for 5 h of flying, what is the maximum distance south the pilot can travel so that the fuel remaining will allow a safe return to the carrier?

7. How many natural numbers less than 200 leave a remainder of 3 when divided by 7 and a remainder of 4 when divided by 5?

Chapter 9
Trigonometric Equations and Identities

Interpreting facts and making decisions are challenges common in everyday life that demand an ability to consider situations, apply knowledge, and justify a position. Solving trigonometric equations involves combining skills for algebraic solutions of linear and quadratic equations with coordinate geometry and with trigonometric ratios. Learning to approach trigonometry with this organization of skills develops strategies for creatively solving problems in other areas of mathematics as well as in everyday life and at work. Proving a trigonometric identity provides experience presenting an argument based on selecting and applying mathematical relationships. Whether you are solving equations or proving identities, analyzing applications of mathematics extends your reasoning skills.

IN THIS CHAPTER, YOU CAN . . .

- prove identities, using the Pythagorean identity, $\sin^2 x + \cos^2 x = 1$, and the quotient relation $\tan x = \dfrac{\sin x}{\cos x}$;
- solve linear trigonometric equations;
- solve quadratic trigonometric equations.

9.1 Simple Trigonometric Equations

We have solved several types of equations. Here we examine equations involving trigonometric ratios and functions. Are there differences from equations met earlier? Yes, there are, but they do not create major difficulties if care is taken. The first example illustrates some of the differences between trigonometric equations and others we have dealt with.

EXAMPLE 1

Solve $5 \sin \theta + 2 = 3 \sin \theta + 1$, $0° \le \theta \le 360°$.

SOLUTION

$$5 \sin \theta + 2 = 3 \sin \theta + 1$$
$$5 \sin \theta - 3 \sin \theta = 1 - 2$$
$$2 \sin \theta = -1$$
$$\sin \theta = -\frac{1}{2}$$

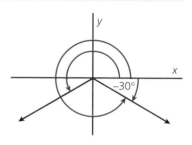

$\theta = \sin^{-1}\left(-\frac{1}{2}\right) = -30°$, from a calculator.

Since $0° \le \theta \le 360°$, there are two values for θ: $\theta = 210°$ and $\theta = 330°$.

This example illustrates two things we find in **trigonometric equations.** First, we have two steps to take, because we must find a value for the ratio itself before we can determine values for the angle defined. If we solve $2x = -1$, we obtain $x = -\frac{1}{2}$, but in solving $2 \sin \theta = -1$ to obtain $\sin \theta = -\frac{1}{2}$, we are only partly done. Second, we obtain a number of solutions, depending on the range of values allowed for the angle. Indeed, if the restriction $0° \le \theta \le 360°$ is removed, there are an infinite number of solutions, as the next example shows.

EXAMPLE 2

Solve $5 \tan \theta = 2 \tan \theta + 7$, correct to one decimal, where θ is measured in degrees.

SOLUTION

$$5 \tan \theta = 2 \tan \theta + 7$$
$$5 \tan \theta - 2 \tan \theta = 7$$
$$3 \tan \theta = 7$$
$$\tan \theta = \frac{7}{3}$$

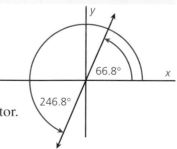

$\theta = \tan^{-1}\left(\frac{7}{3}\right) = 66.8°$, correct to one decimal, from a calculator.

For $0° \le \theta \le 360°$, there are two values for θ: $\theta = 66.8°$ or in quadrant III, $\theta = 180° + 66.8° = 246.8°$.

In addition, all angles coterminal with these angles have the same tangent ratio. The solution is $\theta = (66.8 + 360k)°$, k an integer, and $\theta = (246.8 + 360k)°$, k an integer.

This means that when trigonometric equations are posed, *the domain must be clearly defined,* and from the domain given we determine solutions as required.

EXAMPLE 3

Solve $7 \cos x + 3 = 5 \cos x + 4$, $0 \le x \le 2\pi$.

SOLUTION

$7 \cos x + 3 = 5 \cos x + 4$

$\qquad 2 \cos x = 1$

$\qquad \cos x = \dfrac{1}{2}$

$\qquad x = \dfrac{\pi}{3}$ or $x = \dfrac{5\pi}{3}$

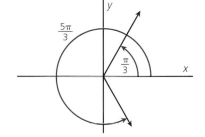

Note that we can consider this solution to give x as angles in radian measure or simply as real numbers. Note also that the answers given are in terms of π. We can of course use a calculator, obtaining the solution

$\qquad \cos x = \dfrac{1}{2}$

$\qquad x = \cos^{-1}(0.5) \doteq 1.047.$

Now $x \doteq 1.047$ or $x = 2\pi \quad 1.047 \doteq 5.236.$

Is there a preference for one form of answer over another? The situation in which the problem arises usually dictates. In this text it will be assumed that if a decimal approximation is expected, you will be told to solve for x, correct to some decimal value. Otherwise, you should give exact answers in terms of π.

It has been noted that we first consider the ratio itself as the variable in the equation. How do we indicate that $\sin \theta$ is to be squared or cubed? There are two ways. It is optional to write $(\sin \theta)^2$ or $\sin^2 \theta$. The first of these is easier to deal with, but the second is by far the more common, so that is the notation we use here.

We refer to $6 \sin^2 \theta + \sin \theta - 1 = 0$ as a quadratic equation in $\sin \theta$.

EXAMPLE 4

Solve $6 \sin^2 x + \sin x - 1 = 0$, $0 \leq x \leq 2\pi$, correct to two decimals.

$$6 \sin^2 x + \sin x - 1 = 0$$
$$(3 \sin x - 1)(2 \sin x + 1) = 0$$
$$3 \sin x - 1 = 0 \quad \text{or} \quad 2 \sin x + 1 = 0$$
$$\sin x = \frac{1}{3} \qquad \sin x = -\frac{1}{2}$$

For $\sin x = \frac{1}{3}$, $x = \sin^{-1}\left(\frac{1}{3}\right)$

$$\doteq 0.340^r$$
$$\text{or } x = \pi - 0.340$$
$$\doteq 2.80^r.$$

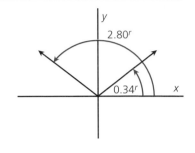

For $\sin x = -\frac{1}{2}$, $x = \sin^{-1}\left(-\frac{1}{2}\right)$

$$\doteq -0.52^r.$$

Since $0 \leq x \leq 2\pi$, $x = \pi + 0.52$

$$\doteq 3.67^r$$
$$\text{or } x = -0.52 + 2\pi$$
$$\doteq 5.76^r.$$

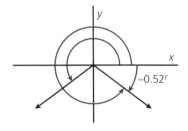

There are four values for x: 0.34^r, 2.80^r, 3.67^r, and 5.76^r.

EXAMPLE 5

Solve $3 \cos^2 \theta - 5 \cos \theta - 1 = 0$, $0° \leq \theta \leq 360°$, correct to one decimal.

$3 \cos^2 \theta - 5 \cos \theta - 1 = 0$

Using the quadratic formula

$$\cos \theta = \frac{5 \pm \sqrt{25 + 12}}{6}$$
$$= \frac{5 \pm \sqrt{37}}{6}.$$

Then $\cos \theta \doteq 1.847$ and there is no solution, or $\cos \theta = -0.1805$ and $\theta = 100.4°$ or $\theta = 259.6°$.

EXAMPLE 6

Determine exact values for all intersection points of the function $f(x) = 4 \cos^2 x$ and the line $y = 3$, $0 \le x \le 2\pi$.

SOLUTION

Use your graphing calculator so that you have a good idea of what to expect. From the graph there seem to be four points of intersection. Equating the given expressions,

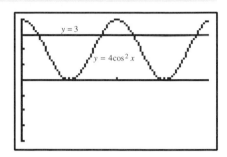

$$4 \cos^2 x = 3$$
$$(\cos x)^2 = \frac{3}{4}$$
$$\cos x = \pm \frac{\sqrt{3}}{2}$$

For $\cos x = \frac{\sqrt{3}}{2}$, $x = \cos^{-1}\left(\frac{\sqrt{3}}{2}\right)$.

Since a right-angled triangle with hypotenuse 2 and one side $\sqrt{3}$ has angles of $\frac{\pi}{6}, \frac{\pi}{3}, \frac{\pi}{2}$, we conclude that

$$x = \frac{\pi}{6} \text{ or } x = 2\pi - \frac{\pi}{6} = \frac{11\pi}{6}.$$

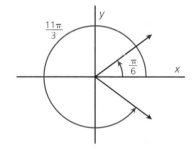

For $\cos x = -\frac{\sqrt{3}}{2}$, using the same argument,

$$x = \pi - \frac{\pi}{6} = \frac{5\pi}{6}$$
$$\text{or } x = \pi + \frac{\pi}{6} = \frac{7\pi}{6}.$$

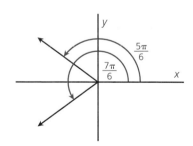

The intersection points are $\left(\frac{\pi}{6}, 3\right)$, $\left(\frac{5\pi}{6}, 3\right)$, $\left(\frac{7\pi}{6}, 3\right)$, and $\left(\frac{11\pi}{6}, 3\right)$.

PART A

1. Solve each of the following equations to the nearest degree, $0 \le \theta \le 360°$.

   a. $\sin \theta = 0.1234$ **b.** $\cos \theta = 0.3472$

   c. $\tan \theta = 1.5732$ **d.** $\sin \theta = -0.7512$

   e. $\cos \theta = -0.7851$ **f.** $\tan \theta = -0.3125$

2. Solve each of the following equations to two decimal places, $0 \le \theta \le 2\pi$.

   a. $\sin \theta = 0.3124$ **b.** $\cos \theta = 0.7315$

   c. $\tan \theta = 3.1571$ **d.** $\sin \theta = -0.8135$

   e. $\cos \theta = -0.1476$ **f.** $\tan \theta = -0.3541$

3. Solve each of the following equations. Give exact answers only, $0° \le \theta \le 360°$.

   a. $\sin \theta = \dfrac{\sqrt{3}}{2}$ **b.** $\cos \theta = \dfrac{1}{\sqrt{2}}$

   c. $\tan \theta = \sqrt{3}$ **d.** $\cos \theta = -\dfrac{1}{2}$

   e. $\sin \theta = -\dfrac{1}{2}$ **f.** $\tan \theta = -1$

4. Solve each of the following equations. Give exact answers only, $0 \le x \le 2\pi$.

   a. $\sin x = \dfrac{1}{\sqrt{2}}$ **b.** $\cos x = \dfrac{\sqrt{3}}{2}$

   c. $\tan x = \dfrac{1}{\sqrt{3}}$ **d.** $\sin x = 1$

   e. $\cos x = 0$ **f.** $\tan x = -1$

PART B

5. Solve each of the following equations. Write your answer to the nearest degree.

   a. $5 \sin \theta + 3 = \sin \theta + 2, \; 0 \le \theta \le 360°$

   b. $7 \sin \theta - 11 = 4 \sin \theta - 13, \; 0 \le \theta \le 360°$

   c. $7 - 11 \cos \theta = 8 - 8 \cos \theta, \; 0 \le \theta \le 360°$

   d. $25 + 15 \tan \theta = 12 \tan \theta + 28, \; 0 \le \theta \le 360°$

6. Solve each of the following equations. Write your answers correct to one decimal.

a. $6 \sin x - 4 = 2 \sin x + 1, 0 \le x \le 2\pi$

b. $7(\cos x + 5) = 13 + 3(\cos x + 8), 0 \le x \le 2\pi$

c. $5 \tan x - 3 = 3 \tan x + 7, 0 \le x \le 2\pi$

d. $2 \sin^2 x + 5 \sin x - 3 = 0, 0 \le x \le 2\pi$

e. $6 \sin^2 x - 6 \sin x - 5 = 0, 0 \le x \le 2\pi$

f. $\tan^2 x = 3, 0 \le x \le 2\pi$

g. $3 \cos^2 x - \cos x = 0, 0 \le x \le 2\pi$

h. $6 \cos^2 x + 7 \cos x - 3 = 0, 0 \le x \le 2\pi$

7. Solve each of the following trigonometric equations. Write your answers correct to two decimals.

a. $5 + 2 \sin x = 4, -\dfrac{\pi}{2} \le x \le \dfrac{3\pi}{2}$

b. $7 + 3 \tan x = 4, 0 \le x \le \pi$

c. $3\sqrt{2} \cos x - 11 = \sqrt{2} \cos x - 9, 0 \le x \le 2\pi$

d. $5(\tan x + 7) = 2(\tan x + 18) - 7, 0 \le x \le 2\pi$

8. If $3 \sin \beta + 2 \cos \alpha = 4$ and $7 \sin \beta - 3 \cos \alpha = 3$, determine the values for $\sin \beta$ and $\cos \alpha$.

9. Determine exact values for all intersection points of the function $f(x) = 2 \sin^2 x$ and the line $y = 1$.

PART C

10. If $\cos \beta + \sin \beta = \dfrac{1 + \sqrt{5}}{\sqrt{6}}$ and $\cos \beta - \sin \beta = \dfrac{1 - \sqrt{5}}{\sqrt{6}}$, determine the exact values of $\sin \beta$ and $\cos \beta$ for $0° \le \beta \le 180°$.

11. Solve the equation $(3 \sin x - 5)(\sin x + 3) = 2 \sin^2 x - 5 \sin x - 11$ for $0 \le x \le 3\pi$.

12. An isosceles triangle ABC has $\angle A = 36°$ and $AB = AC = 1$ unit. Determine the exact value of $\cos 36°$.

13. Determine all solutions of the equation $5 \sin 3x - 11 = 3 \sin 3x - 12$ in the domain $0 \le x \le 2\pi$.

In previous sections, we have seen that there is a connection between the algebraic problem of solving a trigonometric equation and the geometric problem of finding the intersection points between two functions. For example, the solution of the equation $4 \cos^2 x = 3$ produces the x-values for the intersection points between the functions $f(x) = 4 \cos^2 x$ and $g(x) = 3$. To know the number of values for x, we must know the size of the domain. If we decide on a domain of 0 to 2π radians, then there are four values for x, as found in Example 6 of the previous section. On the other hand, if the domain is changed from 0 to $2k\pi$ radians where k is an integer, there will be $4k$ values for x.

Equations of this type are called **conditional trigonometric equations** because the constraints built into the equation require you to find a small number of values for x in one or more periods of the function.

> A **conditional trigonometric equation** is true for some subset of the values for which the equation is defined.

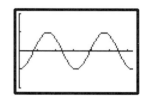

In some situations it appears that two trigonometric functions, which look different on the surface, seem to coincide exactly when graphed. For example, the graphs of $f(x) = \sin^2 x - \cos^2 x$ and $g(x) = 2 \sin^2 x - 1, 0 \le x \le 2\pi$, coincide exactly, suggesting an infinite number of common points. On your graphing calculator, graph the functions $f(x) = \sin^2 x - \cos^2 x$ and $g(x) = 2 \sin^2 x - 1, 0 \le x \le 2\pi$. Observe that this seems to verify that the equation $\sin^2 x - \cos^2 x = 2 \sin^2 x - 1$ is true for all values of x. This type of equation is different from the conditional trigonometric equation and is called an **identity equation,** or simply an **identity.**

> An **identity equation** (or **identity**) is an equation that is true for all defined values of the variable.

One procedure for solving an equation that we suspect is an identity is to split the equation into two functions and then use a graphing calculator to see if the two graphs appear to be the same. This approach is very appealing because it is efficient and quickly gives two graphs that match, seemingly for all points. This, however, is not a proof and there exists the possibility that of the billions of points not included in the calculator calculation, there will be some that will fail to coincide. In this section, we discuss techniques that will determine whether or not a trigonometric equation is true for all values of the variable; in other words, whether or not it is an identity.

We begin with a proof for one of the most important relationships involving trigonometric functions, the **Pythagorean identity** for sine and cosine.

> **Pythagorean Identity Theorem:** If θ is a real angle, the equation
> $\sin^2 \theta + \cos^2 \theta = 1$ is true for all θ.

Proof

Rather than graphing the two functions, we recall the basic definitions of the sine and cosine.

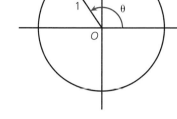

Let any rotation angle $\theta = \angle POX$ be defined in a circle with radius 1 unit, where P is (x, y). By definition $\sin \theta = \dfrac{y}{1}$ and $\cos \theta = \dfrac{x}{1}$.

Separate the given equation into left and right sides and simplify each side.

$$\text{L.S.} = \sin^2 \theta + \cos^2 \theta \qquad\qquad \text{R.S.} = 1$$
$$= y^2 + x^2$$

Now the circle with centre the origin and radius 1 has equation $x^2 + y^2 = 1$.

Thus L.S. $= x^2 + y^2 = 1 =$ R.S.

This statement is true for all positions of point P. It is therefore true for all possible values of θ.

This type of proof is called a proof by **First Principles** because the strategy in constructing the proof makes use of the coordinate definitions. This approach can be used in many situations. Another way of proving identities is to make use of a previously proven theorem. The next example shows how the Pythagorean identity can be used to prove that an equation is an identity.

EXAMPLE 1

Prove $\sin^2 x - \cos^2 x = 2 \sin^2 x - 1$ is true for all real x.

SOLUTION

Consider the two parts of the equation. We designate the left side by L.S. and the right side by R.S.

$$\text{L.S.} = \sin^2 x - \cos^2 x \qquad\qquad\qquad \text{R.S.} = 2 \sin^2 x - 1$$
$$= \sin^2 x - (1 - \sin^2 x), \text{ since } \sin^2 x + \cos^2 x = 1$$
$$= 2 \sin^2 x - 1$$

Since the left side is shown to be identical with the right side, the equation is true for all values of x, and $\sin^2 x - \cos^2 x = 2 \sin^2 x - 1$ is an identity.

The Reciprocal Identities

In trigonometric expressions, we frequently find fractional expressions involving the trigonometric ratios or functions. For convenience, we define the reciprocal functions of the primary functions.

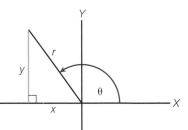

The Cosecant function for angle θ is $\csc \theta = \frac{r}{y}$, $y \neq 0$ (csc being the abbreviation for cosecant).

The Secant function for a variable angle θ is $\sec \theta = \frac{r}{x}$, $x \neq 0$ (sec being the abbreviation for secant).

The Cotangent function for a variable angle θ is $\cot \theta = \frac{x}{y}$, $y \neq 0$ (cot being the abbreviation for cotangent).

The Reciprocal Identities

$$\csc \theta = \frac{1}{\sin \theta}; \qquad \sec \theta = \frac{1}{\cos \theta}; \qquad \cot \theta = \frac{1}{\tan \theta}.$$

Quotient Identities

If θ is a real angle, $\theta \neq \frac{\pi}{2} + k\pi$, $k \in I$, then $\tan \theta = \frac{\sin \theta}{\cos \theta}$.

If θ is a real angle, $\theta \neq k\pi$, $k \in I$, then $\cot \theta = \frac{\cos \theta}{\sin \theta}$.

EXAMPLE 1

Prove that $\tan \theta = \frac{\sin \theta}{\cos \theta}$.

SOLUTION

Let any $\angle POX = \theta$ be defined in a circle with radius 1, centre the origin, where P is (x, y). Separate the equation into left and right sides.

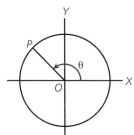

$$\text{L.S.} = \tan \theta$$
$$= \frac{y}{x}, x \neq 0$$

$$\text{R.S.} = \frac{\sin \theta}{\cos \theta} = \frac{\left(\frac{y}{1}\right)}{\left(\frac{x}{1}\right)}$$
$$= \frac{y}{x}, x \neq 0$$

Therefore L.S. = R.S., provided $x \neq 0$.

This means the angles $\frac{\pi}{2}, \frac{3\pi}{2}$, and all multiples of these angles must be excluded because neither L.S. nor R.S. is defined for these angles.

The proof of the second identity is identical. It is helpful to use the following identities as the basic building blocks for identity proofs.

Reciprocal Identities

$$\csc \theta = \frac{1}{\sin \theta}$$

$$\sec \theta = \frac{1}{\cos \theta}$$

$$\cot \theta = \frac{1}{\tan \theta}$$

Pythagorean Identities

$$\sin^2 \theta + \cos^2 \theta = 1$$

$$\tan^2 \theta + 1 = \sec^2 \theta$$

$$\cot^2 \theta + 1 = \csc^2 \theta$$

Quotient Identities

$$\tan \theta = \frac{\sin \theta}{\cos \theta}$$

$$\cot \theta = \frac{\cos \theta}{\sin \theta}$$

Reflection Identities

$$\sin(-\theta) = -\sin \theta$$

$$\cos(-\theta) = +\cos \theta$$

In simplifying trigonometric expressions, we need to recall the algebraic skills of adding rational terms, factoring, and so on, that we use on a regular basis.

EXAMPLE 2

Prove $\sin^4 \theta + 2 \cos^2 \theta - \cos^4 \theta = 1$ for all θ.

SOLUTION

L.S. $= \sin^4 \theta + 2 \cos^2 \theta - \cos^4 \theta$ R.S. $= 1$

$= \sin^4 \theta - \cos^4 \theta + 2 \cos^2 \theta$

$= (\sin^2 \theta + \cos^2 \theta)(\sin^2 \theta - \cos^2 \theta) + 2 \cos^2 \theta$

$= 1(\sin^2 \theta - \cos^2 \theta) + 2 \cos^2 \theta$

$= \sin^2 \theta + \cos^2 \theta$

$= 1$

Therefore L.S. = R.S., and the statement is true for all real values of θ.

ALTERNATE SOLUTION

L.S. $= \sin^4 \theta + 2 \cos^2 \theta - \cos^4 \theta$ R.S. $= 1$

$= (1 - \cos^2 \theta)^2 + 2 \cos^2 \theta - \cos^4 \theta$

$= 1 - 2 \cos^2 \theta + \cos^4 \theta + 2 \cos^2 \theta - \cos^4 \theta$

$= 1$

Therefore L.S. = R.S., and the statement is true for all real values of θ.

EXAMPLE 3

Prove that $\tan^2 x + 1 = \sec^2 x$ for all x, $x \neq \frac{\pi}{2} + k\pi$, $k \in I$.

SOLUTION

L.S. $= \tan^2 x + 1$

$= \left(\dfrac{\sin x}{\cos x}\right)^2 + 1$

$= \dfrac{\sin^2 x}{\cos^2 x} + 1$

$= \dfrac{\sin^2 x + \cos^2 x}{\cos^2 x}$

$= \dfrac{1}{\cos^2 x}$

R.S. $= \sec^2 x$

$= \left(\dfrac{1}{\cos x}\right)^2$

$= \dfrac{1}{\cos^2 x}$

Therefore L.S. $=$ R.S., and the statement is true for all real values of x; $x \neq \frac{\pi}{2} + k\pi$, $k \in I$.

Exercise 9.2

PART A

1. Use the primary trigonometric ratios to show that $\cot \theta = \frac{\cos \theta}{\sin \theta}$ is true for all θ, except those θ that make $\sin \theta = 0$.

2. Express each of the following expressions in a simpler form, by using a known trigonometric identity.

   a. $\cos \theta \sec \theta$

   b. $\tan \theta \cos \theta$

   c. $1 - \sin^2 \theta$

   d. $\tan \theta + \cot \theta$

   e. $\sqrt{1 - \cos^2 \theta}$

   f. $\tan^2 \theta - \sec^2 \theta$

   g. $\dfrac{\cos \theta}{1 + \sin \theta} + \dfrac{\cos \theta}{1 - \sin \theta}$

   h. $\sin^2 \theta + 2 \cos^2 \theta$

3. Use some type of factoring to help simplify each of the following:

   a. $\sin^4 \theta + (\sin^2 \theta)(\cos^2 \theta)$

   b. $(\sin x)^2(\sec x)^2 + (\sin x)^2$

   c. $4 \cos^2 \theta + 8 \cos \theta \sin \theta + 4 \sin^2 \theta$

   d. $\cos^4 x - \sin^4 x$

4. Prove that each of the following equations is an identity by simplifying the left side of the equation.

   a. $\cot A \sin A = \cos A$

   b. $\sec \theta(1 - \cos \theta) = \sec \theta - 1$

   c. $\sin^2 \beta(1 + \cot^2 \beta) = 1$

   d. $\dfrac{1 + \tan \theta}{1 + \cot \theta} = \tan \theta$

PART B

5. Use the analytic definition on a circle diagram to prove that $\sin(-\theta) = -\sin\theta$.

6. Prove that each of the following is an identity.

a. $\tan\theta + \cot\theta = \sec\theta\csc\theta$

b. $\sin^2\theta\sec^2\theta = \sec^2\theta - 1$

c. $\dfrac{\sin x}{\csc x} + \dfrac{\cos x}{\sec x} = 1$

d. $\sin^2\theta + \tan^2\theta = \sec^2\theta - \cos^2\theta$

e. $(\cos x - \sin x)^2 = 1 - 2\sin x\cos x$

f. $\dfrac{1 + \sin\theta}{1 - \sin\theta} = \dfrac{\csc\theta + 1}{\csc\theta - 1}$

g. $\tan\theta = \dfrac{\sin(\theta + 2\pi)}{\cos(\theta - 2\pi)}$

h. $\dfrac{1}{\sec\theta + \tan\theta} = \sec\theta - \tan\theta$

7. Prove $2\sin(-\theta) - \cot\theta\sin\theta\cos\theta = (\sin\theta - 1)^2 - 2$ is an identity equation.

8. Simplify the function $f(\theta) = \sin^2\theta + \dfrac{1 + \cot^2\theta}{1 + \tan^2\theta} + \cos^2\theta$.

9. Prove that each of the following is an identity.

a. $\dfrac{\tan^2 A}{1 + \tan^2 A} = \sin^2 A$

b. $\dfrac{\cos\theta}{1 - \sin\theta} - \sec\theta = \tan\theta$

c. $\dfrac{2\sin^2 x - 1}{\sin x\cos x} = \tan x - \cot x$

d. $\dfrac{1 + \cos(x + 2\pi)}{1 - \cos x} = \dfrac{1 + \sec x}{\sec(x + 2\pi) - 1}$

e. $\sin^4\theta - \cos^4\theta = 1 - 2\cos^2\theta$

f. $\dfrac{\cot\beta}{\csc\beta - 1} + \dfrac{\cot\beta}{\csc\beta + 1} = 2\sec\beta$

10. a. Show that $\sin(45 + 60)° \neq \sin 45° + \sin 60°$.

b. Show that $\sin(45 + 45)° \neq \sin 45° + \sin 45°$.

c. Find another pair of angles A and B to show that $\sin(A + B) \neq \sin A + \sin B$.

PART C

11. Prove that $\dfrac{1}{1 + \sin\theta} = \sec^2\theta - \dfrac{\tan\theta}{\cos\theta}$ is an identity.

12. Prove the identity $\dfrac{\tan^3 A}{1 + \tan^2 A} + \dfrac{\cot^3 A}{1 + \cot^2 A} = \dfrac{1 - 2\sin^2 A\cos^2 A}{\sin A\cos A}$.

13. Simplify $g(\theta) = \dfrac{1}{4\sin^2\theta\cos^2\theta} - \dfrac{(1 - \tan^2\theta)^2}{4\tan^2\theta}$.

In Section 9.1, we solved quadratic equations that involved a single trigonometric function. Sometimes equations are given that contain more than one trigonometric function. When this situation arises, it is usually necessary to find a substitution that will reduce the equation to a single trigonometric function. The identities of the previous section provide the required substitutions.

EXAMPLE 1

Solve $2 \cos x + 3 \tan x = 0$, $0 \le x \le 2\pi$.

SOLUTION

Since $\tan x = \frac{\sin x}{\cos x}$, $\cos x \ne 0$, the equation becomes

$$2 \cos x + \frac{3 \sin x}{\cos x} = 0.$$

Multiply by $\cos x$:
$$2 \cos^2 x + 3 \sin x = 0$$
$$2(1 - \sin^2 x) + 3 \sin x = 0$$
$$2 \sin^2 x - 3 \sin x - 2 = 0$$
$$(2 \sin x + 1)(\sin x - 2) = 0$$
$$\sin x = -\frac{1}{2} \quad \text{or} \quad \sin x = 2$$

There is no value of x for which $\sin x = 2$.

If $\sin x = -\frac{1}{2}$, $x = \sin^{-1}\left(-\frac{1}{2}\right) = -\frac{\pi}{6}$.

For $0 \le x \le 2\pi$, we obtain $x = \frac{7\pi}{6}$ or $x = \frac{11\pi}{6}$.

EXAMPLE 2

Solve $3 \sin x - 2 \cos x = 0$, $0 \le x \le 2\pi$, correct to three decimals.

SOLUTION

$$3 \sin x - 2 \cos x = 0$$

Then $\frac{\sin x}{\cos x} = \frac{2}{3}$, $\cos x \ne 0$

$$\tan x = \frac{2}{3}$$

$$x = \tan^{-1}\left(\frac{2}{3}\right) = 0.588, \text{ correct to three decimals.}$$

Then $x = 0.588$ or $x = \pi + 0.588 = 3.730$.

EXAMPLE 3

Find the intersection points of the functions $f(x) = 2 \cos^2 x$ and $g(x) = 5 \sin x - 1$, on the domain 0 to 2π.

SOLUTION

From the graph, there appear to be two intersection points. For intersection points, $f(x) = g(x)$,

$$2 \cos^2 x = 5 \sin x - 1. \qquad ①$$

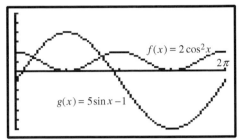

From the Pythagorean identity,

$$\cos^2 x = 1 - \sin^2 x.$$

Then equation ① becomes

$$2 - 2 \sin^2 x = 5 \sin x - 1$$
$$2 \sin^2 x + 5 \sin x - 3 = 0$$
$$(2 \sin x - 1)(\sin x + 3) = 0.$$
$$\sin x = \tfrac{1}{2} \text{ or } \sin x = -3$$

There is no value of x for which $\sin x = -3$.

If $\sin x = \tfrac{1}{2}$, $x = \tfrac{\pi}{6}$ or $x = \tfrac{5\pi}{6}$.

If $x = \tfrac{\pi}{6}$, $g\left(\tfrac{\pi}{6}\right) = 5 \sin\left(\tfrac{\pi}{6}\right) - 1$

$$= 5\left(\tfrac{1}{2}\right) - 1$$
$$= \tfrac{3}{2}.$$

If $x = \tfrac{5\pi}{6}$, $g\left(\tfrac{5\pi}{6}\right) = \tfrac{3}{2}.$

There are two intersection points: $\left(\tfrac{\pi}{6}, \tfrac{3}{2}\right)$ and $\left(\tfrac{5\pi}{6}, \tfrac{3}{2}\right)$.

Exercise 9.3

PART B

1. Solve $\sin \theta(\cos \theta - 1) = 0$ for $0 \leq \theta \leq 2\pi$.

2. Solve $(\tan x - \sqrt{3})(\sqrt{2} \sin x + 1) = 0$ for $0° \leq x \leq 360°$.

3. In each of the following, use a substitution for $\tan \theta$ or $\cot \theta$ in order to simplify. Solve the resulting equation for $0° \leq \theta \leq 360°$.

   a. $2 \cos \theta - \cot \theta = 0$ **b.** $2 \sin \theta \tan \theta = 5 - \dfrac{1}{\cos \theta}$

4. Solve each of the following for $0 \le x \le 2\pi$.

a. $\cos^2 x + \sin x + 1 = 0$ **b.** $\sqrt{2} \sin^2 x = \sqrt{2} - \sin x$

c. $4 \sin^2 x + \sin x + 3 = 6 \cos^2 x$ **d.** $\sec^2 x = 3 \tan^2 x + \tan x$

5. Find the points of intersection of the functions $f(x) = \sin^2 x - 3 \cos^2 x$ and $g(x) = 8 \cos x - 4$, $0 \le x \le 2\pi$.

6. Solve the equations for $-\pi \le x \le 2\pi$.

a. $6 \cos x + 5 \tan x = 0$ **b.** $5 \tan x \cos^2 x - 3 \sin^2 x = 0$

7. a. Find the exact points of intersection of the functions
$f(x) = 4 \sin^2 x + 7 \sin x + 6$ and $g(x) = 2 \cos^2 x - 4 \sin x + 11$ on domain $0 \le x \le 3\pi$.

    b. Check the work of part **a** on a graphing calculator.

8. Solve the following equations for $0 \le x \le 2\pi$.

a. $10 - 7 \cos x = 6 \cos^2 x + 9$ **b.** $2 \sin^2 x = 1 - \cos x$

c. $5 \cos x = \sin^2 x$ **d.** $2 \tan x + \sec x = 1$

9. If $-\dfrac{\pi}{2} < x < \dfrac{\pi}{2}$, solve and verify $1 + 2 \tan x = \sqrt{\tan^2 x + 2 \tan x + 6}$.

10. The function $S = -3 \cos\left(\dfrac{\pi}{12}t\right) + 20$ represents the surface temperature S of a pond, in degrees Celsius, and t represents the number of hours since sunrise at 06:00. At what time is the surface temperature of the water 20°C?

11. The function $h = 3 \sin(0.523t - 3.141) + 4$ represents the depth of water h, in metres, in a river seaport, where t is the time of day. Note at midnight $t = 0$ and at 01:00, $t = 1$. A ship finishes loading at 12:00. If the ship requires a minimum depth of 5 m to clear the harbour, what is the earliest time the ship can set sail? Express your answer to the nearest minute.

12. Determine the points of intersections of the graphs of the functions
$f(x) = 3 \sin^2 x - 1 \cos x$ and $g(x) = 2 \cos x + 1$, $0 \le x \le 2\pi$.

PART C

13. For domain $0 \le \alpha \le 2\pi$, solve $\sin \alpha + \cos \alpha = 1$.

14. Solve $\cos^2 x(6 \tan^2 x \sin x - 1) = 4 \sin x - 1$ for $0 \le x \le 2\pi$.

15. If $m = \csc \theta - \sin \theta$ and $n = \sec \theta - \cos \theta$, prove that $m^{\frac{2}{3}} + n^{\frac{2}{3}} = (mn)^{-\frac{2}{3}}$. Discuss any restrictions.

9.4 Compound Angle Identities

There are many instances in which we are required to work with compound angles (angles that are the combination of two or more angles). In determining the trigonometric ratios of such angles, the calculations are trivial if the angles are known specifically. For example, if $\angle A = \angle B + \angle C$, $\angle B = 50°$, and $\angle C = 23°$, then $\cos A = \cos 73° \doteq 0.292\,37$. We note that $\cos B + \cos C \doteq 1.563\,29$, so we are sure that $\cos (B + C) \neq \cos B + \cos C$ in general. By trying a few further examples with your calculator, you can convince yourself that this is so. You might be able to find a few instances in which it is true, however! Given that there are times when we must work with compound angles (they occur frequently in identities and are of great importance in analysis), we develop the Addition Identities. The first addition identity is as follows:

Theorem: $\cos (A + B) = \cos A \cos B - \sin A \sin B$

Proof

Consider a circle, with centre $(0, 0)$ and radius 1, containing angles of measure A, $A + B$, and $-B$, as drawn. By definition of sine and cosine, the points M, N, P, and Q have coordinates as shown in the diagram.

Since $\angle MOP$ and $\angle NOQ$ have equal measure $(A + B)$ by construction, $\triangle NOQ$ is congruent to $\triangle MOP$. Then $|MP| = |NQ|$.

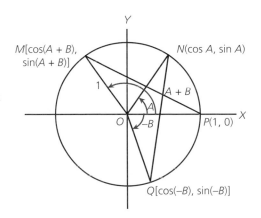

Using the distance formula, we obtain

$$\sqrt{[\cos(A + B) - 1]^2 + \sin^2(A + B)} = \sqrt{[\cos A - \cos(-B)]^2 + [\sin A - \sin(-B)]^2}.$$

On squaring both sides and expanding the brackets, we obtain

$$\cos^2(A + B) - 2 \cos(A + B) + 1 + \sin^2(A + B)$$
$$= \cos^2 A - 2 \cos A \cos(-B) + \cos^2(-B) + \sin^2 A - 2 \sin A \sin(-B) + \sin^2(-B).$$

Since $\sin^2 x + \cos^2 x = 1$, $\cos(-x) = \cos x$, and $\sin(-x) = -\sin x$, the equation can be simplified to $1 - 2 \cos(A + B) + 1 = 1 - 2 \cos A \cos B + 1 + 2 \sin A \sin B$.
Hence, $\cos(A + B) = \cos A \cos B - \sin A \sin B$.

Here is a listing of the Addition Identities and others that follow easily from them. In the examples following, proofs are provided for some. Once you have seen how one or two are developed, you will see how others can be done by the same methods, and you will be asked to do so in the exercises.

Addition Identities

$$\cos(A + B) = \cos A \cos B - \sin A \sin B$$

$$\cos(A - B) = \cos A \cos B + \sin A \sin B$$

$$\sin(A + B) = \sin A \cos B + \cos A \sin B$$

$$\sin(A - B) = \sin A \cos B - \cos A \sin B$$

$$\tan(A + B) = \frac{\tan A + \tan B}{1 - \tan A \tan B}$$

Complementary Identities

$$\cos\left(\frac{\pi}{2} - A\right) = \sin A$$

$$\sin\left(\frac{\pi}{2} - A\right) = \cos A$$

Double-Angle Identities

$$\cos 2A = \cos^2 A - \sin^2 A$$

$$\sin 2A = 2 \sin A \cos A$$

$$\tan 2A = \frac{2 \tan A}{1 - \tan^2 A}$$

EXAMPLE 1

Show that $\cos(A - B) = \cos A \cos B + \sin A \sin B$.

SOLUTION

$$\cos(A - B) = \cos[A + (-B)]$$
$$= \cos A \cos(-B) - \sin A \sin(-B)$$
$$= \cos A \cos B + \sin A \sin B$$

EXAMPLE 2

Show that $\cos\left(\dfrac{\pi}{2} - A\right) = \sin A$.

$$\cos\left(\frac{\pi}{2} - A\right) = \cos\frac{\pi}{2}\cos A + \sin\frac{\pi}{2}\sin A$$
$$= 0 \times \cos A + 1 \times \sin A$$
$$= \sin A$$

EXAMPLE 3

Show that $\sin(A + B) = \sin A \cos B + \cos A \sin B$.

We know that $\sin A = \cos\left(\dfrac{\pi}{2} - A\right)$.

Then $\quad \sin(A + B) = \cos\left[\dfrac{\pi}{2} - (A + B)\right]$

$$= \cos\left[\frac{\pi}{2} - A - B\right]$$
$$= \cos\left[\left(\frac{\pi}{2} - A\right) - B\right]$$
$$= \cos\left(\frac{\pi}{2} - A\right)\cos B + \sin\left(\frac{\pi}{2} - A\right)\sin B$$
$$= \sin A \cos B + \cos A \sin B$$

EXAMPLE 4

Show that $\tan(A + B) = \dfrac{\tan A + \tan B}{1 - \tan A \tan B}$.

L.S. $= \tan(A + B) = \dfrac{\sin(A + B)}{\cos(A + B)}$

$\qquad\qquad = \dfrac{\sin A \cos B + \cos A \sin B}{\cos A \cos B - \sin A \sin B}$

R.S. $= \dfrac{\tan A + \tan B}{1 - \tan A \tan B}$

$\qquad = \dfrac{\dfrac{\sin A}{\cos A} + \dfrac{\sin B}{\cos B}}{1 - \dfrac{\sin A}{\cos A}\dfrac{\sin B}{\cos B}}$

$\qquad = \dfrac{\dfrac{\sin A \cos B + \cos A \sin B}{\cos A \cos B}}{\dfrac{\cos A \cos B - \sin A \sin B}{\cos A \cos B}}$

$\qquad = \dfrac{\sin A \cos B + \cos A \sin B}{\cos A \cos B - \sin A \sin B}$

Then L.S. = R.S. and the identity is true.

EXAMPLE 5

If $\tan A = \frac{3}{2}$, where $0° < A < 90°$, and $\cos B = \frac{-3}{5}$, where $180° < B < 270°$, determine the exact value of $\sin(A + B)$.

Since $\tan A = \frac{y}{x} = \frac{3}{2}$, let point P be $(2, 3)$.

Then $OP = \sqrt{4 + 9} = \sqrt{13}$.

Therefore $\sin A = \frac{3}{\sqrt{13}}$ and $\cos A = \frac{2}{\sqrt{13}}$.

Since $\cos B = \frac{x}{r} = \frac{-3}{5}$,

$$9 + y^2 = 25$$
$$y = \pm 4.$$

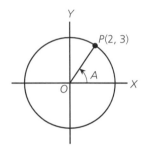

Since $180° < B < 270°$, Q is $(-3, y)$ and $y = -4$.

Then $\sin B = \frac{-4}{5}$.

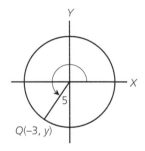

Using the addition identity $\sin(A + B) = \sin A \cos B + \cos A \sin B$

$$= \left(\frac{3}{\sqrt{13}}\right)\left(\frac{-3}{5}\right) + \left(\frac{2}{\sqrt{13}}\right)\left(\frac{-4}{5}\right)$$

$$= \frac{-9 - 8}{5\sqrt{13}}$$

$$= \frac{-17}{5\sqrt{13}}.$$

EXAMPLE 6

Prove $\tan \frac{\theta}{2} + \cot \frac{\theta}{2} = 2 \csc \theta$.

SOLUTION

Whenever half angles appear in a problem, it is a good strategy to use a substitution to remove them. In this case, let $\frac{\theta}{2} = W$ and separate the problem into two parts.

L.S. $= \tan W + \cot W$

$= \dfrac{\sin W}{\cos W} + \dfrac{\cos W}{\sin W}$

$= \dfrac{\sin^2 W + \cos^2 W}{\sin W \cos W}$

$= \dfrac{1}{\sin W \cos W}$

R.S. $= 2 \csc 2W$

$= \dfrac{2}{\sin(W + W)}$

$= \dfrac{2}{\sin W \cos W + \cos W \sin W}$

$= \dfrac{2}{2 \sin W \cos W}$

$= \dfrac{1}{\sin W \cos W} =$ L.S.

Exercise 9.4

PART A

1. a. Show that $\sin(90° + W) = \cos W$.

   b. Prove $\cos(90° + W) = -\sin W$.

2. a. Use the identity for $\sin(A + B)$ to show $\sin 2A = 2 \sin A \cos A$.

   b. Record this identity in your journal.

3. a. Expand and simplify $\cos\left(\frac{3\pi}{2} - R\right)$.

   b. Show that $\sin(270° + R) = -\cos R$.

4. If $\sin W = \frac{3}{5}$ and $\cos T = \frac{12}{13}$, where both W and T are acute angles, then

   a. find the exact value of $\sin(W + T)$;

   b. show that $\cos(W - T) > \sin(W + T)$.

PART B

5. Prove that $\sin(A - B) = \sin A \cos B - \cos A \sin B$ by replacing B by $-B$.

6. a. Prove $\cos 2A = \cos^2 A - \sin^2 A$.

b. Prove $\cos 2A = 2 \cos^2 A - 1$.

c. Show that $\cos 2A$ can be expressed as an identity in terms of $\sin A$.

d. Record the three different identities for $\cos 2A$. Write a short commentary on why it could be useful to have different identities for $\cos 2A$.

7. Show that $\sin(45° + x) + \cos(45° + x) = \sqrt{2} \cos x$.

8. Determine the exact value for each of the following:

a. $\cos 75°$ **b.** $\sin 15°$ **c.** $\cos 105°$ **d.** $\sin 255°$

9. Express each quantity in terms of $\sin x$ and $\cos x$.

a. $\sin\left(\dfrac{\pi}{3} + x\right)$ **b.** $\cos\left(x + \dfrac{3\pi}{4}\right)$

c. $\cos\left(\dfrac{\pi}{4} - x\right)$ **d.** $\sin(2\pi - x)$

10. If $\cos A = \dfrac{1}{3}$, with $0 < A < \dfrac{\pi}{2}$, and $\sin B = \dfrac{1}{4}$, with $\dfrac{\pi}{2} < B < \pi$, calculate each quantity.

a. $\cos(A + B)$ **b.** $\sin(A + B)$

c. $\cos 2A$ **d.** $\sin 2B$

11. If $\tan A = \dfrac{1}{3}$ and $\pi < A < \dfrac{3\pi}{2}$, calculate each quantity. In what quadrant does the angle $2A$ lie?

a. $\sin 2A$ **b.** $\cos 2A$

12. Prove that each of the following is an identity.

a. $\cos^4 A - \sin^4 A = \cos 2A$

b. $1 + \sin 2\alpha = (\sin \alpha + \cos \alpha)^2$

c. $\sin(A + B) \bullet \sin(A - B) = \sin^2 A - \sin^2 B$

d. $\dfrac{\cos W - \sin 2W}{\cos 2W + \sin W - 1} = \cot W$

e. $\dfrac{\sin 2\theta}{1 - \cos 2\theta} = 2 \csc 2\theta - \tan \theta$

f. $\tan \dfrac{\theta}{2} = \dfrac{\sin \theta}{1 + \cos \theta}$

13. Simplify the function $f(x) = \sin 3x \csc x - \cos 3x \sec x$.

14. Prove that each of the following is an identity.

a. $\dfrac{1 + \sin \theta - \cos \theta}{1 + \sin \theta + \cos \theta} = \tan \dfrac{\theta}{2}$

b. $\dfrac{\cos 2W}{1 + \sin 2W} = \dfrac{\cot W - 1}{\cot W + 1}$

c. $\sin 3\theta = 3 \cos^2 \theta \sin \theta - \sin^3 \theta$

d. $\cos 3\theta = \cos^3 \theta - 3 \cos \theta \sin^2 \theta$

PART C

15. For $0 \le \beta \le 2\pi$, determine all solutions of $\sin \beta + \cos \beta = \sin \beta \cos \beta$.

16. In any acute-angled $\triangle ABC$, prove that $b^2 \sin 2C + c^2 \sin 2B = 2\, bc \sin A$.

17. a. Use the identity $\tan Q = \dfrac{\sin Q}{\cos Q}$ to help prove the identity

$\tan(A - B) = \dfrac{\tan A - \tan B}{1 + \tan A \tan B}$.

b. Prove that $\dfrac{1 + \tan \dfrac{\theta}{2}}{1 - \tan \dfrac{\theta}{2}} = \sec \theta + \tan \theta$.

c. Find an identity for $\tan(2A)$.

18. Use the identities for $\sin(A + B)$ and $\sin(A - B)$ to help prove each of the following:

a. $\sin x + \sin y = 2 \sin\left(\dfrac{x + y}{2}\right)\cos\left(\dfrac{x - y}{2}\right)$

b. $\sin x - \sin y = 2 \cos\left(\dfrac{x + y}{2}\right)\sin\left(\dfrac{x - y}{2}\right)$

19. In any $\triangle ABC$, prove that $\dfrac{a - b}{b} = \dfrac{2 \sin\left(\dfrac{C}{2}\right)\sin\left(\dfrac{A - B}{2}\right)}{\sin B}$.

PART B

1. Solve each of the following equations to the nearest degree, $0 \leq \theta \leq 360°$.

   a. $\sin \theta = 0.7134$ **b.** $\cos \theta = 0.5173$

   c. $\tan \theta = 3.1435$ **d.** $\sin \theta = -0.1351$

   e. $\cos \theta = -0.1031$ **f.** $\tan \theta = -0.4315$

2. Solve each of the following equations to two decimal places, $0 \leq \theta \leq 2\pi$.

   a. $\sin \theta = 0.1513$ **b.** $\cos \theta = 0.8415$

   c. $\tan \theta = 0.3157$ **d.** $\sin \theta = -0.6139$

   e. $\cos \theta = -0.5314$ **f.** $\tan \theta = -2.3145$

3. Solve each of the following equations. Write your answers to the nearest degree, $0 \leq \theta \leq 360°$.

   a. $3 \sin \theta + 5 = \sin \theta + 6$ **b.** $5 - 3 \cos \theta = 2 - 6 \cos \theta$

   c. $2 \cos^2 \theta + 4 = 5$ **d.** $3 \sin^2 \theta - 8 \sin \theta + 4 = 0$

   e. $9 \tan \theta - 5 = 4 \tan \theta + 3$ **f.** $3 \sin^2 \theta + \sin \theta = 0$

4. Solve each of the following equations. Write your answers correct to one decimal, $0 \leq \theta \leq 2\pi$.

   a. $5 \cos \theta - 2 = 3 \cos \theta - 1$ **b.** $7 - 2 \sin \theta = 4 - 5 \sin \theta$

   c. $4 \sin^2 \theta + 1 = 4$ **d.** $6 \cos^2 \theta - 5 \cos \theta + 1 = 0$

   e. $3 \tan \theta - 2 = 2 \tan \theta + 3$ **f.** $2 \cos^2 \theta - \cos \theta = 0$

5. Prove that each of the following is an identity.

   a. $\cos x \tan x = \sin x$ **b.** $\csc \theta - \sin \theta = \cot \theta \csc \theta$

   c. $\cos x + \tan x \sin x = \sec x$ **d.** $\dfrac{\cos \theta - \sin \theta}{\cos \theta} = 1 - \tan \theta$

   e. $\dfrac{\sin x}{1 - \sin^2 x} = \tan x \sec x$ **f.** $(1 - \cos \theta)^2 + \sin^2 \theta = 2(1 - \cos \theta)$

   g. $\dfrac{\sin x + \tan x}{1 + \sec x} = \sin x$ **h.** $(1 - \cos^2 x)\csc x = \sin x$

   i. $\dfrac{\sin x}{1 - \cos x} = \dfrac{1 + \cos x}{\sin x}$ **j.** $\cos^4 \theta - \sin^4 \theta = 1 - 2 \sin^2 \theta$

6. Solve each of the following equations, $0 \le x \le 2\pi$.

a. $2 \cos^2 x + 5 \cos x - 3 = 0$

b. $2 \sin^2 x - \sin x - 2 = 0$

c. $2 \cos^2 x - \cos x - 1 = 0$

d. $3 \sin x = \cos x$

e. $3 - 3 \sin x - 2 \cos^2 x = 0$

f. $2 \sin^2 x - 3 \cos x = 0$

g. $5 - 5 \cos x = 4 \sin^2 x$

h. $3(\tan \theta + \sin^2 \theta) - 4 = 4 \tan \theta - 3 \cos^2 \theta$

7. Determine the exact values of the intersection points for the functions $f(x) = \sin^2 x$ and $g(x) = 3 \cos^2 x$, $-\pi \le x \le 2\pi$.

8. If $\sin A = \frac{1}{5}$ with $0 < A < \frac{\pi}{2}$ and $\cos B = -\frac{1}{3}$, $\frac{\pi}{2} < B < \pi$, calculate each quantity.

a. $\sin(A + B)$ **b.** $\cos(A + B)$

c. $\sin 2A$ **d.** $\cos 2B$

9. Prove that $\tan 2A = \dfrac{2}{\cot A - \tan A}$.

10. Prove that $\sin B = \sqrt{\dfrac{1 - \cos 2B}{2}}$ for $0 \le B \le \frac{\pi}{2}$.

11. If $0 < A < \frac{\pi}{2}$ and $\sin A + 1 = 2\sqrt{1 - \sin^2 A}$, determine the value of $\sin A$.

1. In a math class of 25 students, the numbers 1 to 25 were assigned, one to each student. The teacher then wrote a large number on the chalkboard and asked each student whether or not the number they were assigned was a divisor of it. Every student answered "yes," but two were wrong, and these two, surprisingly, had been assigned consecutive numbers.

    a. Which two students were wrong?

    b. What was the number the teacher wrote on the board?

2. If $Q + \dfrac{R}{x-3} = \dfrac{3x+5}{x-3}$ for all values of x, determine values for Q and R.

3. If $a, a + d, a + 9d, d > 0$, are the sides of a right-angled triangle, determine the ratio $a{:}d$.

4. Determine values for p and q so that $\dfrac{-5 + 9\cos^2 x}{2 + 3\sin x} = p + q\sin x$ for all values of x.

5. In triangle ABC, $AB = AC$ and $\dfrac{BC}{AB} = r$. Determine the value of
$\cos A + \cos B + \cos C$ in terms of r.

6. Determine all values of x, $0 \le x \le 2\pi$, such that
$$\sin 2x + \cos 2x + \sin x + \cos x + 1 = 0.$$

7. In triangle ABC, the angles A, B, and C satisfy the equation
$\cos(A - B) = 1 + \sin A \sin B(1 - \sin C)$. Determine the values of the three angles. (This is an exercise in thinking. Do not expand the expressions. Think about the range of values of the functions involved.)

Chapter 10

Locus

Designs created by architects, engineers, mechanics, and technical artists often require an understanding of a locus, which is a path traced by a point as it follows geometric conditions. Whether you are planning an invention, designing an object, or visualizing a path in a sport that involves a ball or a jump, knowing how to set and vary these paths can be essential for success. You can also use dynamic geometry software to experiment with such paths to make and test your hypotheses to form generalizations. Understanding the path of a locus will advance your reasoning process for many areas of your life.

IN THIS CHAPTER, YOU CAN . . .

- construct a geometric model to represent a described locus of points, using geometry software and drawing by hand;
- determine properties of the geometric model constructed, and use the properties to interpret the locus;
- explain the process used in constructing a geometric model;
- determine equations for conics.

10.1 What Is a Locus?

The ancient Greeks used deductive reasoning and Euclidean geometry to study lines, circles, and the **conic sections** (parabolas, circles, ellipses, and hyperbolas). In the seventeenth century, conic sections took on an important role in scientific applications when Kepler and Newton used conic sections to describe the orbit of the planets. In the same century, Descartes and Fermat developed analytic geometry, which gives a correspondence between an equation and the set or **locus** of all those points in the plane with coordinates that satisfy the equation. If we have a geometric locus or curve, we can use analytic geometry to algebraically obtain the properties from the defining equation of the curve.

Kepler   Descartes   Fermat   Newton

Photos: Kepler, Descartes, and Fermat: Bettmann/CORBIS; Newton: Leonard de Selva/CORBIS

We will start our study of the conic sections by finding the set of points or locus of these curves.

> A **locus** is a set of points that satisfy a given condition or the path traced out by a point that moves according to a stated geometric condition.

You are very familiar with some loci.

For example: Find the locus of points P that is 6 cm from a point C.

You know that these points lie on a circle with a radius of 6 cm.

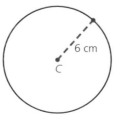

Some loci are more difficult to determine. To find the locus of points P that satisfy a given condition or conditions, you need to find the location of some of the points and then sketch the locus using these points as a guide. Dynamic geometry software, such as *Geometer's Sketchpad,* makes this process easier.

INVESTIGATION 1

Using *Geometer's Sketchpad*, determine the locus of points equidistant from two points *A* and *B*.

Here are some steps to help you get started.

Step 1: Construct and label points *A* and *B*.

Step 2: Construct a line segment of arbitrary length. Label the end points *M* and *N*.

Step 3: Construct a circle with centre *A* and radius *MN*.

Step 4: Construct a circle with centre *B* and radius *MN*.

Step 5: Select the points of intersection of the two circles. You may need to adjust the length of line segment *MN* so that the circles intersect.

Step 6: Choose *Trace* from the *Display* menu and vary the length of line segment *MN*.

The two points of intersection of the circles are equidistant from *A* and *B* and thus are two points on the locus. The trace function gave you many more points on the locus.

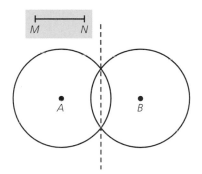

a. Describe the locus of points equidistant from two points.

b. Hide the circles and points of intersection. Use *Geometer's Sketchpad* to construct the locus you identified in part **a.**

INVESTIGATION 2

Using *Geometer's Sketchpad*, determine the locus of points at a constant distance from a line *l*.

Draw line *l* and a line segment *MN* representing the distance of the points from line *l*. If you know where the locus is located, construct the locus.

If not, follow the steps below.

Step 1: Hide the control points on line *l*.

Step 2: Construct a point *P* on line *l* and use the features of *Geometer's Sketchpad* to find point(s) *A* at the perpendicular distance *MN* from point *P*.

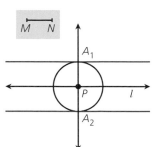

Step 3: Select point *A* and choose *Trace* from the *Display* menu.

Step 4: Draw the locus by moving point *P* along line *l*.

a. Describe the locus of points equidistant to line *l*.

b. When you click anywhere, the locus disappears. Construct the locus of points at a distance *MN* from line *l* without using the *Trace* function.

INVESTIGATION 3

Using *Geometer's Sketchpad*, determine the locus of points that are twice as far from point A as they are from point B.

Here are some steps to help you get started.

Step 1: Construct and label points A and B.

Step 2: Construct a line segment of arbitrary length. Label the end points M and N.

Step 3: Think about how you obtain a line segment that is twice as long as another line segment.

Step 4: Complete the diagram.

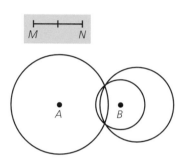

a. Describe the locus.

b. Change the location of point A. Describe how the locus changes

    i) when points A and B are closer together;

    ii) when points A and B are farther apart.

INVESTIGATION 4

Chords of equal length are drawn in a circle. Using *Geometer's Sketchpad*, determine the locus of midpoints of the chords.

Here are some steps to help you get started.

Step 1: Construct a line segment MN. This will be the length of the chord.

Step 2: Construct a circle with centre A.

Step 3: Place a point P on the circle.

Step 4: Locate a second point Q on the circle, at a distance MN from point P.

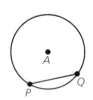

Step 5: Construct line segment PQ. As you move point P around the circle, the length of line segment PQ remains a constant length.

Step 6: Construct the midpoint of line segment PQ.

Step 7: Determine the locus of the midpoints of the chords.

a. Describe the locus of midpoints of the chords.

b. Change the length of the chord. Describe how the locus changes

    i) when the length of the chord is decreased;

    ii) when the length of the chord is increased.

INVESTIGATION 5

Using *Geometer's Sketchpad*, determine the locus of points equidistant from a point *F* and a line *d*.

Here are some steps to help you get started.

Step 1: Construct and label line *d*. Hide its control points.

Step 2: Construct and label point *F*. Point *F* is not on line *d*.

Step 3: Construct line segment *MN*. You will be looking for points at a distance *MN* from point *F* and line *d*.

Step 4: Construct a circle with radius *MN* and centre *F*.

Step 5: Construct points at a distance *MN* from line *d*. Recall what you learned in Investigation 2.

Step 6: Draw the locus of all points equidistant from point *F* and line *d*.

a. Describe the locus.

    Hint: Hiding all the construction lines and control points will make it easier to see how the locus changes when you change the position of point *F*.

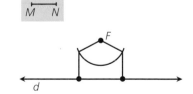

b. Change the location of point *F*. Describe how the locus changes

    i) when point *F* is closer to line *d*;

    ii) when point *F* is farther from line *d*.

A wonderful application of the combination of geometry and algebra is illustrated through locus problems. We have discussed the concept of locus and the determination of a locus using the technology of the computer. We can also use algebra to describe the locus of a point moving under a stated condition. When doing so, we frequently must choose points in the *xy* plane as references on which to build the algebra. We can frequently make our work much easier if we choose points with care. For example, if we wish to determine the locus of points four times as far from one point as from another, choosing the points $(0, 0)$ and $(5, 0)$ will probably be advantageous. Why do we choose one point 5 units from the other? That is because the sum of the distances 4 and 1 is 5.

Determine the equation of the locus of a point that moves so that it is four times as far from point $A(0, 0)$ as from point $B(5, 0)$.

SOLUTION

Let the point tracing the locus be $P(x, y)$.

Then $PA = 4PB$

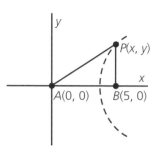

$PA = \sqrt{x^2 + y^2}$ and $PB = \sqrt{(x - 5)^2 + y^2}$

$\sqrt{x^2 + y^2} = 4\sqrt{(x - 5)^2 + y^2}$

Squaring both sides,

$$x^2 + y^2 = 16[(x - 5)^2 + y^2]$$
$$x^2 + y^2 = 16[x^2 - 10x + 25 + y^2]$$
$$x^2 + y^2 = 16x^2 - 160x + 400 + 16y^2$$
$$15x^2 + 15y^2 - 160x + 400 = 0$$

This is the equation of the locus. Note that while we have the equation, it is not clear that we can say with certainty what the shape of the locus is. This issue is addressed in the next chapter.

Exercise 10.1

1. In Investigation 1 you found that the locus of points equidistant from two points A and B appeared to be the right bisector of the line segment AB. Although the line drawn by *Geometer's Sketchpad* looks like the right bisector, you have not proved that the locus is the right bisector. Use your knowledge of Euclidean geometry to prove that all points on the right bisector of a line segment are equidistant from the ends of the line segment.

2. Analytic geometry can be used to give an alternate proof that points equidistant from the ends of a line segment are on the right bisector of the line segment.

   a. For points $A(3, 8)$ and $B(7, 14)$, find the equation satisfied by point $P(x, y)$ such that $PA = PB$.

   b. Show that the equation found in part **a** is the equation of the right bisector of line segment AB.

3. In Investigation 3 you found the locus of points that are twice as far from point A as they are from point B.

    a. If the coordinates of A are $(2, 0)$ and the coordinates of B are $(5, 0)$, find an equation of the locus.

    b. Does this equation agree with your finding in the investigation? Explain.

4. In Investigation 5 you found the locus of points equidistant from a point F and a line d. Coordinates have been added to a sketch drawn using *Geometer's Sketchpad*.

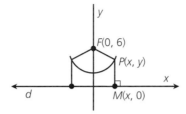

    a. Find an equation of the locus of points P.

    b. Does this equation agree with your finding in the investigation? Explain.

5. As in Question 4, coordinates have been added to a sketch drawing using *Geometer's Sketchpad*.

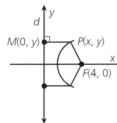

    a. Find an equation of the locus of points P.

    b. Does this equation agree with your finding in the investigation? Explain.

10.2 Drawing the Conic Sections

In the last section, you used *Geometer's Sketchpad* to draw a **parabola**.

> A **parabola** is the set of points equidistant from a fixed point, the **focus**, and a fixed line, the **directrix**.

You will be using *Geometer's Sketchpad* to draw the locus of a parabola, as you did in the last section. Then, by making changes to this diagram, you will create the other conic sections.

INVESTIGATION 1

Using *Geometer's Sketchpad*, draw a horizontal line and hide the control points. Construct a point on the line and label the point *F*. This will be the **focus**.

Construct a line perpendicular to the horizontal line and label this line **directrix.**

Construct a line segment. Label the end points *A* and *B*. Construct two points on the line segment *AB*. Label these points *C* and *D*. Construct line segments *AC* and *AD*. Practise selecting line segments *AB*, *AC*, and *AD*. Notice that dragging control point *B* changes the length of all the line segments.

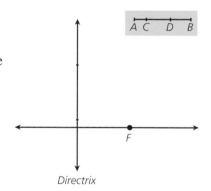

Construct lines parallel to the directrix, at a distance *AD* from the directrix, using the method from Section 10.1, Investigation 2.

To simplify the drawing, hide the circle, the perpendicular line, and the control points on the parallel lines. Do not hide the control points on the directrix.

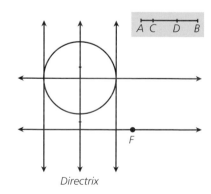

For any point on a parabola, the distance from the directrix to the point is the same as the distance from the focus to the point.

Construct a circle with centre F and radius AD.

Construct the points equidistant from the directrix and the focus.

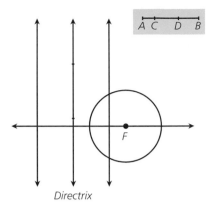

Directrix

Draw the locus of the parabola by dragging control point B.

a. Experiment with different locations for the focus F on the horizontal line. Trace the locus of the parabola and observe the changes in the appearance of the curve. Describe your findings.

b. Describe what happens to the parabola when you rotate the directrix.

c. Describe any symmetries of the parabola.

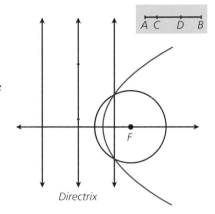

Directrix

**Keep this sketch in *Geometer's Sketchpad*.
You will be using it in the next two investigations.**

A number called the **eccentricity** of a conic will help us. For a point P on the conic, the eccentricity is the distance from the focus to P divided by the distance from the directrix to P.

For a parabola,
$$\text{eccentricity} = \frac{\text{distance from focus to } P}{\text{distance from directrix to } P} = 1.$$

Let's see what happens if we change the eccentricity.

Using your sketch in *Geometer's Sketchpad*, delete the circle with centre F. If the directrix is not vertical, rotate the directrix to a vertical position.

Construct a circle with centre F and radius AC.

Construct the intersection points of the circle centre F and the line parallel to the directrix.

For a point P on the conic, the eccentricity is the distance from the focus to P divided by the distance from the directrix to P.

◆ The distance from the focus to P is AC.
  Measure AC.

◆ The distance from the directrix to P is AD.
  Measure AD.

◆ Calculate the eccentricity.

$$\text{eccentricity} = \frac{\text{distance from focus to } P}{\text{distance from directrix to } P}$$

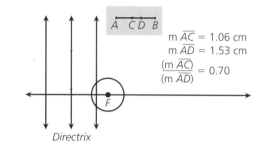

m $\overline{AC}$ = 1.06 cm
m $\overline{AD}$ = 1.53 cm
$\dfrac{(\text{m } \overline{AC})}{(\text{m } \overline{AD})}$ = 0.70

Directrix

Notice that the eccentricity is less than 1.

Draw the locus by dragging control point B.

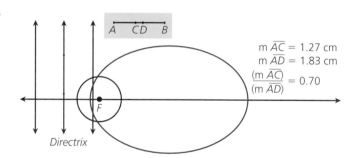

m $\overline{AC}$ = 1.27 cm
m $\overline{AD}$ = 1.83 cm
$\dfrac{(\text{m } \overline{AC})}{(\text{m } \overline{AD})}$ = 0.70

Directrix

The locus is no longer a parabola. This conic shape is called an **ellipse.**

Notice that as the locus is drawn, the length of line segments AC and AD change, but the eccentricity remains constant.

a. Changing the position of point C between A and D changes the eccentricity. Experiment with different eccentricities. Trace the locus of the ellipse and observe the changes in the appearance of the curve. Describe your findings.

b. Experiment with different locations for the focus F on the horizontal line. Trace the locus of the ellipse and observe the changes in the appearance of the curve. Describe your findings.

c. Describe what happens to the ellipse when you rotate the directrix.

d. Describe any symmetries of the ellipse.

e. Describe the ellipse when the eccentricity is 0.5.

If the directrix is not vertical, rotate the directrix to a vertical position.

Change the position of point C so that C is between D and B. Notice that the value of the eccentricity is now greater than 1.

m $\overline{AC}$ = 3.06 cm
m $\overline{AD}$ = 2.48 cm
$\dfrac{(\text{m } \overline{AC})}{(\text{m } \overline{AD})}$ = 1.24

Draw the locus by dragging control point B.

The appearance of the locus has changed again.

Since the distance from a point on the curve to the focus is greater than the distance from the point to the directrix, there will also be points of intersection between the circle and the second parallel line.

m $\overline{AC}$ = 4.66 cm
m $\overline{AD}$ = 1.39 cm
$\dfrac{(\text{m } \overline{AC})}{(\text{m } \overline{AD})}$ = 3.36

Create these intersection points and redraw the locus. Describe how the second branch can be obtained.

This conic section is called a **hyperbola.**

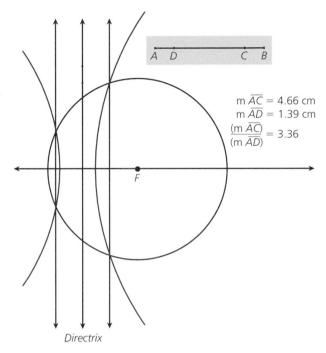

Directrix

a. Changing the position of point C between D and B changes the eccentricity. Experiment with different eccentricities. Trace the locus of the hyperbola and observe the changes in the appearance of the curve. Describe your findings.

m $\overline{AC}$ = 1.14 cm
m $\overline{AD}$ = 1.14 cm
$\dfrac{(\text{m } \overline{AC})}{(\text{m } \overline{AD})}$ = 1.00

b. Experiment with different locations for the focus F on the horizontal line. Trace the locus of the hyperbola and observe the changes in the appearance of the curve. Describe your findings.

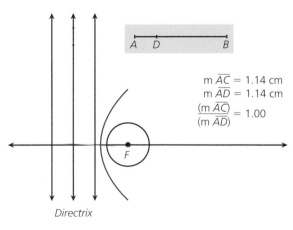

Directrix

c. Describe what happens to the hyperbola when you rotate the directrix.

d. Describe any symmetries of the hyperbola.

e. Change the position of point C until the eccentricity is 1. Trace the locus. What curve have you drawn?

Here is a summary of the eccentricity of the conic sections.

> For a point P on the conic, the eccentricity e is the distance from the focus to P divided by the distance from the directrix to P.
>
> If $e = 1$, the conic is a parabola.
>
> If $0 < e < 1$, the conic is an ellipse.
>
> If $e > 1$, the conic is a hyperbola.

We can define an ellipse and a hyperbola in a different way that is sometimes more useful.

We are familiar with the following definition of a circle.

A **circle** is the set of points such that the distance from each point to a fixed point (the centre) is constant.

Any line segment joining center C to a point on the circle is called the **radius.**

Now assume the centre point of the circle splits into two points and the points move away from one another. As the points, called the foci, move, the shape of the circle changes. The circle elongates and now, from any point on the circumference, instead of a radius we have two **focal radii** (one to each focus).

For a circle, the radius is constant. For an ellipse, the sum of the focal radii is constant.

> An **ellipse** is the set of points such that the sum of the distances from each point to two fixed points, the foci, is constant.

INVESTIGATION 4

Using *Geometer's Sketchpad*, construct two points. Label the points F_1 and F_2. These points will be the foci of the ellipse.

Construct a line and hide its control points. Construct three points on the line. Label the points A, B, and C. Point B is between A and C. Hide the line. Construct line segments AB and BC.

Moving point B between A and C will change the length of line segments AB and BC, but the sum of the line segments will be constant.

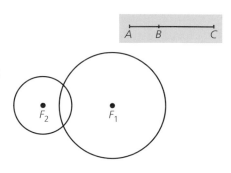

Construct a circle centre F_2 with radius AB. Construct a circle centre F_1 with radius BC.

Construct the two intersection points of the circles. You may need to adjust the length of line segment AC so that the circles intersect.

If P is a point on the locus,

$$PF_1 + PF_2 = BC + AB$$
$$= AC$$

If the length of the line segment AC is kept constant, we have satisfied the conditions for an ellipse.

Draw the locus by dragging point B between A and C.

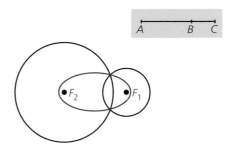

a. How do you make a long, thin ellipse?

b. How do you make an ellipse that is nearly circular?

c. Describe how the shape of the ellipse changes as the foci move

   i) closer together;

   ii) farther apart.

**Keep this sketch in *Geometer's Sketchpad*.
You will be using it in the next investigation.**

A **hyperbola,** like an ellipse, has two points that are called **focal points,** or foci.

For an ellipse, the *sum* of the distances from a point on the curve to the two foci remains constant. For a hyperbola, it is the *difference* of the two distances that remains constant. The difference of two numbers is either positive or negative, depending on the order of subtraction. Thus, it is the absolute value of the differences that is constant.

A **hyperbola** is the set of points such that the absolute value
of the difference of the distances from each point
to two fixed points, the foci, is constant.

We can draw a hyperbola using the sketch from Investigation 4.

Drag point B beyond C.

Drag point C toward A, until $AC < F_1F_2$.

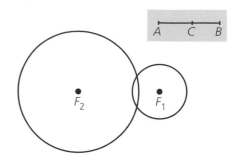

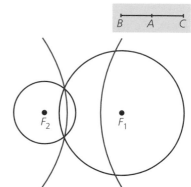

Draw the locus by dragging point B back and forth, extending beyond points A and C.

a. Explain how this construction method satisfies the definition of a hyperbola.

b. Describe how the shape of the hyperbola changes as the foci move

  i) closer together;

  ii) farther apart.

Exercise 10.2

1. An ellipse can be drawn using a piece of string. Mark points F_1 and F_2 on a piece of paper. Tape the two ends of the string at the points you marked. Using your pencil, pull the string taut and trace the curve.

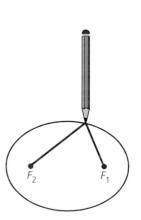

a. Explain how this construction method satisfies the definition of an ellipse given in Investigation 3.

b. Describe how the shape of the ellipse would change if

  i) the ends of the string were taped closer together;

  ii) the ends of the string were taped farther apart.

c. If the two ends of the string were taped at the same point, what curve would your pencil trace? Explain.

d. If the two ends of the string were taped as far apart as possible so that the string was stretched straight, what curve would the pencil trace? Explain.

2. A hyperbola can be drawn using a piece of string and a small ring. Select a point on the string at a short distance from the centre of the string. Tie a knot at this point. Mark points F_1 and F_2 on a piece of paper. Tape the two ends of the string at F_1 and F_2. Place the ring over the string, as shown in the diagram. Place the end of your pencil in the ring. Hold the knot in the string. Pull the string taut and trace the curve with your pencil by sliding the ring down the string.

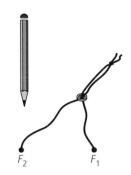

a. What adjustment do you need to make to draw the second branch of the hyperbola?

b. Explain how this construction method satisfies the definition of a hyperbola from Investigation 4.

c. Describe how the shape of the hyperbola would change if

   i) the ends of the string were taped closer together;

   ii) the ends of the string were taped farther apart.

d. Do you think a hyperbola has asymptotes? Explain.

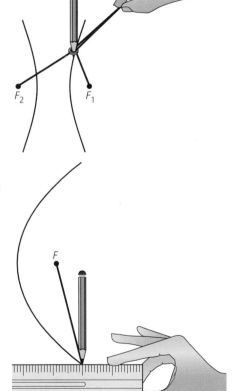

3. A parabola can be drawn using a ruler and a piece of string the same length as the ruler. Tape the string to one end of the ruler. On a piece of paper, mark a point F (focus). Tape the other end of the string to the point F. Place the ruler with one end at the edge of the paper and with the ruler perpendicular to the edge of the paper. Use the tip of your pencil to hold the string taut along the ruler. Mark this point. Slide the ruler, keeping the end at the edge of the paper and the ruler perpendicular to the edge of the paper. Keeping the string taut, trace the parabola.

a. Explain how this construction method satisfies the definition of a parabola.

b. Describe how the shape of the parabola would change if the point F were closer to the edge of the paper against which the end of the ruler was placed.

c. How would you change your construction to draw a parabola that looks narrower than the one drawn?

d. Do you think a parabola has asymptotes? Explain.

An ellipse has two important lines, called the major and minor axes.

A_1A_2 is the major axis.
B_1B_2 is the minor axis.

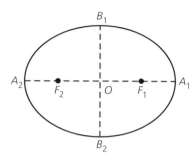

4. Draw two perpendicular lines intersecting at O. On the horizontal line, locate F_1 and F_2 so that $OF_1 = OF_2 = 6$ cm. Draw an ellipse using a piece of string 20 cm long, as you did in Question 1.

    a. What is the length of A_1A_2?

    b. What is the length of B_1B_2?

    c. What is the length of B_1F_1?

    d. What is the relationship connecting OA_1, OB_1, and OF_1?

5. Repeat Question 4 with $OF_1 = OF_2 = 8$.

6. Repeat Question 4 with $OF_1 = OF_2 = 4$.

Chapter 11

Equations of Conic Sections

Conics—circles, parabolas, ellipses, and hyperbolas—are model shapes seen in nature and in design. Circles and approximations are found in examples from antiquity such as coins, wheels, and the arrangement of stones at Stonehenge. More recent examples include cross-sections of pipes, balls, tree trunks, planets, and Olympic rings. A parabolic shape is frequently chosen for supporting bridges and arches, for picking up sound with satellite dishes, and for reflecting light. An ellipse describes orbits of planets, moons, and comets. In navigation, hyperbolas are used for planning locations of stations that send signals to guide aircraft. Understanding conic sections is essential for planning, creating, and using objects whose conic shapes contribute to achieving specific purposes.

IN THIS CHAPTER, YOU CAN . . .

- illustrate the conics as intersections of planes with cones;
- identify standard forms for the equations of conic sections: parabolas, circles, ellipses, and hyperbolas;
- identify the type of conic, and determine the key features of a conic;
- sketch the graph of a conic;
- solve problems from a variety of applications involving conics;
- solve problems involving the intersections of lines and conics.

Appolonius of Perga (c. 225 B.C.) discovered that by intersecting a right circular cone with a plane, he could form a number of different shapes. A right circular cone is the surface generated by a straight line that passes through a fixed point (vertex) and touches a circle whose plane is perpendicular to the axis joining the centre to a fixed point. The cone consists of two parts or nappes, one on either side of the vertex. If cones of this type are intersected by plane surfaces, the resulting curves are called **conic sections** or **conics.**

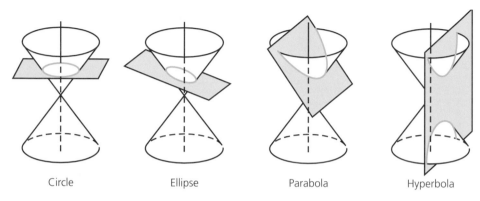

| Circle | Ellipse | Parabola | Hyperbola |

Position of Plane	Resulting Curve
perpendicular to the axis of the cone	circle
cutting one nappe only and not parallel to the generating line	ellipse
parallel to the generating line	parabola
cutting both nappes	hyperbola

Pappus of Alexandria (c. A.D. 300) redefined each of these curves in two dimensions as the locus of a point such that the ratio of its distance from a fixed point, the focus, to a fixed line, the directrix, is constant. In the seventeenth century, analytic geometers, Descartes in particular, developed two-variable equations from these definitions and thus represented the conic sections as a set of second-degree relations.

11.1 The Circle

The **circle** is the set of points in a plane that are equidistant from a fixed point, the centre. The distance from the centre to any point on the circle is called the **radius.**

EXAMPLE 1

Find the equation of a circle with radius 6 units and centre the origin $O(0, 0)$.

SOLUTION

If $P(x, y)$ is a point on the circle, then $OP = 6$.

Using the distance formula, $\sqrt{(x - 0)^2 + (y - 0)^2} = 6$.

Square both sides of this equation.

$$x^2 + y^2 = 36$$

The required equation is $x^2 + y^2 = 36$.

Note that the circle has x-intercepts at ± 6, y-intercepts at ± 6, domain $-6 \le x \le 6$, $x \in R$, and range $-6 \le y \le 6$, $y \in R$.

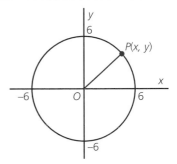

EXAMPLE 2

Find the equation of a circle with radius 5 units and centre $C(1, -2)$.

SOLUTION

If $P(x, y)$ is a point on the circle, then $PC = 5$.

Therefore $\sqrt{(x - 1)^2 + (y + 2)^2} = 5$.

Square both sides of this equation.

$$(x - 1)^2 + (y + 2)^2 = 25$$

The equation of the circle is $(x - 1)^2 + (y + 2)^2 = 25$.

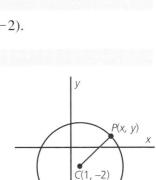

If we expand and simplify this equation we get

$$x^2 + y^2 - 2x + 4y - 20 = 0.$$

Using the methods outlined in Examples 1 and 2, we can determine the equation of the circle with centre $C(h, k)$ and radius r units.

Therefore $CP = r$ and $\sqrt{(x - h)^2 + (y - k)^2} = r$.

Square both sides of the equation to obtain

$$(x - h)^2 + (y - k)^2 = r^2.$$

This is referred to as the standard form of the equation of a circle.

If we expand and simplify the equation above, we get

$$x^2 - 2xh + h^2 + y^2 - 2yk + k^2 = r^2$$

or $\quad x^2 + y^2 - 2hx - 2ky + h^2 + k^2 - r^2 = 0, r \ge 0.$

This is referred to as the general form of the equation of a circle.

> The **equation of a circle** with centre $C(h, k)$ and radius r units is given by
> $$(x - h)^2 + (y - k)^2 = r^2.$$

If the centre of the circle is $(0, 0)$ and the radius is r, the equation is $x^2 + y^2 = r^2$.

EXAMPLE 3

Determine the centre and radius for each of the following circles.

a. $x^2 + y^2 = 10$

b. $x^2 + y^2 - 6x - 4y - 3 = 0$

SOLUTION

a. By comparing to the standard form, $(x - h)^2 + (y - k)^2 = r^2$, $h = 0$, $k = 0$, and $r^2 = 10$. The centre is $(0, 0)$ and the radius $\sqrt{10}$.

b. To find the centre and radius, we must complete the squares on the x- and y-terms to write the equation in standard form.

$$x^2 + y^2 - 6x + 4y - 3 = 0$$
$$x^2 - 6x + y^2 + 4y = 3$$

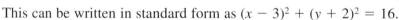

Completing the squares gives

$$x^2 - 6x + 9 + y^2 + 4y + 4 = 3 + 9 + 4.$$

This can be written in standard form as $(x - 3)^2 + (y + 2)^2 = 16$.

By comparing to the standard form, $h = 3$, $k = -2$, and $r^2 = 16$.
The centre is $(3, -2)$ and the radius is 4.

EXAMPLE 4

Determine the equation of the circle with diameter from $(1, 4)$ to $(5, -4)$.

SOLUTION

The centre is the midpoint of the diameter.
The coordinates of the centre are $\left(\dfrac{1 + 5}{2}, \dfrac{4 - 4}{2}\right)$ or $(3, 0)$.
The radius is the distance from the centre to either point; therefore,

$$r = \sqrt{(3 - 1)^2 + (0 - 4)^2}$$
$$= \sqrt{20}.$$

The required equation is $(x - 3)^2 + y^2 = 20$.

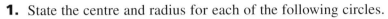

PART A

1. State the centre and radius for each of the following circles.

   a. $x^2 + y^2 = 100$ **b.** $x^2 + y^2 = 50$

   c. $x^2 + y^2 = \dfrac{1}{4}$ **d.** $(x + 3)^2 + y^2 = 81$

   e. $x^2 + (y - 4)^2 = 24$ **f.** $(x - 3)^2 + (y - 1)^2 = 9$

   g. $2x^2 + 2y^2 = 32$ **h.** $9x^2 + 9y^2 = 4$

   i. $2(x - 2)^2 + 2(y + 2)^2 = 4$ **j.** $5(x + 3)^2 + 5(y - 9)^2 = 100$

2. If the following points are on the circle given by $x^2 + y^2 = 50$, determine the values for the missing coordinates.

   a. $(-1, y)$ **b.** $(x, 5)$ **c.** $(7, y)$

   d. $(0, y)$ **e.** $(x, -4)$ **f.** $(-3, y)$

3. For the circle given by $x^2 + y^2 = 100$, determine whether each of the following points is on the circle, inside the circle, or outside the circle.

   a. $(7, -7)$ **b.** $(-6, 8)$ **c.** $(-5, -9)$

   d. $(4, 4\sqrt{5})$ **e.** $(2\sqrt{5}, 4\sqrt{5})$ **f.** $(-5, -2\sqrt{5})$

4. For the circle given by $x^2 + y^2 = 20$,

   a. show that the point $(2, 4)$ is on the circle;

   b. use symmetry to determine the coordinates of three other points on the circle.

5. Explain why each of the following is not the equation for a circle.

   a. $x^2 - y^2 = 16$ **b.** $x^2 + 4y^2 = 16$ **c.** $4x^2 + 4y^2 = -4$

6. Determine the equation for each of the following circles.

   a. centre $(0, 0)$, radius 7   **b.** centre $(0, 0)$, radius $\sqrt{3}$

   c. centre $(0, 0)$, radius $\dfrac{3}{4}$ **d.** centre $(1, 2)$, radius 5

   e. centre $(-4, 0)$, radius 4   **f.** centre $(2, -4)$, radius $2\sqrt{2}$

7. What is the domain and range for the circle defined by $(x - 3)^2 + (y - 4)^2 = 16$?

8. Determine the centre and radius for each of the following circles.

   a. $x^2 + y^2 + 6x - 27 = 0$ **b.** $x^2 + y^2 - 2x - 2y = 0$

   c. $x^2 + y^2 - 2x + 6y - 6 = 0$ **d.** $x^2 + y^2 = 8y$

   e. $x^2 + y^2 + 6x + 8y + 17 = 0$ **f.** $4x^2 + 4y^2 - 8x + 16y + 11 = 0$

9. Determine the equation in standard form for each of the following circles.

   a. centre $(0, 0)$, through $(-2, 3)$ **b.** centre $(0, 0)$, through $(\sqrt{2}, -\sqrt{2})$

   c. centre $(0, 0)$, x-intercepts at ± 8 **d.** centre $(3, 4)$, through $(0, 0)$

   e. centre $(-1, 3)$, through $(1, -1)$ **f.** centre $(-2, -2)$, y-intercept -2

10. Determine the equations of the circles with the given diameters.

   a. from $(-3, 5)$ to $(3, -5)$ **b.** from $(-1, 2)$ to $(5, 8)$

11. a. Show that the points $P(-2, 4)$ and $Q(2, -4)$ are both on the circle $x^2 + y^2 = 20$.

    b. Show that PQ is a diameter of the circle.

12. Determine the x- and y-intercepts for each of the following circles.

   a. $(x - 5)^2 + (y + 1)^2 = 50$ **b.** $(x + 4)^2 + (y - 3)^2 = 25$

   c. $x^2 + y^2 - 6x + 10y + 9 = 0$

13. A circle with its centre on the y-axis passes through $(-3, 0)$ and $(5, 4)$. Determine its centre and radius.

14. A circle passes through $(-2, 2)$, $(6, 2)$, and $(-2, -4)$.

   a. Determine the equation of the circle.

   b. Determine the coordinates of the highest and lowest points on the circle.

15. Describe the graph of $x^2 + y^2 = 0$. What would this indicate according to Appolonius?

16. For the circle given by $x^2 + y^2 = 34$,

   a. show that the line segment from $P(-5, 3)$ to $Q(3, 5)$ is a chord of the circle;

   b. find the midpoint M of the chord;

   c. show that $MO \perp PQ$.

17. Given the circle $x^2 + y^2 + 8x - 6y = 0$,

    a. show that the circle passes through $O(0, 0)$;

    b. if O is one end of a diameter of the circle, find the other end P;

    c. pick any other point Q on the circle and show that $\angle OQP = 90°$.

Refracting and Reflecting Telescopes

Two types of telescopes have been used for looking great distances—refracting and reflecting telescopes.

Lenses are used in refracting telescopes. The rays of light are concentrated at the focus of the lens. There is a loss in the intensity of the light ray and there is some change in the colour of the light. Thus, refracting telescopes are not very suitable for looking at the dim light from the stars.

The reflecting telescope uses both parabolic and hyperbolic mirrors. With a parabolic mirror, rays parallel to the axis of symmetry are reflected to the focus. With a hyperbolic mirror, rays from one focus are reflected to the other focus. Both of these principals are used in a reflecting telescope.

The incoming rays of light from space are parallel. They strike a parabolic mirror, which reflects the light to the focus of the parabola. Because of the design of the reflecting telescope, this is also the focus of the hyperbola. The rays are now reflected to the other focus of the hyperbola. The eyepiece is located at this focus.

Reflecting telescopes have many advantages over refracting telescopes. Mirrors are easier to make in large sizes than are lenses. Mirrors can concentrate the light of the dim images from deep space. Reflecting telescopes are currently used for space exploration.

Photos: Roger Ressmeyer/CORBIS

Refracting Telescope

Reflecting Telescope

We have seen the parabola as the graph of a quadratic function in our previous work. Parabolas occur in many applications such as mirrors on telescopes, satellite dishes, and suspension bridge cables.

A **parabola** is the set of points equidistant from a fixed line, the directrix, and a fixed point, the focus.

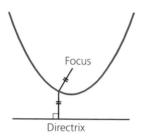
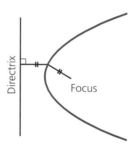

EXAMPLE 1

Determine the equation of a parabola whose focus is (2, 0), vertex (0, 0), and directrix the line $x = -2$.

SOLUTION

Let $P(x, y)$ be a point on the parabola.

$$PF = PD$$
$$\sqrt{(x - 2)^2 + (y - 0)^2} = x + 2$$

Square both sides of the equation.

$$(x - 2)^2 + (y - 0)^2 = (x + 2)^2$$
$$x^2 - 4x + 4 + y^2 = x^2 + 4x + 4$$
$$y^2 = 8x$$

The equation of the parabola is $y^2 = 8x$.

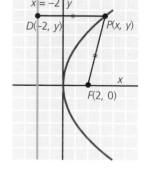

We can follow the same steps to arrive at a general equation for a parabola in **standard position;** that is, with its vertex at the origin and its axis of symmetry along the y-axis.

We first note that from the symmetry of the shape and the description of the locus, the vertex will be on the axis of symmetry halfway between the focus and the directrix. Therefore, we will let the focus be $F(0, c)$ and the directrix the line $y = -c$.

Let $P(x, y)$ be a point on the parabola.

$$PF = PD$$
$$\sqrt{x^2 + (y - c)^2} = y + c$$
$$x^2 + y^2 - 2cy + c^2 = y^2 + 2cy + c^2$$
$$x^2 = 4cy$$

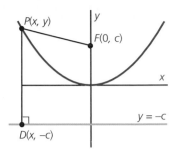

> The general **equation of a parabola** with the y-axis as the axis of symmetry and focus at $(0, c)$ is $x^2 = 4cy$.

Similarly:

> The general **equation of a parabola** with the x-axis as the axis of symmetry and focus at $(c, 0)$ is $y^2 = 4cx$.

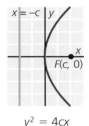

$y^2 = 4cx$

Focus $(c, 0)$

Directrix $x = -c$

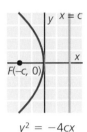

$y^2 = -4cx$

Focus $(-c, 0)$

Directrix $x = c$

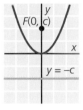

$x^2 = 4cy$

Focus $(0, c)$

Directrix $y = -c$

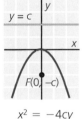

$x^2 = -4cy$

Focus $(0, -c)$

Directrix $y = c$

Note that the parabola opens up or to the right if $c > 0$, down or to the left if $c < 0$.

The distance c, from the vertex to the focus, is called the **focal length.**

If the vertex is not at the origin, we can apply the appropriate translation to arrive at the equation.

EXAMPLE 2

Determine the equation for the parabola with vertex at $(-1, 4)$ and focus at $(-1, 2)$.

SOLUTION

Since the vertex is at $(-1, 4)$ and the focus is $(-1, 2)$, $c = 2$.

In standard position the equation would be $x^2 = -4cy$ or $x^2 = -4(2)y$.

Thus, $x^2 = -8y$.

Since the vertex is at $(-1, 4)$, the equation becomes
$(x + 1)^2 = -8(y - 4)$.

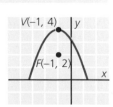

EXAMPLE 3

Determine the coordinates of the focus for the parabola $y^2 - 4y = 4x$, and sketch the graph.

SOLUTION

Complete the square on the left side.

$$y^2 - 4y = 4x$$
$$y^2 - 4y + 4 = 4x + 4$$
$$(y - 2)^2 = 4(x + 1)$$

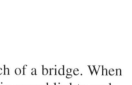

Since the equation is in the form $y^2 = 4cx$, the parabola opens to the right with vertex $(-1, 2)$ and $4c = 4$ or $c = 1$.

To find the focus, we must move 1 unit right from the vertex. The focus is $(0, 2)$.

The parabola occurs in many common structures such as the arch of a bridge. When rotated about its axis, we have the surface of a paraboloid used in searchlights and sound-receiving dishes. In these latter cases, the focal point is very important as the incoming waves all reflect to this point.

EXAMPLE 4

The arch of a bridge across a stream is parabolic. The bridge is 24 m wide and 9 m high at the centre of the arch. A horizontal support beam is placed 4 m below the vertex. What is the length of the beam?

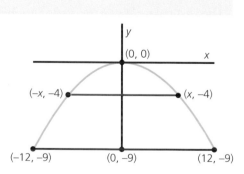

SOLUTION

Let the vertex of the parabolic arch be $(0, 0)$.

Let the equation of the parabola be $x^2 = -4cy$.

Since $(-12, -9)$ lies on the curve, then

$$(-12)^2 = -4c(-9)$$
$$144 = 36c$$
$$c = 4.$$

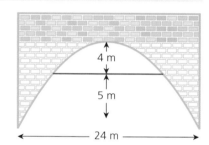

The equation is $x^2 = -16y$.

Since one point at the end of the beam is $(x, -4)$, then

$$x^2 = -16(-4)$$
$$x^2 = 64$$
$$x = \pm 8.$$

The length of the beam is 2×8 or 16 m.

Parabolic Mirrors

Giant Parabolic Mirror

A parabolic mirror is obtained by revolving a parabola around its axis of symmetry. Because of the law of reflection for light rays, a beam of incoming rays parallel to the axis will converge at the focus of the parabola.

The reflective property is used in reverse when rays emanating from the focus are reflected in a beam parallel to the axis. This reflection preserves the intensity of the light beam.

Parabolic mirrors are used in visual and radio telescopes, radar antennas, searchlights, automobile headlights, microphone systems, and solar heating devices.

In the early 1970s, energy conservation became very important. Using solar power instead of conventional sources was a major topic for articles in many publications. There were many plans published for "Solar Hot Dog Cookers." To build such a device, a large paraboloid was first built. A paraboloid is a three-dimensional object obtained by revolving a parabola about its axis of symmetry. Articles frequently showed pictures of someone with a piece of plywood cut in the shape of a parabola patiently moving the shape around a large pile of plaster. As the plaster hardened it took the shape of a paraboloid. A new plaster coating was made around this paraboloid. When the casting was removed, the entire inside was painstakingly tiled with tiny pieces of mirror. The hot dog was suspended at the focus and the solar cooker aimed at the sun. This was a lot of work just to cook a hot dog—but the solar energy used was free.

Photo: Adam Woolfitt/CORBIS

Exercise 11.2

PART A

1. State the focal length c for each of the following parabolas.

   a. $x^2 = 8y$ **b.** $y^2 = -4x$ **c.** $x^2 = -16y$

   d. $y^2 = -12x$ **e.** $x^2 = -y$ **f.** $y^2 = 5x$

   g. $x^2 = 5y$ **h.** $y^2 = -2x$

2. State the coordinates of the focus for each of the parabolas in Question 1.

3. State the equations for each of the parabolas drawn below.

a.

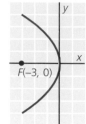

$F(-3, 0)$

b.

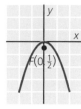

$F(0, \frac{1}{2})$

c.

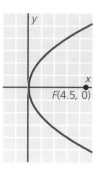

$F(4.5, 0)$

PART B

4. Determine the intercepts, the coordinates of the vertices, and then sketch the graph for each of the following:

a. $x^2 = 4(y + 1)$

b. $y^2 = -4(x - 4)$

c. $(y - 2)^2 = 8(x + 2)$

d. $(x - 2)^2 = -5(y - 5)$

5. Determine the coordinates of the vertex and focus for each of the following:

a. $(x - 3)^2 = 4y$

b. $(y + 2)^2 = 4x$

c. $(x + 1)^2 = -8y$

d. $(y + 1)^2 = -12x$

e. $x^2 - 6x - 4y - 3 = 0$

f. $y^2 - 4y + 8x + 52 = 0$

g. $y^2 + 8y - 2x + 10 = 0$

h. $x^2 - 6x + 4y + 21 = 0$

6. Determine the equation for each of the following parabolas.

a. vertex $(0, 0)$, focus $(-2, 0)$

b. vertex $(0, 0)$, focus $(0, 0.5)$

c. vertex $(4, 0)$, focus $(0, 0)$

d. vertex $(1, 2)$, focus $(1, 2.25)$

e. vertex $(-2, 0)$, focus $(-2, -1)$

f. vertex $(0, -3)$, focus $(0, 3)$

7. Determine the equation for each of the following parabolas.

a. vertex $(4, 0)$, y-intercepts ± 2

b. vertex $(-2, 4)$, x-intercepts 2 and -6

c. focus $(0, 2)$, directrix $y = -4$

d. focus $(4, 0)$, directrix $x = 1$

8. A parabolic arch is used to support a bridge. The arch is 80 m wide at the base and the height of the vertex is 20 m. How high is the arch at a point 20 m in from either end of the base?

9. A parabolic dish for picking up sound has a diameter of 1.5 m and is 20 cm deep.

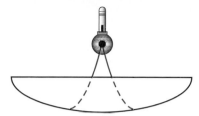

How high above the centre of the dish should the microphone be placed to pick up the sound that is reflected to the focus?

10. The cross-section of a parabolic reflector is as shown. The bulb is located at the focus, F.

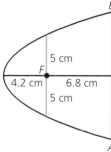

Find the diameter AB of the opening.

PART C

11. Find the equation for a parabola whose vertex is $(0, b)$ and whose directrix is the line $y = -b$.

12. A parabola has its vertex at $(1, 3)$ and its directrix is the line $y = x$. Where is its focus?

13. The **latus rectum** for a conic section is defined as the chord through the focus perpendicular to the axis.

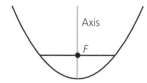

Find the length of the latus rectum for each of the following:

a. $x^2 = 4y$ **b.** $y^2 = 12x$ **c.** $x^2 = -2y$

Make a general conclusion.

11.3 The Ellipse

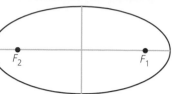

The path followed by a planet as it orbits about the sun is an ellipse with the sun at one focal point. An ellipse looks like a flattened circle with symmetry about two perpendicular axes. These are referred to as the **major axis** (the longer one) and the **minor axis.** It has two **focal points** or **foci** F_1 and F_2 that are equidistant from the centre on the major axis. The points where the curve crosses the major axis are the **vertices of the ellipse.**

The **ellipse** can be defined as a conic section whose eccentricity is less than 1; that is, using a focus-directrix definition in the same way we did for the parabola,

$$PF = ePD \text{ where } e < 1.$$

Varying the value of e changes the amount of the flattening or elongation so that the closer the value of e comes to 1, the flatter the ellipse becomes.

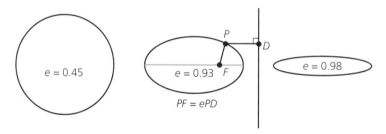

Ellipses and Ellipsoids

The Whispering Gallery is found within the Capitol building in Washington, DC.

The beautiful, simple shape of an ellipse makes it a natural shape for consideration in the design of buildings. If an ellipse is revolved about an axis, we get a three-dimensional object called an ellipsoid. Just like an ellipse, an ellipsoid has two foci.

For any point on an ellipse, the tangent line makes equal angles with the two focal radii; therefore, any sound wave emanating from one focus of an ellipse is reflected to the other focus. The same type of reflection occurs with an ellipsoid.

In the U.S. Senate there is a room referred to as the Whispering Gallery. The ceiling is shaped like half an ellipsoid. A word, whispered at one focus, can be heard at the second focus, which is 15 m away. Diplomats must be careful that their private conversation, quietly shared at one focus, is not overheard by a reporter standing at the other focus.

Architects must be aware of such phenomena when they design buildings. How sound is reflected from a surface of a room is of utmost importance in the design of concert halls and theatres.

Photo: George Lepp/CORBIS

In developing the equation of an ellipse, we will use the locus definition.

> An **ellipse** is the set of points such that the sum of the distances from each point to two fixed points, the foci, is constant.

The two distances from a point on the ellipse to its foci are the **focal radii.**

EXAMPLE 1

Determine the equation for an ellipse with foci at $(-4, 0)$ and $(4, 0)$, given that the sum of the focal radii is 10.

SOLUTION

If $P(x, y)$ is a point on the ellipse,

$$PF_1 + PF_2 = 10$$
$$\sqrt{(x + 4)^2 + y^2} + \sqrt{(x - 4)^2 + y^2} = 10$$
$$\sqrt{(x - 4)^2 + y^2} = 10 - \sqrt{(x + 4)^2 + y^2}.$$

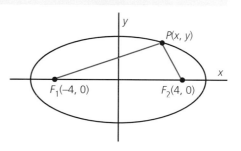

Squaring both sides,

$$(x^2 - 8x + 16) + y^2 = 100 - 20\sqrt{(x + 4)^2 + y^2} + (x^2 + 8x + 16) + y^2$$
$$5\sqrt{(x + 4)^2 + y^2} = 4x + 25.$$

Squaring again,

$$25[(x + 4)^2 + y^2] = (4x + 25)^2$$
$$25x^2 + 200x + 400 + 25y^2 = 16x^2 + 200x + 625$$
$$9x^2 + 25y^2 = 225.$$

This curve has x-intercepts ± 5 and y-intercepts ± 3. The major axis has length 10 and the minor axis has length 6. If we divide both sides of the equation by 225, we get the equation in a more convenient form,

$$\frac{x^2}{25} + \frac{y^2}{9} = 1 \text{ or } \frac{x^2}{5^2} + \frac{y^2}{3^2} = 1.$$

From this equation, we can read the intercepts and hence determine the lengths of the axes.

As well, the eccentricity is defined as the ratio of the focal length to the semi-major axis $e = \frac{c}{a}$.

Thus, the eccentricity for this ellipse is $e = \frac{4}{5}$ or 0.8.

An ellipse is in **standard position** if its axes lie along the x- and y-axes of the plane. We can use the method from above to develop the general equation for an ellipse in standard position.

Let the foci be $(c, 0)$ and $(-c, 0)$, and let the sum of the focal radii be $2a$, $a > c$.

If $P(x, y)$ is any point on the ellipse, then

$$PF_1 + PF_2 = 2a$$

$$\sqrt{(x + c)^2 + y^2} + \sqrt{(x - c)^2 + y^2} = 2a$$

$$\sqrt{(x + c)^2 + y^2} = 2a - \sqrt{(x - c)^2 + y^2}.$$

Squaring both sides,

$$(x^2 + 2cx + c^2) + y^2 = 4a^2 - 4a\sqrt{(x - c)^2 + y^2} + (x^2 - 2cx + c^2) + y^2$$

$$4a\sqrt{(x - c)^2 + y^2} = 4a^2 - 4cx$$

$$a\sqrt{(x - c)^2 + y^2} = a^2 - cx.$$

Squaring again,

$$a^2[(x - c)^2 + y^2] = (a^2 - cx)$$

$$a^2(x^2 - 2cx + c^2) + a^2y^2 = a^4 - 2a^2cx + c^2x^2$$

$$a^2x^2 - c^2x^2 + a^2y^2 = a^4 - a^2c^2.$$

Factoring

$$(a^2 - c^2)x^2 + a^2y^2 = a^2(a^2 - c^2).$$

For the x-intercepts, let $y = 0$.

Then $(a^2 - c^2)x^2 = a^2(a^2 - c^2)$

and $\qquad x^2 = a^2$

$$x = \pm a.$$

The x-intercepts are $\pm a$ and the length of the major axis is $2a$.

For the y-intercepts of the curve, we let $x = 0$.

Then $a^2y^2 = a^2(a^2 - c^2)$

$$y^2 = a^2 - c^2.$$

Since $a^2 > c^2$, we let $a^2 - c^2 = b^2$.

Then $\quad y^2 = b^2$

and $\quad y = \pm b.$

The y-intercepts are $\pm b$ and the length of the minor axis is $2b$.

But the y-intercepts are $\pm b$.

Then $b^2 = a^2 - c^2$ or $a^2 = b^2 + c^2$.

Our equation becomes

$$b^2x^2 + a^2y^2 = a^2b^2$$

or $\qquad \dfrac{x^2}{a^2} + \dfrac{y^2}{b^2} = 1.$

> The **equation of the ellipse** centred at the origin,
> with foci $(c, 0)$ and $(-c, 0)$ and major axis of length $2a$, is
>
> $$\dfrac{x^2}{a^2} + \dfrac{y^2}{b^2} = 1$$
>
> where $a^2 = b^2 + c^2$ and the minor axis has length $2b$.

Similarly:

> The **equation of the ellipse** centred at the origin,
> with foci $(0, c)$ and $(0, -c)$ and major axis of length $2a$, is
>
> $$\dfrac{x^2}{b^2} + \dfrac{y^2}{a^2} = 1$$
>
> where $a^2 = b^2 + c^2$ and the minor axis has length $2b$.

Note that we can tell the orientation of the major and minor axes from the relative sizes of the denominators.

EXAMPLE 2

For the ellipse given by $\dfrac{x^2}{9} + \dfrac{y^2}{16} = 1$, state

a. the intercepts,

b. the lengths of the axes,

c. the foci,

and sketch the graph of the ellipse.

SOLUTION

a. The x-intercepts are ± 3; the y-intercepts are ± 4.

b. The length of the major axis is 8 and the length of the minor axis is 6.

c. Since $a^2 = 16$, $b^2 = 9$, and $a^2 = b^2 + c^2$,

$$c^2 = 16 - 9$$
$$c^2 = \sqrt{7}$$
$$c = \pm\sqrt{7}.$$

The foci are $(0, \sqrt{7})$ and $(0, -\sqrt{7})$.

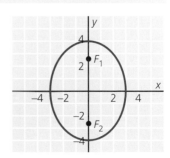

EXAMPLE 3

Determine the equation of an ellipse with foci (2, 0) and (−2, 0) and major axis of length 8.

SOLUTION

Since the foci are on the x-axis, the equation is in the form $\frac{x^2}{a^2} + \frac{y^2}{b^2} = 1$.

Since $2a = 8$, $a = 4$.

Since the foci are (2, 0) and (−2, 0), $c = 2$.

Then $b^2 = a^2 - c^2$

$\qquad = 16 - 4$

$b^2 = 12$.

The required equation is $\frac{x^2}{16} + \frac{y^2}{12} = 1$.

EXAMPLE 4

Determine the equation of the ellipse having foci $F_1(1, 3)$ and $F_2(1, 7)$ and major axis of length 10. Sketch its graph.

SOLUTION

The centre is the midpoint of F_1F_2.

The coordinates of the centre are $\left(\frac{1 + 1}{2}, \frac{3 + 7}{2}\right)$ or (1, 5).

Since c is the distance from the centre to F_1, then $c = 2$.

Since the major axis has length 10 units, and is parallel to the y-axis, the equation is in the form $\frac{x^2}{b^2} + \frac{y^2}{a^2} = 1$.

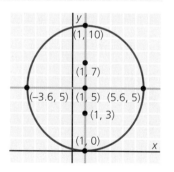

$\qquad 2a = 10$

$\qquad a = 5$

$\qquad a^2 = b^2 + c^2$

or $\qquad b^2 = a^2 - c^2$

$\qquad\qquad = 25 - 4$

$\qquad\qquad = 21$.

Since the centre is at (1, 5), the equation of the ellipse is

$$\frac{(x - 1)^2}{21} + \frac{(y - 5)^2}{25} = 1.$$

To sketch the graph, we determine the ends of the axes.

Since $a = 5$ and the centre of the ellipse is $(1, 5)$, the ends of the major axis are $(1, 10)$ and $(1, 0)$.

Since $b = \sqrt{21}$, the ends of the minor axis are $(1 - \sqrt{21}, 5)$ and $(1 + \sqrt{21}, 5)$ and are approximately $(-3.6, 5)$ and $(5.6, 5)$.

Exercise 11.3

PART A

1. State the lengths of the major and minor axes for each of the following ellipses.

   a. $\dfrac{x^2}{64} + \dfrac{y^2}{16} = 1$ **b.** $\dfrac{x^2}{16} + \dfrac{y^2}{25} = 1$ **c.** $\dfrac{x^2}{3} + \dfrac{y^2}{2} = 1$

2. State the coordinates of the vertices for each of the ellipses of Question 1.

3. State the coordinates of the foci for each of the ellipses of Question 1.

PART B

4. Determine the coordinates of the vertices and foci for each of the following ellipses.

   a. $x^2 + 9y^2 = 36$

   b. $25x^2 + 16y^2 = 400$

   c. $2x^2 + 3y^2 = 12$

5. Determine the vertices and foci for each of the following ellipses.

   a. $\dfrac{(x - 2)^2}{25} + \dfrac{y^2}{24} = 1$

   b. $\dfrac{x^2}{12} + \dfrac{(y - 4)^2}{16} = 1$

   c. $\dfrac{(x + 1)^2}{100} + \dfrac{(y - 3)^2}{36} = 1$

6. Sketch graphs of each of the following ellipses.

   a. $4x^2 + y^2 = 36$

   b. $x^2 + 9y^2 = 9$

   c. $9(x + 2)^2 + 4(y + 3)^2 = 36$

   d. $(x - 3)^2 + 4(y + 5)^2 = 16$

7. Determine the equation for each of the following ellipses.

   a. vertices $(6, 0)$ and $(-6, 0)$, foci $(4, 0)$ and $(-4, 0)$

   b. foci $(0, 3)$ and $(0, -3)$, minor axis length 8

   c. vertices $(0, 8)$ and $(0, -8)$, ends of minor axis $(4, 0)$ and $(-4, 0)$

   d. foci $(2, 0)$ and $(-2, 0)$, major axis length 16

   e. centre $(0, 0)$, major axis 10 units long, foci on the x-axis, $c = \dfrac{1}{2}$

8. In each of the following, a set of properties is given that defines an ellipse. Determine the equation of the ellipse and sketch the graph.

   a. vertices $(\pm 20, 0)$, passes through $(10, 10)$

   b. vertices $(2, 0)$ and $(2, 8)$, ends of minor axis $(0, 4)$ and $(4, 4)$

   c. vertices $(0, 0)$ and $(10, 0)$, foci $(1, 0)$ and $(9, 0)$

   d. foci $(-5, 2)$ and $(3, 2)$ and minor axis length 12

   e. major axis length 16, centre $(2, -1)$, and one focus $(2, 5)$

9. A bubble over a set of tennis courts is semi-elliptical in cross-section, 20 m wide at the ground, and 8 m high in the centre. How high is the bubble at a point 2 m in from the outer edge?

10. Halley's comet appears to move around the sun, one focus of its elliptical orbit, every 76 years. The major axis of this orbit is 5370 million km and the eccentricity of the orbit is 0.97, where eccentricity $e = \dfrac{c}{a}$.

   a. Find an equation for the orbit.

   b. Find the minimum distance between the comet and the sun (the perigee of the orbit).

11. A spacecraft is in a circular orbit 800 km above Earth. To transfer the craft to a lower circular orbit 150 km above Earth, the spacecraft must be placed in an elliptical orbit as shown with the centre of Earth at one focus.

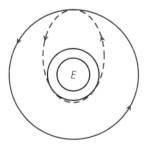

Find an equation of the transfer orbit if the radius of Earth is 6336 km.

12. Given the equation $4x^2 + 9y^2 - 16x + 54y + 61 = 0$, do the following:

  a. Show that the equation represents an ellipse by writing it in the form
  $$\frac{(x - h)^2}{a^2} + \frac{(y - k)^2}{b^2} = 1.$$

  b. Locate the centre, vertices, and focal points for the ellipse.

  c. Sketch the graph of the ellipse.

13. A scale model of the elliptical orbit of a satellite about a planet is built on a scale where 1 cm = 10 km. If the equation for the model orbit is $5x^2 + 9y^2 = 4500$, what are the least and greatest distances between the satellite and the planet (assuming the planet is at one focus of the orbit)?

14. a. Find the length of the latus rectum for $9x^2 + 25y^2 = 225$.

  b. How can you use this information to help you sketch the ellipse?

Conic Sections and Comets

Astronomers are very familiar with conic sections. The planets and comets travel in elliptical paths with the sun at one focus.

Before it was realized that comets followed an elliptical path around the sun, comets were regarded as harbingers of evil. They were thought to predict wars, plagues, or earthquakes. The large comet that appeared before the battle of Hastings in 1066 was interpreted by King Harold as an indication of his forthcoming defeat by William the Conqueror.

Edmund Halley (1656–1742), seeing a great comet in 1682, studied the records of earlier comets. He concluded that the comet he had seen had also been sighted in 1607, 1531, 1456, and in 1066. He predicted the comet would return in 1759, 1835, and 1910. Because of these correct predictions, this comet is now called Halley's comet.

The period of Halley's comet is about 76 years. However, you cannot just add 76 years to the date of the last sighting to predict when the comet will appear again. The gravitational pull of the major planets alters the orbital period from revolution to revolution. Halley's comet was most recently seen in 1986. It will return again in 2061.

Internet search engines will provide you with many sites where you can learn more about Halley's comet.

Photo: NASA

Halley's comet

11.4 The Hyperbola

The hyperbola, used extensively in navigation, is defined in much the same way as the ellipse. It has two branches that are symmetric relative to two perpendicular axes. The **vertices** are the end points of the **transverse axis** and the **foci** lie on the extensions of this transverse axis. The **conjugate axis,** although not touching the curve, helps to define the **asymptotes.** The hyperbola is bounded by a pair of asymptotes that are the diagonals of the rectangle formed about the centre with its sides at the ends of the axes.

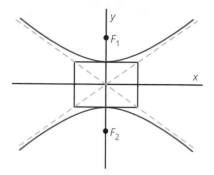

The hyperbola is a conic section with eccentricity greater than 1, so that its focus-directrix equation could be written $PF = ePD$ where $e > 1$. In developing the equation of a hyperbola, we will use the focus definition.

> A **hyperbola** is the set of points such that the absolute value of the difference between the distances from any point on the curve to two fixed points, the foci, is constant. The constant is the length of the transverse axis.

EXAMPLE 1

Find the equation for the hyperbola with foci at $F_1(-5, 0)$ and $F_2(5, 0)$ and where the constant difference between the focal radii is 8.

SOLUTION

If $P(x, y)$ is a point on the curve,
$$PF_1 - PF_2 = 8.$$

Note that due to the symmetry, the order of subtraction does not matter.

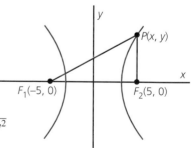

$$\sqrt{(x + 5)^2 + y^2} - \sqrt{(x - 5)^2 + y^2} = \pm 8$$
$$\sqrt{(x + 5)^2 + y^2} = \pm 8 + \sqrt{(x - 5)^2 + y^2}$$

Squaring both sides and simplifying,

$$(x^2 + 10x + 25) + y^2 = 64 \pm 16\sqrt{(x-5)^2 + y^2} + (x^2 - 10x + 25) + y^2$$

$$20x - 64 = \pm 16\sqrt{(x-5)^2 + y^2}$$

$$5x - 16 = \pm 4\sqrt{(x-5)^2 + y^2}$$

Squaring again,

$$(5x - 16)^2 = 16[(x-5)^2 + y^2]$$

$$25x^2 - 160x + 256 = 16(x^2 - 10x + 25 + y^2)$$

$$= 16x^2 - 160x + 400 + 16y^2$$

$$9x^2 - 16y^2 = 144.$$

Dividing both sides by 144,

$$\frac{x^2}{16} - \frac{y^2}{9} = 1.$$

From this equation, we can see that the x-intercepts are ± 4, and there are no y-intercepts, since $\frac{y^2}{9} = -1$ has no solution.

If we isolate the y in the equation,

$$y^2 = \frac{9(x^2 - 16)}{16}$$

$$y = \frac{\pm 3\sqrt{x^2 - 16}}{4}.$$

Since there are real values for y only if $x^2 - 16 \geq 0$, there are no values for y if $-4 < x < 4$, and hence no points on the curve having x-values for $-4 < x < 4$.

The diagonals of the rectangle formed by the lines $x = 4$, $x = -4$, $y = 3$, and $y = -3$ are the asymptotes of the hyperbola having equations $y = \frac{3}{4}x$ and $y = -\frac{3}{4}x$.

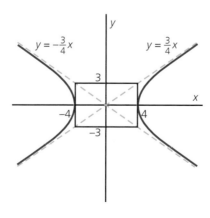

As we did for the ellipse, we can follow this same method to develop the general equation for a hyperbola in standard position. For the hyperbola with foci $(c, 0)$ and $(-c, 0)$ and transverse axis $2a$, $PF_1 - PF_2 = \pm 2a$. (The sign depends on which branch P is located.)

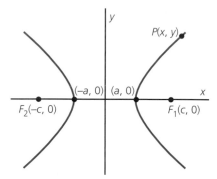

For any point $P(x, y)$ on the hyperbola we write

$$PF_1 - PF_2 = \pm 2a$$
$$\sqrt{(x - c)^2 + y^2} - \sqrt{(x + c)^2 + y^2} = \pm 2a$$
$$\sqrt{(x - c)^2 + y^2} = \pm 2a + \sqrt{(x + c)^2 + y^2}$$

Square both sides.

$$x^2 - 2cx + c^2 + y^2 = 4a^2 \pm 4a\sqrt{(x + c)^2 + y^2} + x^2 + 2cx + c^2 + y^2$$
$$-4cx - 4a^2 = \pm 4a\sqrt{(x + c)^2 + y^2}$$
$$-(cx + a^2) = \pm a\sqrt{(x + c)^2 + y^2}$$
$$c^2x^2 + 2a^2cx + a^4 = a^2x^2 + 2a^2cx + a^2c^2 + a^2y^2$$
$$(c^2 - a^2)x^2 - a^2y^2 = a^2(c^2 - a^2).$$

For the x-intercepts, let $y = 0$.

Then $(c^2 - a^2)x^2 = a^2(c^2 - a^2)$

and $\qquad\qquad x^2 = a^2$

$$x = \pm a.$$

The x-intercepts are $\pm a$ and the length of the transverse axis is $2a$.

By setting $x = 0$, the y-intercepts are obtained from

$$-a^2y^2 = a^2(c^2 - a^2)$$
$$y^2 = -(c^2 - a^2).$$

Since $c > a$, there are no real y-intercepts. Allowing for consistency with the equation of the ellipse, we designate an imaginary conjugate of length $2b$ and define

$$b^2 = c^2 - a^2$$

or $\qquad c^2 = a^2 + b^2$.

Then the equation of the hyperbola is

$$b^2x^2 - a^2y^2 = a^2b^2$$

or

$$\frac{x^2}{a^2} - \frac{y^2}{b^2} = 1.$$

This equation can be written

$$\frac{y^2}{b^2} = \frac{x^2}{a^2} - 1$$

$$y^2 = \frac{b^2}{a^2}x^2 - b^2.$$

As x becomes very large, y^2 is very close to $\frac{b^2}{a^2}x^2$. For example, if $a = 5$, $b = 3$, $y^2 = \frac{9}{25}x^2 - 9$, and if $x = 100$, $y = \pm\sqrt{3591} \doteq \pm59.92$ if we use the complete expression, while $y = \pm\sqrt{3600} = \pm60$ if we use only the first term in the right side.

Although there is no y-value exactly equal to $\pm\frac{b}{a}x$, we see that $y = \pm\frac{b}{a}x$ is a very close approximation. This tells us that the graph of the curve approaches increasingly closer to the straight lines $y = \frac{b}{a}x$ and $y = -\frac{b}{a}x$, and we call these lines the asymptotes of the hyperbola. They are valuable in sketching the graph, as shown in the diagram.

> The **equation of the hyperbola** centred at the origin, with foci $(c, 0)$ and $(-c, 0)$ and transverse axis $2a$ on the x-axis, is
>
> $$\frac{x^2}{a^2} - \frac{y^2}{b^2} = 1$$
>
> where $c^2 = a^2 + b^2$.
> Its asymptotes are $y = \frac{b}{a}x$ and $y = -\frac{b}{a}x$.
> Similarly, the **equation of the hyperbola** centred at the origin, with foci $(0, c)$ and $(0, -c)$ and transverse axis on the y-axis, is
>
> $$\frac{x^2}{b^2} - \frac{y^2}{a^2} = -1$$
>
> where $c^2 = a^2 + b^2$.
> Its asymptotes are $y = \frac{a}{b}x$ and $y = -\frac{a}{b}x$.

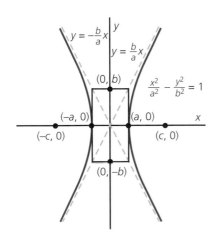

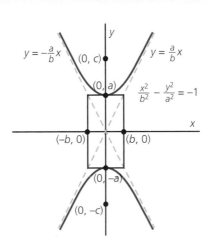

EXAMPLE 2

For the hyperbola $\frac{x^2}{16} - \frac{y^2}{25} = 1$, state

a. the coordinates of the vertices,

b. the coordinates of the foci,

c. the length of the transverse axis,

d. the equations of the asymptotes. Sketch the graph of the hyperbola.

SOLUTION

Since the equation is $\frac{x^2}{16} - \frac{y^2}{25} = 1$, for this hyperbola,

$$a^2 = 16, \ b^2 = 25.$$

For a hyperbola, $c^2 = a^2 + b^2$,

then $\quad c^2 = 16 + 25$

$$= 41.$$

a. The vertices are $(4, 0)$ and $(-4, 0)$.

b. The foci are $(\sqrt{41}, 0)$ and $(-\sqrt{41}, 0)$.

c. The length of the transverse axis is 8.

d. The asymptotes are $y = \frac{5}{4}x$ and $y = -\frac{5}{4}x$.

To sketch the graph, use the vertices and the asymptotes, recalling that the graph approaches, but never touches, them.

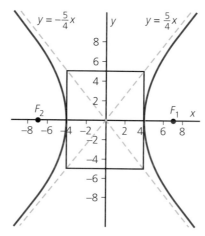

EXAMPLE 3

Determine the equation of the hyperbola having foci $(-1, 6)$ and $(-1, -2)$ and transverse axis of length 4. Sketch its graph.

SOLUTION

From the given foci, $2c = 8$ so $c = 4$.

The length of the transverse axis is $2a$ so $2a = 4$ and $a = 2$. The transverse axis lies on the vertical line connecting the foci.

From $c^2 = a^2 + b^2$, we obtain $b^2 = 16 - 4 = 12$.

Since the foci are $(-1, 6)$ and $(-1, -2)$, the centre of the hyperbola is the midpoint or $\left(\dfrac{-1 + (-1)}{2}, \dfrac{6 + (-2)}{2}\right)$ or $(-1, 2)$.

If the hyperbola was centred at the origin, its equation would be $\dfrac{x^2}{b^2} - \dfrac{y^2}{a^2} = -1$ or $\dfrac{x^2}{12} - \dfrac{y^2}{4} = -1$.

The vertices of the graph in standard position would be $(0, 2)$ and $(0, -2)$, but since the centre is at $(-1, 2)$ rather than $(0, 0)$, the vertices are $(-1, 4)$ and $(-1, 0)$.

Then the equation of the hyperbola is $\dfrac{(x + 1)^2}{12} - \dfrac{(y - 2)^2}{4} = -1$.

Its asymptotes in standard position would be $y = \dfrac{2}{\sqrt{12}}x$ and $y = -\dfrac{2}{\sqrt{12}}x$.

Since $\dfrac{2}{\sqrt{12}} = \dfrac{2}{2\sqrt{3}} = \dfrac{1}{\sqrt{3}}$, these simplify to $y = \dfrac{1}{\sqrt{3}}x$ and $y = -\dfrac{1}{\sqrt{3}}x$.

For the translated hyperbola, the asymptotes are $y - 2 = \dfrac{1}{\sqrt{3}}(x + 1)$ and $y - 2 = -\dfrac{1}{\sqrt{3}}(x + 1)$.

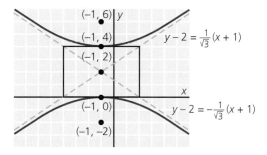

Two "Special" Cases

If $a = b$, then $x^2 - y^2 = k$. This hyperbola is called a rectangular hyperbola.

If a rectangular hyperbola is rotated so the x- and y-axes are the asymptotes, the equation takes the form $xy = k$, $k \neq 0$.

PART A

1. State the length of the transverse axis for each of the following hyperbolas.

a. $\dfrac{x^2}{36} - \dfrac{y^2}{64} = 1$

b. $\dfrac{x^2}{9} - \dfrac{y^2}{16} = -1$

c. $\dfrac{x^2}{2} - \dfrac{y^2}{2} = 1$

d. $\dfrac{x^2}{8} - y^2 = -1$

2. State the foci for each of the hyperbolas of Question 1.

3. State the equations for the asymptotes for each hyperbola of Question 1.

PART B

4. Determine the foci and vertices for each of the following hyperbolas.

a. $5x^2 - 4y^2 = -20$

b. $x^2 - 2y^2 = 4$

c. $20x^2 - 16y^2 = 320$

d. $49x^2 - y^2 = -49$

5. Determine the vertices and foci for each of the following hyperbolas.

a. $\dfrac{(x-2)^2}{4} - \dfrac{y^2}{12} = 1$

b. $\dfrac{x^2}{7} - \dfrac{(y+4)^2}{9} = -1$

c. $\dfrac{(x+2)^2}{25} - \dfrac{(y-3)^2}{75} = 1$

6. Sketch graphs for the following hyperbola showing vertices and asymptotes.

a. $4x^2 - y^2 = 36$

b. $50y^2 - 8x^2 = 200$

c. $9(x-2)^2 - 4y^2 = 36$

7. In each of the following, a set of properties is given that defines a hyperbola. Determine the equation of the hyperbola.

a. vertices $(5, 0)$ and $(-5, 0)$, foci $(6, 0)$ and $(-6, 0)$

b. foci $(0, 8)$ and $(0, -8)$, transverse axis of length 10

c. vertices $(3, 0)$ and $(-3, 0)$, asymptotes $y = 2x$ and $y = -2x$

d. foci $(0, 5)$ and $(0, -5)$, length of conjugate axis 2

8. Hyperbolas such as $9x^2 - 4y^2 = 36$ and $9x^2 - 4y^2 = -36$ are known as **conjugate hyperbolas.**

   a. Sketch these hyperbolas on the same axes.

   b. What is true about their asymptotes?

9. Determine the equation for each of the following hyperbolas and sketch the graphs.

   a. vertices (2, 1) and (6, 1), foci (0, 1) and (8, 1)

   b. foci (3, 3) and (3, −3), $a = 2$

10. Sketch graphs for each of the following:

   a. $x^2 - y^2 = 16$

   b. $y^2 - x^2 = 16$

   What observations can you make about the asymptotes of a rectangular hyperbola?

11. Two LORAN (LOng RAnge Navigation) stations at A and B simultaneously send electronic signals to aircraft. The on-board computer can guide the aircraft by converting the time difference in receiving the signals into a distance difference $PA - PB$. This locates the aircraft on one branch of a hyperbola.

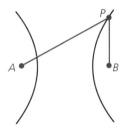

   a. Determine an equation of a hyperbola described by the flight path of an aircraft 20 km closer to station B than to station A, if the stations are 100 km apart.

   b. Determine an equation of the hyperbola described by a position 80 km from B and 140 km from A, if the stations are 100 km apart.

   c. Determine an equation of the hyperbola described by a position 120 km from B and 100 km from A, if the stations are 80 km apart.

PART C

12. Write the equation $7x^2 - 2y^2 - 14x + 12y = 101$ in the form $\dfrac{(x - h)^2}{a^2} - \dfrac{(y - k)^2}{b^2} = 1$ and sketch the graph indicating centre, vertices, and asymptotes.

13. a. Sketch the graph of $xy = 4$.

   b. State the equations of the axes for this curve.

   c. Find the vertices and foci for this curve.

14. Sketch graphs for each of the following:

   a. $x^2 - y^2 = 0$

   b. $xy = 0$

What would this indicate according to Appolonius?

The LORAN Navigation System

LORAN Station

Ships need to know their exact position when they are entering a harbour, approaching a rocky shore, or calling for assistance from the Coast Guard.

The LORAN Navigation System, which covers the coastal waters of North America, the Great Lakes, and the Bering Sea, permits ships to identify their position within 50 m of their exact location. Aircraft also use this system. There are LORAN stations in the United States, Canada, and Russia that cooperate in providing electronic coverage.

LORAN stations are situated along the coasts. Electronic signals are sent out from the stations. The ship or aircraft that wants to know its location locks onto two stations. As the signals come from different stations, there is a time difference that is converted into a distance difference. The ship or aircraft carries a map showing systems of hyperbolic curves. Knowing the difference in the distance from the two stations, the ship can identify the hyperbolic curve on which it is located. The ship then locks onto two different stations and repeats the process. The intersection of the two hyperbolic curves on the map gives the current position of the ship.

You can learn more about the LORAN Navigation System by visiting the web site of the United States Coast Guard.

Photo: United States Coast Guard

11.5 The General Quadratic Equation

We have seen that the conic sections are all represented by quadratic equations. In this section, we address the question of identifying the locus if given a quadratic equation of the form $ax^2 + by^2 + 2gx + 2fy + c = 0$.

EXAMPLE 1

Determine the conic represented by the equation $x^2 - 4x + y^2 + 8y - 5 = 0$.

SOLUTION

Consider $x^2 - 4x + y^2 + 8y = 5$.

Complete the squares on x and y.

$$(x^2 - 4x + 4) + (y^2 + 8y + 16) = 5 + 4 + 16$$
$$(x - 2)^2 + (y + 4)^2 = 25$$

This equation represents a circle with centre $(2, -4)$ and radius 5.

Any quadratic equation in two variables can always be treated in the same way. By expressing the equation as the combination of two squares, we can identify the conic and its properties. Do we always get a conic section from a quadratic equation in two variables? The answer is no, and in the exercise you will be asked to examine this.

EXAMPLE 2

Sketch the graph of $x^2 - 6x - 8y - 23 = 0$.

SOLUTION

We complete the square only on x.

$$x^2 - 6x - 8y = 23$$
$$(x^2 - 6x + 9) - 8y = 23 + 9$$
$$(x - 3)^2 - 8y = 32$$
$$(x - 3)^2 = 8y + 32$$
$$(x - 3)^2 = 8(y + 4)$$

This is the equation of a parabola. Comparing it with $x^2 = 4cy$, we see that the vertex is $(3, -4)$ and that $c = 2$. Its intercepts are obtained from

$$\frac{1}{8}(x - 3)^2 - 4 = 0$$
$$(x - 3)^2 = 32$$
$$x - 3 = \pm 4\sqrt{2}$$
$$x = 3 \pm 4\sqrt{2} \doteq 8.7 \text{ and } -2.7.$$

The parabola opens upward.

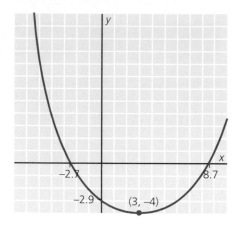

EXAMPLE 3

Determine the type of conic represented by $x^2 - 4y^2 - 4x - 8y = 0$.

State its centre, the length of its axes, and its foci. What translation has the conic undergone from standard position?

SOLUTION

Completing the square,

$$x^2 - 4y^2 - 4x - 8y - 16 = 0$$
$$(x^2 - 4x) - 4(y^2 + 2y) = 16$$
$$(x^2 - 4x + 4) - 4(y^2 + 2y + 1) = 16 + 4 - 4$$
$$(x - 2)^2 - 4(y + 1)^2 = 16$$
$$\frac{(x - 2)^2}{16} - \frac{(y + 1)^2}{4} = 1.$$

This is the equation of a hyperbola. Its centre is $(2, -1)$.

The translation from standard position is 2 units to the right and 1 unit down or a $(2, -1)$ translation.

Since the equation is in the form $\frac{x^2}{a^2} - \frac{y^2}{b^2} = 1$, $a = 4$ and the length of the transverse axis is $2a = 8$. The conjugate axis is $2b$, where $b = 2$, so its length is 4.

$$c^2 = a^2 + b^2$$
$$= 4^2 + 2^2$$
$$= 20$$
$$c = 2\sqrt{5} \text{ (Remember that } c > 0.)$$

The foci are $2\sqrt{5} \doteq 4.5$ units to the right and to the left of the centre, so the foci are $(6.5, -1)$ and $(-2.5, -1)$.

PART A

1. For $ax^2 + by^2 + 2gx + 2fy + c = 0$, state the conditions on a and b for this equation to represent the following:

   a. a circle   **b.** a parabola   **c.** an ellipse   **d.** a hyperbola

PART B

2. Determine the centre and radius of each circle.

   a. $x^2 + y^2 - 2x + 4y - 4 = 0$ **b.** $4x^2 + 4y^2 - 16x + 32y + 55 = 0$

   c. $9x^2 + 9y^2 + 36y + 31 = 0$ **d.** $(x + 3)^2 + (y - 4)^2 = 12$

   e. $x^2 + y^2 + 4x + 4y = 0$ **f.** $x^2 + y^2 - 8x - 6y + 21 = 0$

3. Identify the conic represented by each of the following. For each, state the translation it has undergone from standard position and identify the values of a, b, and c, if they exist.

   a. $4x^2 - y^2 + 16x - 4y + 16 = 0$ **b.** $x^2 + 9y^2 - 6x - 54y + 81 = 0$

   c. $4x^2 + 9y^2 - 8x + 54y + 49 = 0$ **d.** $4x^2 - y^2 + 24x - 6y + 31 = 0$

   e. $x^2 - 10x - 8y + 1 = 0$ **f.** $5x^2 + 20x + 2y + 22 = 0$

   g. $y^2 - 8x + 6y + 17 = 0$ **h.** $4x^2 - y^2 - 8x - 4y + 16 = 0$

4. Sketch the graph represented by each of the following equations, labelling the centre, where applicable, and the vertices.

   a. $x^2 - 4x - 2y - 6 = 0$ **b.** $25x^2 + 4y^2 - 250x + 24y + 561 = 0$

   c. $x^2 + y^2 - 10x + 6y + 34 = 0$ **d.** $16x^2 + 16y^2 + 16x - 24y - 3 = 0$

   e. $3y^2 - 4x + 18y + 31 = 0$ **f.** $4x^2 + y^2 + 40x - 4y + 100 = 0$

5. a. For the hyperbola with equation $x^2 - y^2 = 100$, state the asymptotes and vertices.

   b. Repeat part **a** for the hyperbola with equation $x^2 - y^2 = k$,

      where $k = 25, 9, 1, \frac{1}{4}, \frac{1}{25}, \frac{1}{100}, 0$.

   c. What happens if $k = 0$?   **d.** What happens if $k = -\frac{1}{100}, -\frac{1}{4}, -1, -25$?

6. Describe the graph of each of the following:

   a. $x^2 - 4y^2 = 0$ **b.** $(x - 3)^2 - y^2 = 0$

   c. $4(x - 2)^2 - (y - 1)^2 = 0$ **d.** $x^2 + y^2 = 0$

   e. $(x - 1)^2 + 9(y + 3)^2 = 0$ **f.** $16x^2 - 25(y + 1)^2 = 0$

11.6   Linear-Quadratic Systems

In this section we consider the intersection of two loci, where one is represented by a linear equation and the other by a quadratic. In other words, find the point(s) where a given straight line crosses the graph of one of the conic sections. There are three possibilities: the line may not cross the curve; it may touch at one point only (a tangent); or it may cross at two distinct points (a secant).

In all cases the method is the same.

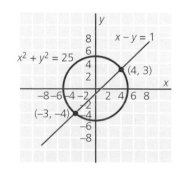

Step 1: Solve the linear equation for one of its variables.

Step 2: Substitute for this result for the variable in the quadratic equation.

Step 3: Solve the resulting quadratic equation.

Step 4: Substitute back into the linear equation to find the other coordinate.

EXAMPLE 1

Find the points of intersection of the line $x - y = 1$ and the circle $x^2 + y^2 = 25$. Illustrate with a graph.

SOLUTION

From the linear equation $y = x - 1$.

Substituting into the circle equation,
$$x^2 + (x - 1)^2 = 25.$$

Simplifying,
$$2x^2 - 2x + 1 = 25$$
$$x^2 - x - 12 = 0$$
$$(x - 4)(x + 3) = 0$$
$$x = 4 \text{ or } x = -3.$$

Substituting in $y = x - 1$,

if $x = 4$ then $y = 3$;

if $x = -3$ then $y = -4$.

The points of intersection are $(4, 3)$ and $(-3, -4)$.

Exercise 11.6

PART B

1. Solve each of the following systems and illustrate with a graph.

a. $y = x + 1$, $y = x^2 - 2x - 3$ **b.** $y = 5x - 5$, $y = x^2 + 4x - 5$

c. $x - 2y + 3 = 0$, $y^2 = 4x$ **d.** $x + 2y = 4$, $y^2 = -4x$

e. $x + y = 5$, $x^2 + y^2 = 25$ **f.** $3x - 4y = 25$, $x^2 + y^2 = 25$

g. $2x + 3y = 6$, $4x^2 + 9y^2 = 36$ **h.** $x - y - 2 = 0$, $x^2 - y^2 = 8$

i. $2x = 3y$, $4x^2 + y^2 = 100$ **j.** $x - y = 3$, $xy = 4$

2. a. Determine the value(s) of k making $y = 4x + k$ tangent to the circle $x^2 + y^2 = 25$.
(A tangent is a line that touches a circle or a conic at exactly one point.)

    b. Determine the value(s) of k such that $y = 4x + k$ intersects the circle $x^2 + y^2 = 25$ at two points.

    c. Discuss with your class how these results tie in with the character of the roots of a quadratic equation.

PART C

3. Determine the equation of the tangent to the circle $x^2 + y^2 = 25$ at the point $(4, 3)$.

4. Determine the equation of the tangent to the circle $x^2 + y^2 = r^2$ at the point (x_1, y_1).

1. Determine the centre and radius of each of the following:

   a. $x^2 + y^2 = 144$ **b.** $4x^2 + 4y^2 = 9$

   c. $(x - 3)^2 + (y - 2)^2 = 13$ **d.** $x^2 + (y + 3)^2 = 9$

2. Determine the x- and y-intercepts for each circle of Question 1.

3. Determine the centre, radius, and intercepts for $x^2 + y^2 + 8x - 6y = 0$.

4. Show that the line segment from $P(1, 3)$ to $Q(9, -3)$ is a diameter of the circle $x^2 + y^2 = 10x$.

5. Determine the equation for each of the following circles.

   a. centre $(0, 0)$, radius $\frac{2}{3}$ **b.** centre $(0, 0)$, through $(1, \sqrt{3})$

   c. centre $(-1, 2)$, radius 3 **d.** centre $(2, 0)$, through $(0, 2)$

   e. with diameter from $(-2, 4)$ to $(4, -2)$

6. Determine the focal point for each of the following:

   a. $y = 12x^2$ **b.** $y^2 = 12x$

7. Determine the vertex and focal point for each of the following:

   a. $y = 16 - x^2$ **b.** $4x + y^2 = 16$ **c.** $y^2 - 4y = 8x$

8. Determine the equation for each of the following parabolas.

   a. vertex $(0, 0)$, focus $(0, 8)$

   b. vertex $(-2, 0)$, focus $(0, 0)$

   c. vertex $(0, 4)$, x-intercepts ± 2

   d. vertex $(1, -2)$, through $(0, 0)$, horizontal axis of symmetry

9. Determine the equations of two parabolas that have vertex $(0, 1)$ and pass through $(2, 2)$.

10. Determine the vertices and foci for each of the following:

a. $\dfrac{x^2}{100} + \dfrac{y^2}{64} = 1$

b. $x^2 + \dfrac{y^2}{25} = 1$

c. $\dfrac{9x^2}{16} + \dfrac{9y^2}{25} = 1$

d. $8x^2 + 9y^2 = 72$

e. $4x^2 + y^2 = 8$

f. $\dfrac{(x-3)^2}{36} + \dfrac{(y+2)^2}{20} = 1$

11. Determine the equation for each of the following ellipses.

a. vertices $(\pm 6, 0)$, length of minor axis is 4

b. vertices $(0, \pm 15)$, foci $(0, \pm 12)$

c. foci $(\pm 4, 1)$, length of major axis is 16

d. vertices $(0, 0)$, $(0, -12)$, length of minor axis is 10

e. foci $(1, 3)$ and $(9, 3)$, sum of the focal radii is 10

12. Determine the vertices, foci, and the equations of the asymptotes for each of the following:

a. $\dfrac{x^2}{16} - \dfrac{y^2}{9} = 1$

b. $\dfrac{x^2}{5} - \dfrac{y^2}{4} = -1$

c. $4x^2 - 4y^2 = 25$

d. $4y^2 - 3x^2 = 84$

e. $\dfrac{(x+4)^2}{9} - \dfrac{(y-2)^2}{16} = 1$

f. $24(x-1)^2 - (y-2)^2 = 24$

13. Determine the equation for each of the following hyperbolas.

a. vertices $(0, \pm 6)$, foci $(0, +12)$

b. vertices $(\pm 4, 0)$, asymptotes $y = \pm \dfrac{5}{2}x$

c. foci $(\pm 10, 0)$, $a = 4$

d. foci $(0, 0)$ and $(8, 0)$, difference of focal radii 2

14. The arch of a bridge is 80 m wide at the base and 20 m high in the centre. A horizontal support beam is 10 m below the highest point.

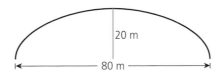

Find the length of the beam

a. if the arch is parabolic,

b. if the arch is elliptical.

15. A toy rocket follows a parabolic path after it is launched from the ground.

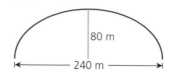

Find the equation for this path, using $(0, 0)$ as its starting point, if it reaches a height of 80 m and travels 240 m down range.

16. A satellite in an elliptical orbit about the earth has a minimum height 215 km and a maximum height 939 km. If the diameter of the earth is 12 672 km, find the equation of the orbit

a. if the earth is at the centre of the orbit,

b. if the earth is at a focal point of the orbit.

17. Determine the equation of a hyperbola described by the flight path of an aircraft 40 km closer to one of the two LORAN stations that are 120 km apart.

18. Sketch the graph represented by each of the following, labelling the centre and vertices.

a. $x^2 + y^2 - 4x + 6y - 12 = 0$

b. $25x^2 + 4y^2 - 100x + 24y + 36 = 0$

c. $y^2 + 4x - 4y - 4 = 0$

d. $2x^2 + y^2 + 12x - 2y + 15 = 0$

e. $x^2 - 4y^2 - 6x - 40y - 116 = 0$

19. Solve the following systems.

a. $x - 2y = 2$
$x^2 + 4y = 0$

b. $x + 2y = 10$
$x^2 + y^2 = 20$

c. $y = 2x + 4$
$4x^2 + y^2 = 16$

20. Show that $y = x - 2$ intersects $x^2 - y^2 = 8$ at only one point but is not a tangent to the curve.

1. Find all three-digit numbers that are 25 times the sum of their digits.

2. If $2^{x+3} + 2^x = 3^{y+2} - 3^y$, where x and y are integers, determine all possible values of x and y.

3. The interior angles of a convex polygon, measured in degrees, form an arithmetic sequence in which the smallest term is $120°$ and the common difference is $5°$. Determine the number of sides of the polygon.

4. Determine the equation of the locus of the vertex of a triangle, given that the triangle base is the line joining $(5, 2)$ and $(-1, -6)$ and that the vertex moves so that the triangle area is always 20.

5. The graphs of $y = \sqrt{3}x^2$ and $x^2 + y^2 = 4$ intersect at points A and B.

   a. Determine the coordinates of A and B.

   b. Determine the length of the minor arc AB of the circle.

6. A machine gives out three bronze discs when one silver disc is inserted and three silver discs when one bronze disc is inserted. Can Donald, who starts out with one bronze disc, play the machine repeatedly and end up with an equal number of bronze and silver discs?

7. Ten friends send greeting cards to each other, with each sending five cards. Must there be two who sent cards to each other? If so, what is the minimum number of such pairs?

1. Find the value of θ and x in each of the following, correct to one decimal place.

a.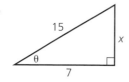

b.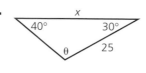

c.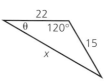

2. Solve each of the following for $0° \leq θ \leq 360°$ to one decimal place.

a. $\cos θ = -0.7934$

b. $5(\tan θ - 2) = 3 \tan θ + 7$

c. $6 \sin^2 θ - \sin θ - 2 = 0$

3. Determine the exact value of θ for each of the following, $0 \leq θ \leq 2π$.

a. $5(\sin θ - 2) = 3(\sin θ - 4)$

b. $7 \cos^2 θ + 5 = 3 \cos^2 θ + 8$

c. $2 \sin^2 θ - \sin θ = 1$

d. $4 \tan θ(3 \tan θ - 3) = 6(2 \tan^2 θ - \tan θ + 1)$

4. Given the functions $f(x) = 3 \cos x - 1$ and $g(x) = \cos x - 3$, do the following:

a. Graph the two functions on the same set of axes for $0 \leq x \leq 3π$.

b. Determine the exact points of intersection of the two functions for $0 \leq x \leq 3π$.

5. a. In each of the diagrams below, the sine function $y = \sin x$ has been transformed into the given curve. Determine an equation for each function.

i)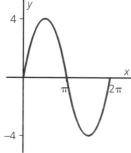

ii)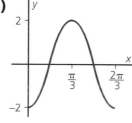

iii)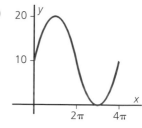

b. In each of the diagrams above, the cosine function $y = \cos x$ has been transformed into the given curve. Determine an equation for each function.

6. State the period, amplitude, equation of the sinusoidal axis, and the phase shift for each of the following functions, and then sketch two complete periods of each.

a. $y = 3 \sin 2x$

b. $y = 4 \cos\left(\frac{x}{2}\right) + 2$

c. $y = 5 \sin 2\left(x - \frac{\pi}{6}\right)$

d. $y = 3 \cos 2\left(x + \frac{\pi}{4}\right) - 3$

e. $h = -2 \cos 6\left(t - \frac{\pi}{3}\right) + 5$

f. $y = 0.5 \sin(5\pi x) - 0.25$

7. Express each of the following in a simpler form by using known identities.

a. $\sec\theta \csc\theta \cot\theta + \dfrac{1}{\cos^2\theta}$

b. $\sin^3\theta + \sin\theta \cos^2\theta - \cos\theta \tan\theta$

c. $\dfrac{1 - \sin^2\theta}{\csc^2\theta - 1} - \sin^2\theta$

d. $\cot\theta - \dfrac{\sin\theta}{1 - \cos\theta}$

8. Prove each of the following identities.

a. $\cos\theta \cot\theta = \dfrac{1}{\sin\theta} - \sin\theta$

b. $(\csc^2\theta - 1)(1 - \cos^2\theta) = \cos^2\theta$

c. $\dfrac{\sin\theta + \cos\theta}{\sin\theta - \cos\theta} = \dfrac{\sec\theta + \csc\theta}{\sec\theta - \csc\theta}$

d. $\dfrac{1 + \cos\theta}{1 - \cos\theta} = \dfrac{1 + \sec\theta}{\sec\theta - 1}$

e. $\dfrac{\sqrt{1 - \cos^2\theta}\,\sqrt{\sec^2\theta - 1}}{\cos\theta} = \tan^2\theta$

f. $\dfrac{1 + \tan^2\theta}{1 + \cot^2\theta} = \dfrac{(1 - \cos^2\theta)(\tan\theta)}{\sin\theta \cos\theta}$

g. $\csc^2\theta - \csc\theta \cot\theta = \dfrac{1}{1 + \cos\theta}$

h. $\dfrac{\tan^2\theta - 1}{\cot^2\theta - 1} = 1 - \sec^2\theta$

9. Two tangents from a point A are drawn to a circle with centre C and radius 12 cm.

a. If the angle between the tangents at point A is $40°$, then find the length of each tangent.

b. Determine the area of quadrilateral $ABCD$, where B and D are the points of contact of the tangents with the circle.

10. The roof for an airplane hangar is to be constructed with steel girders, as shown. If $AB = 12$ m, $BT = 9$ m, and $AC = 52$ m, then determine the following:

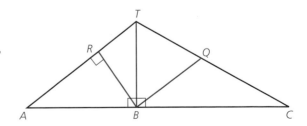

a. the length of the steel girders AT and TC,

b. the roof angle at point A,

c. the length of the cross-brace BR,

d. the length of the cross-brace BQ, if Q is the midpoint of TC.

11. Radar at the control tower T of an airport is tracking two aircraft. Airplane A is on a bearing of $N35°E$ at a distance of 10 km from the tower. Aircraft B is on a bearing of $N20°W$ at a distance of 6 km from T. Determine the distance AB.

12. Sunset Point coast guard station is 40 km SW from Rock Point Station. At midnight both stations pick up an SOS radio signal from a ship in distress at point D. The bearing from Sunset Point is $N10°E$ and the bearing from Rock Point is $S70°W$.

a. Determine the distances from each station to D, correct to two decimal places.

b. If the rescue boat from Rock Point can make 12 km/h and the boat from Sunset Point can only make 8 km/h, which boat will get to point D first? What is the difference in the times?

13. The graph describes the height of a bicycle pedal in centimetres above the road as a cyclist rides around a flat track.

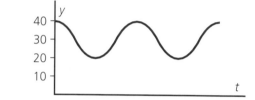

a. Determine the closest distance from the pedal to the road.

b. What is the amplitude of this periodic function?

c. If the cyclist is pedalling at the rate of 15 cycles per minute, then determine the period for the function.

d. Find an equation to represent the height of the pedal in terms of time.

e. What is the height of the pedal when $t = 2.5$ min?

14. In late June at Red Lake, Ontario, the maximum angle of the sun at 12:00 is 65°. The graph shows that the angle of elevation for the sun at different times of the day is part of a sinusoidal function. Explain what time of day is represented by point A.

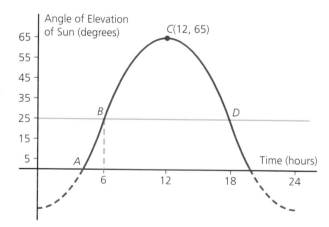

a. Find the equation of the sinusoidal axis BD.

b. Find the equation for the function.

c. Explain why this function could not be used to describe the angle of elevation of the sun in other months of the year.

15. One complete cycle of the moon takes 29 days and nights. The percentage of the moon's surface, as viewed from mid-Ontario cottage country, forms a sinusoidal function when measured against time. At the new moon, the percentage visible is 0% and at full moon the percentage visible is 100%.

  a. Make a sketch to illustrate one complete period of the cycle, starting with the new moon.

  b. Determine the equation for this function.

  c. Use the function of part **b** to determine the percentage of the moon that should be visible on night 12 in the cycle (to one decimal place).

  d. Explain why this function is not continuous.

16. To determine if a tree-covered mountain is suitable for development, a survey crew set up a base line $RQ = 850$ m. From point R, the top of the mountain was visible and $\angle TRQ = 140.5°$, $\angle TRB = 31.5°$, where line $RB \perp TB$. From point Q, $\angle RQT = 29.5°$.

  a. Determine the height of the mountain, to the nearest metre.

  b. From point R to the foot of the mountain, distance $RW = 250$ m. Determine the length of the ski run, to the nearest metre, from T to W (assume a smooth straight line).

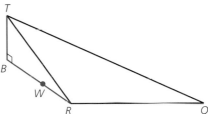

17. Determine the centre and radius of each circle.

  a. $(x \quad 1)^2 + y^2 = 49$

  b. $x^2 + y^2 - 4x - 4y - 8 = 0$

18. Determine the equation of each of the following circles.

  a. centre $C(-3, 5)$ and radius of 6

  b. centre on the x-axis and passing through the points $(2, 4)$ and $(6, 4)$

19. For each of the following parabolas, determine the coordinates of the vertex and the focus, the equation of the directrix, and the direction of opening. Sketch each graph.

  a. $x^2 = -8y$

  b. $y^2 + 4y - 4x - 6 = 0$

20. Write the equation of the parabola with focus at $(-3, 1)$ and directrix $x - 1 = 0$.

21. For each ellipse, find the coordinates of each vertex, the foci, and the lengths of the major and minor axes. Sketch the graph.

  a. $\dfrac{x^2}{16} + \dfrac{y^2}{25} = 1$

  b. $2x^2 + 3y^2 = 6$

22. Write an equation for the ellipse with centre $(0, 0)$, major axis of 6 units, and one focus at $(0, 2)$.

23. For each hyperbola, find the coordinates of the vertices, the foci, and the lengths of the transverse and conjugate axes. Sketch each graph.

a. $\dfrac{x^2}{4} - \dfrac{y^2}{9} = 1$

b. $\dfrac{x^2}{144} - \dfrac{y^2}{25} = -1$

24. Find an equation of the hyperbola with centre at $C(1, -3)$, passing through $P(4, 1)$, and with a vertical transverse axis of 6 units.

25. Identify the conic represented by each of the following. For each conic, state the coordinates of its vertices.

a. $4y = 4x - x^2$

b. $y^2 + 4x - 4y + 8 = 0$

c. $x^2 - y^2 + 12x + 4y + 16 = 0$

d. $3x^2 + 4y^2 + 12x + 32 + 28 = 0$

26. Determine the points of intersection of the line $x + 2y - 12 = 0$ with the conic represented by $x^2 + y^2 - 6x - 4y - 12 = 0$.

27. Simplify the following:

a. $\dfrac{(5x^2y)(4xy^2)^2}{(2xy)^2}$

b. $\dfrac{4^{2n} \times 16^{n-1}}{8^{n-3}}$

28. Solve the following:

a. $x^{\frac{1}{3}} = 3$

b. $4^{2y-3} = 8$

c. $2^{2x} + 4 = 5(2^x)$

29. Evaluate the following:

a. $\left(\dfrac{8}{27}\right)^{\frac{1}{3}}$

b. $4^{\frac{3}{2}}$

c. $\sqrt[3]{125} - (243)^{\frac{3}{5}}$

30. Determine the domain and range of each of the following functions.

a. $y = 2x^2 - 4x + 7$

b. $y = \sqrt{x - 3}$

Glossary

Amortization period The time required to repay a mortgage loan.

Amortization schedule A schedule of mortgage payments.

Amount of an annuity The value at the end of n payment periods if one unit has been deposited at the end of each of the n payment periods.

Amount of an investment The total value at the end of n payment periods.

Amplitude (of a periodic function) The maximum displacement from a reference line in either a positive or a negative direction.

Angle of depression The angle between the horizontal line of sight and an object below the line of sight.

Angle of elevation The angle between the horizontal line of sight and an object above the line of sight.

Angle of rotation The angle created when a line segment is rotated about its vertex to a terminal position.

Annuity A series of equal payments or deposits made at regular intervals of time.

Arithmetic sequence A sequence for which the difference between consecutive terms is a constant. Each term of the sequence equals the previous term increased or decreased by a fixed amount.

Arithmetic series The indicated sum of an arithmetic sequence.

Asymptote (of a given curve) A line that the curve approaches but does not touch.

Axis of symmetry The line associated with a geometric figure such that every point on either side of the line can be reflected onto a point on the other side of the line.

Circle The set of points such that the distance from each point to a fixed point (the centre) is constant.

Clockwise A rotation is clockwise if it moves in the direction of a clock's hands.

Common difference In an arithmetic sequence, the difference between consecutive terms.

Common ratio In a geometric sequence, the ratio of consecutive terms.

Complex number An expression of the form $a + bi$, where a and b are real numbers and i is a symbol with the property that $i^2 = -1$.

Complex number system A method of writing numbers which includes the complex numbers.

Compound interest The interest paid on both the original investment and interest previously earned.

Conditional trigonometric equation A conditional trigonometric equation is true for some subset of the values for which the equation is defined.

Conic sections The curves (circle, ellipse, parabola, and hyperbola) formed by the intersection of a plane and both halves of a circular cone.

Conjugate axis (of a hyperbola) The line segment through the midpoint of the transverse axis and perpendicular to the transverse axis and equal in length to the line segment formed by the points of intersections of the asymptotes of the hyperbola and the line through a vertex and perpendicular to the transverse axis.

Conjugate complex numbers Complex numbers differing only in the sign of their imaginary components. The conjugate of $a + bi$ is $a - bi$.

Conjugate hyperbolas Two hyperbolas with their transverse and conjugate axes reversed.

Contraction A dilatation that shrinks the size of an object.

Coterminal Two angles in standard position are coterminal if they have the same terminal arm. Coterminal angles differ by a multiple of $360°$.

Counterclockwise A rotation is counterclockwise if it moves in the opposite direction of a clock's hands.

Dependent variable The output value of a relation.

Dilatation A transformation that preserves the shape of a figure but allows the size to change.

Directrix The name given to the fixed line in the topic of conics.

Discriminant For the quadratic equation $ax^2 + bx + c = 0$, $b^2 - 4ac$ is called the *discriminant*. It is used to determine the nature of the roots of the quadratic equation.

Divisors (of a number) Those numbers that divide evenly into the given number.

Domain (of a relation) The set of the first coordinates of the ordered pairs of the relation.

Eccentricity (of a conic) The ratio, for a point on the conic, of its distance from the focus to its distance from the directrix.

Effective interest rate The annual rate that would give the same yield as the nominal rate calculated over conversion periods of less than a year.

Ellipse The set of points such that the sum of the distances from each point to two fixed points (the foci) is constant.

Entire radical A radical expressed as a number completely under the radical sign.

Equation of a circle $(x - h)^2 + (y - k)^2 = r^2$, where $C(h, k)$ is the centre and r is the radius.

Equation of an ellipse $\frac{x^2}{a^2} + \frac{y^2}{b^2} = 1$ if the ellipse is centred at the origin and has foci $(c, 0)$ and $(-c, 0)$ or $\frac{x^2}{b^2} + \frac{y^2}{a^2} = 1$ if the ellipse is centred at the origin and has foci $(0, c)$ and $(0, -c)$. For both, the major axis has length $2a$, the minor axis has length $2b$, and $a^2 = b^2 + c^2$.

Equation of a hyperbola $\frac{x^2}{a^2} - \frac{y^2}{b^2} = 1$ if the hyperbola is centred at the origin and has foci $(c, 0)$ and $(-c, 0)$ or $\frac{x^2}{b^2} - \frac{y^2}{a^2} = -1$ if the hyperbola is centred at the origin and has foci $(0, c)$ and $(0, -c)$. For both, the transverse axis has length $2a$, the conjugate axis has length $2b$, and $c^2 = a^2 + b^2$. The asymptotes are $y = \frac{a}{b}x$ and $y = -\frac{a}{b}x$.

Equation of a parabola $x^2 = 4cy$ if the axis of symmetry is the y-axis and the focus is $(0, c)$. $y^2 = 4cx$ if the axis of symmetry is the x-axis and the focus is $(c, 0)$.

Exponent A number placed in a superscript position to the right of another number or variable to indicate repeated multiplication; for example, a^3 is a short form for a times a times a.

Exponential expression An expression containing numbers written in exponential form.

Exponential function A curve with an equation of the form $y = a^x$.

Factor To factor a number or expression means to write it as a product of two or more divisors.

First Principles, Method of A proof by the Method of First Principles is a proof that is developed by using the original definitions.

Focal length (of a parabola) The distance from the vertex to the focus.

Focal point Another name for a focus of a conic.

Focal radii The line segments between the focii of a conic and a point on the conic.

Focal radius The line segment between the focus of a conic and a point on the conic.

Focus The name given to the fixed point in the topic of conics.

Function A function f is a relation that assigns to each element in the domain exactly one element in the range.

Geometric sequence A sequence for which the ratio between consecutive terms is a constant. Each term of the sequence is a constant times the previous term.

Geometric series The indicated sum of a geometric sequence.

Horizontal dilatation A stretch or contraction in the direction of the x-axis.

Horizontal reflection A horizontal reflection reflects all points on the graph of a relation to the opposite side of the y-axis. The point (x, y) is mapped onto $(-x, y)$.

Horizontal translation A horizontal translation moves all points on the graph of a relation left or right a constant amount. The point (x, y) is mapped onto $(x + a, y)$.

Hyperbola The set of points such that the absolute value of the difference of the distances from each point to two fixed points (the foci) is constant.

Identity equation (or identity) An equation that is true for all defined values of the variable.

Independent variable The input value of a relation.

Installment buying Paying for an article with equal payments at predetermined regular times.

Inverse of a function The inverse of a function reverses the coordinates of the ordered pairs defined by the function.

Latus rectum (of a parabola) The chord passing through the focus perpendicular to the axis of symmetry.

(of an ellipse) The chord passing through a focus perpendicular to the major axis.

(of a hyperbola) The chord passing through the focus perpendicular to the transverse axis.

Linear inequality A linear inequality is solved in the same way as a linear equation with the exception that if both sides of an inequation are multiplied or divided by a negative quantity, the sense of the inequality is reversed.

Locus A set of points that satisfy a given condition or the path traced out by a point that moves according to a stated geometric condition.

Logarithm (of a number to a given base) The exponent that must be used with that base to obtain the given number.

Logarithmic function A curve with an equation of the form $y = \log_a x$. This is the inverse function of $y = a^x$.

Major axis (of an ellipse) The longest diameter of the ellipse.

Minor axis (of an ellipse) The shortest diameter of the ellipse.

Mixed radical A radical expressed as the product of a rational number and a number under the radical sign.

Mortgage A loan on a property with the property as collateral.

Nominal interest rate The rate quoted for a year when the interest is calculated over periods of less than a year.

Ordinary annuities Annuities with payments made at the end of the payment period.

Parabola The set of points equidistant from a fixed point (the focus) and a fixed line (the directrix).

Period of an annuity The time between two consecutive payments.

Period (of a function) The time taken to make one complete cycle.

Periodic function A function whose values repeat at regular intervals, measured along the horizontal (or independent) axes.

Phase shift In trigonometry, another name for a horizontal translation.

Present value of an annuity The amount necessary to provide payments of one unit each at the end of n payment periods.

Present value (of an investment) The amount of money that must be invested now to produce a specified amount of money at some time in the future.

Quadratic equation A quadratic equation can be written in the form $ax^2 + bx + c = 0$, where $a \neq 0$.

Quadratic formula The quadratic formula states that the roots of the quadratic equation $ax^2 + bx + c = 0$ are $x = \dfrac{-b \pm \sqrt{b^2 - 4ac}}{2a}$.

Quadratic function A function that can be written in the form $y = ax^2$, where $a \neq 0$.

Radian The measure of an angle subtended at the centre of a circle by an arc equal in length to the radius of the circle.

Radical The root of a quantity as indicated by the sign $\sqrt{}$ (the radical sign). A number placed to the left of the radical sign shows the type of root. For example, $\sqrt[4]{}$ is a fourth root. If there is no number, the root is a square root.

Radicand The number or expression under a radical sign.

Radius (of a circle) Any line segment joining the centre of the circle to a point on the circle.

Range (of a relation) The set of the second coordinates of the ordered pairs of the relation.

Rational expression A rational expression of the form $\dfrac{f(x)}{g(x)}$, where $f(x)$ and $g(x)$ are polynomials, $g(x) \neq 0$.

Recursively A sequence is defined recursively if the first one or two terms is/are given and each term thereafter is given by an expression involving the previous term or terms.

Refinance (a mortgage) To renegotiate the mortgage.

Reflection A reflection in a line l is a transformation that maps any point on the graph of a relation onto an image point such that the line segment joining the point to its image is perpendicular to l and has its midpoint on l.

Relation A rule that associates each element x in a set A with one or more element(s) y in a set B. This rule creates a set of ordered pairs (x, y).

Roots of an equation A number that when substituted for a variable in a given equation satisfies both sides of the equation (i.e., makes both sides equal).

Rotation A transformation which changes the position of a given geometric figure by rotating the figure about a fixed point.

Sequence A function whose domain is the set of natural numbers.

Series The indicated sum of the terms of a sequence.

Simple interest Interest calculated on the principal only.

Sinusoidal axis The line halfway between the maximum and minimum values of the graph of a sinusoidal function.

Sinusoidal function A function expressed in terms of a sine function.

Standard position (of a parabola) If its vertex is at the origin and its axis of symmetry is along either the x-axis or the y-axis.

(of a circle) If its centre is at the origin.

(of an ellipse or a hyperbola) If its centre is at the origin and its axes lie along the x- and y-axes.

Stretch A dilatation that enlarges the size of an object.

Term of a sequence Each element of a sequence is called a *term*.

Term of an annuity The total time involved in completing the annuity.

Term of the mortgage The length of time of the current mortgage agreement.

Theorem A statement that has been proven rather than just assumed.

Transformation A change in size, shape, or position of a geometric figure. Examples include translations, reflections, rotations, and dilatations.

Translation A transformation which only changes the position of a geometric figure. The line segments joining points to their images all have the same magnitude and direction. A translation maps the point (x, y) onto the point $(x + a, y + b)$, where a and b are constants.

Transverse axis (of a hyperbola) The shortest line segment between two points of the hyperbola if one point is on one branch of the hyperbola and the other point is on the other branch of the hyperbola.

Trigonometric equations Equations involving trigonometric ratios and functions.

Trigonometry The branch of mathematics concerned with solving triangles.

Vertex of a parabola The maximum or minimum point of the parabola.

Vertical dilatation A stretch or contraction in the direction of the y-axis.

Vertical line test If no two points on the graph of a relation lie on the same vertical line, the relation represented by the graph is a function.

Vertical reflection A vertical reflection reflects all points on the graph of a relation to the opposite side of the x-axis. The point (x, y) is mapped onto $(x, -y)$.

Vertical translation A vertical translation moves all points on the graph of a relation up or down a constant amount. The point (x, y) is mapped onto $(x, y + a)$.

Vertices of a hyperbola The points where the hyperbola crosses the transverse axis.

Vertices of an ellipse The points where the ellipse crosses the major axis.

Zeros of a function The zeros of a function $f(x)$ are the values of x which make $f(x) = 0$.

Answers

Chapter 1

Exercise 1.1

1. a. (112, 27), (120, 28), (122, 29), (132, 31), (140, 33)
b. domain: {112, 120, 122, 132, 140}; range: {27, 28, 29, 31, 33}
2. a. domain: {0, 1, 3, 4, 5}; range: {2, 5, 11, 14, 15}
b. domain: {−3, 2, 4, 9}; range: {4, 1, 5}; **3. b.** 100 kL
c. 400 min **4. a.** domain: $0 \le t \le 7$, $t \in R$; range: $0 \le d \le 18$,
$d \in R$ **5. a.** domain: {−2, −1, 1, 2}; range: {−1, 0, 1}
b. domain: {0, 1, 2, 4}; range: {0, 1, 2, 4} **c.** domain:
$-3 \le x \le 2$, $x \in R$; range: $-1 \le y \le 3$, $y \in R$ **d.** domain:
$0 \le x \le 3$, $x \in R$; range: $0 \le y < 3$, $y \in R$ **6.** domain: $x \in R$;
range: $y \in R$ **7. a.** {(Henry, Karen), (Henry, Shenif), (Henry,
Hinta), (Hinta, Karen), (Hinta, Shenif), (Shenif, Karen)}
b. domain: {Henry, Hinta, Shenif}; range: {Karen, Shenif,
Hinta} **8. a.** {(1, 1), (2, 1), (2, 2), (3, 1), (3, 3), (4, 1), (4, 2),
(4, 4), (5, 1), (5, 5), (6, 1), (6, 2), (6, 3), (6, 6), (7, 1), (7, 7),
(8, 1), (8, 2), (8, 4), (8, 8), (9, 1), (9, 3), (9, 9), (9, 9)} **b.** domain:
{1, 2, 3, 4, 5, 6, 7, 8, 9}; range: {1, 2, 3, 4, 5, 6, 7, 8, 9}
9. a. {(Vicki, Elizabeth), (Vicki, Chris), (Vicki, David),
(Elizabeth, Chris), (Elizabeth, David), (Chris, David)}
b. domain: {Vicki, Elizabeth, Chris}; range: {Elizabeth, Chris,
David} **10.a.** domain: $x \in R$; range: $y \in R$ **b.** domain: $x \in R$;
range: $y \ge 0$, $y \in R$ **c.** domain: $x \ge -4$, $x \in R$; range: $y \ge 0$,
$y \in R$ **d.** domain: $x \ne 0$, $x \in R$; range: $y \ne 0$, $y \in R$ **e.** domain:
$x \in R$; range: $y \ge 3$, $y \in R$ **f.** domain: $x \ne 2$, $x \in R$; range:
$y \ne 0$, $y \in R$ **11. b.** 19.25 m **c.** yes **12.** Answers will vary.
14. a. domain: $-5 \le x \le 5$, $x \in R$; range: $-5 \le y \le 5$, $y \in R$
b. domain: $-5 \le x \le 5$, $x \in R$; range: $-\frac{5}{2} \le y \le \frac{5}{2}$, $y \in R$
c. domain: $x \le -5$ or $x \ge 5$, $x \in R$; range: $y \in R$ **d.** domain:
$-3 \le x \le 7$, $x \in R$; range: $-5 \le y \le 5$, $y \in R$

Exercise 1.2

1. a, c, c, and f **2. a.** 5 **b.** −10 **c.** 0 **d.** $3a-1$ **3.** a, d, and e
4. $f(0) = 4$, $f(2) = 6$ **5. a.** 1 **b.** 3 **c.** 3 **d.** 2 **6.** a and c are functions.
7. a. domain: $x \in R$; range: $y \ge -3$, $y \in R$ **b.** domain: $x \ge 3$,
$x \in R$; range: $y \ge 0$, $y \in R$ **c.** domain: $x \ne -2$, $x \in R$; range:
$y \ne 0$, $y \in R$ **8. a.** −2 **b.** −17 **c.** −4.0625 **d.** $-a^2 + 4a - 5$
9. a. −2 **b.** 2 **c.** −3 **10. a.** −7 **b.** 5 **c.** −6 **d.** $4x - 5$ **e.** $2x - 3$
f. $2x + 2a - 5$ **g.** $\frac{6}{x} - 5$ **h.** $4x - 15$ **i.** $4x^2 - 2x - 5$ **11. a.** 38
b. 26 **c.** $27a^2 - 6a + 5$ **d.** $12x^2 - 4x + 5$ **e.** $3x^2 + 2x + 5$
f. $3x^2 + 16x + 26$ **g.** $\frac{3}{x^2} - \frac{2}{x} + 5$ **h.** $27x^2 + 30x + 13$
i. $3x^4 - 8x^2 + 10$ **12. a.** 7 **b.** 3 **c.** $x^2 + 6x + 5$ **d.** $2x^2 + 15x + 5$
e. $-x^2 + 3x + 10$ **f.** $4x^2 + x - 6$ **14.** $\frac{-7}{4}$ **15.** 61

Exercise 1.3

1.

x	−3	−2	−1	0	1	2	3
$f(x)$	−6	−2	0	3	5	−4	8
$f(\frac{1}{2}x)$			0		3		5
$f(2x)$			−2	3	−4		

3. a. horizontal translation of +4 **b.** vertical stretch by a factor
of 5 **c.** vertical translation of −5 **d.** horizontal stretch by a
factor of 4 **e.** horizontal translation of +2, vertical translation
of −1 **f.** horizontal translation −1, vertical stretch by a factor
of 3 **g.** reflection in the x-axis followed by a vertical translation
of +2 **h.** horizontal contraction by a factor of 2, reflection
in the y-axis **i.** vertical contraction by a factor of 4, vertical
translation of +2 **j.** horizontal translation of +1, reflection
in the y-axis **k.** horizontal translation of +6, horizontal stretch
by a factor of 3 **l.** horizontal compression by a factor of 2,
vertical stretch by a factor of 3, reflection in the x-axis,
vertical translation of +3

Exercise 1.4

1.		i)	ii)	iii)	iv)
a.		up	(0, −3)	$y \ge -3$	$x = 0$
b.		up	(3, 0)	$y \ge 0$	$x = 3$
c.		down	(0, 4)	$y \le 4$	$x = 0$
d.		up	(−2, 1)	$y \ge 1$	$x = -2$
e.		down	(4, −3)	$y \le -3$	$x = 4$
f.		down	(5, 0)	$y \le 0$	$x = 5$
g.		up	(0, −3)	$y \ge -3$	$x = 0$
h.		down	(−4, 3)	$y \le 3$	$x = -4$
i.		up	(−3, −5)	$y \ge -5$	$x = -3$

4. a. $y = -4(x + 1)^2 + 2$ **b.** $y = 2x^2 - 3$ **c.** $y = -(x - 1)^2 + 3$
d. $y = 5(x + 2)^2 - 2$

Exercise 1.5

1. a. domain: $x \ge 5$, $x \in R$; range: $y \ge 0$, $y \in R$
b. domain: $x \ge -3$, $x \in R$; range: $y \ge 0$, $y \in R$
c. domain: $x \le 2$, $x \in R$; range: $y \ge 0$, $y \in R$
d. domain: $x \ge 0$, $x \in R$; range: $y \ge 0$, $y \in R$
e. domain: $x \le 3$, $x \in R$; range: $y \le 0$, $y \in R$
f. domain: $x \ge -1$, $x \in R$; range: $y \ge 2$, $y \in R$
g. domain: $x \ge 4$, $x \in R$; range: $y \ge -2$, $y \in R$
h. domain: $x \le 0$, $x \in R$; range: $y \le 3$, $y \in R$
2. a. $y = \sqrt{x - 3} + 1$ **b.** $y = -\sqrt{4 - x}$ **c.** $y = \sqrt{2 - x} - 1$
d. $y = \sqrt{x}$ **4.** $x \ge \frac{3}{2}$, $x \in R$ **5.** domain: $x \le \frac{1}{2}$, $x \in R$;
range: $y \le 4$, $y \in R$ **6. b.** $y = \sqrt{x}$, $y = -\sqrt{x}$ **d.** $y = \sqrt{-x}$,
$y = -\sqrt{-x}$

Exercise 1.6

1. ii)	Domain $x \in R$	Range $y \in R$	Equation of the Asymptotes
a.	$x \ne -3$	$y \ne 0$	$x = -3, y = 0$
b.	$x \ne 0$	$y \ne -3$	$x = 0, y = -3$
c.	$x \ne 4$	$y \ne 0$	$x = 4, y = 0$
d.	$x \ne -1$	$y \ne 1$	$x = -1, y = 1$
e.	$x \ne 2$	$y \ne 0$	$x = 2, y = 0$
f.	$x \ne -2$	$y \ne 3$	$x = -2, y = 3$
g.	$x \ne 2$	$y \ne 0$	$x = 2, y = 0$
h.	$x \ne 1$	$y \ne 1$	$x = 1, y = 1$
i.	$x \ne 3$	$y \ne 2$	$x = 3, y = 2$

2. a. $y = 1 + \frac{2}{x + 1}$ **b.** $y = 2 + \frac{2}{x - 1}$

Exercise 1.7

1. no **2. a.** {(5, 2), (6, 3), (7, 4), (8, 5)}; yes **b.** {(2, 4), (3, 5),
(2, −3), (3, 6)}; no **c.** {(3, 1), (3, 2), (3, 4), (3, 7)}; no
d. {(3, −2), (4, −1), (7, 2), (9, 4)}; yes **3. a.** $y = \frac{x + 2}{3}$
b. $y = \frac{10 - 2x}{5}$ **c.** $x = y^2 - 3$ or $y = \pm\sqrt{x + 3}$

d. $4y^2 + 9x^2 = 36$ or $y = \frac{\pm 3\sqrt{4 - x^2}}{2}$ **4.** a and c **5. c.** yes
6. b. yes **d.** no **7. a.** $y = \frac{1}{x}$ **b.** same **8. a.** $g(3) = 9, g(-2) = -1,$
$g(0) = 3$ **b.** $y = \frac{x - 3}{2}$ **c.** $g^{-1}(9) = 3, g^{-1}(-1) = -2,$
$g^{-1}(3) = 0$ **d.** x **e.** x **f.** 11 **9. a.** $y = \frac{x + 5}{2}$; a function; domain:
$x \in R$; range: $y \in R$ **b.** $y = \frac{x - 1}{3}$; a function; domain: $x \in R$;
range: $y \in R$ **c.** $y = \frac{x}{2}$; a function; domain: $x \in R$; range:
$y \in R$ **d.** $y = \pm\sqrt{x + 4}$; not a function **e.** $y = \frac{1 + 3x}{x}$;
a function; domain: $x \neq 0, x \in R$; range: $y \neq 3, y \in R$
10. a. $y = 20\,000 + 0.15x$ **b.** domain: $x \geq 0, x \in R$; range:
$y \geq 20\,000, y \in R$ **c.** $y = \frac{20x - 40\,000}{3}$ **d.** yes; domain:
$x \geq 20\,000, x \in R$; range: $y \geq 0, y \in R$ **11. a.** $y = \frac{x + 3}{2}$
c. (3, 3) **12. a.** $g^{-1}(x) = \frac{x - 4}{3}$ **b.** -11 **15. b.** domain: $x \neq -2$,
$x \in R$; range: $y \neq 0, y \in R$ **c.** $y = \frac{3 - 2x}{x}$ **d.** yes **e.** domain:
$x \neq 0, x \in R$; range: $y \neq -2, y \in R$ **17. b.** No, since two points
on the curve can be joined by a vertical line. **c.** For $y = f(x)$,
domain is $-2 \leq x \leq 0, x \in R$, and the range is $-1 \leq y \leq 3$,
$y \in R$. For the inverse, domain is $-1 \leq x \leq 3, x \in R$, and the
range is $-2 \leq y \leq 0, y \in R$.

1.8 Review Exercise

1. a. domain: $\{-2, -1, 0, 1, 2\}$; range: $\{1, 0, -1\}$ **b.** domain:
$\{-2, -1, 0\}$; range: $\left\{-1, -\frac{1}{2}, 0, \frac{1}{2}, 1\right\}$ **c.** domain: $x \in R$;
range: $y \in R$ **d.** domain: $x \in R$; range: $y \leq 4, y \in R$ **e.** domain:
$x \in R$; range: $y \in R$ **f.** domain: $x \geq -1, x \in R$; range: $y \geq 0$,
$y \in R$ Parts a, c, d, e, and f are functions. **2. a.** $\{(1, -2),$
$(0, -1), (-1, 0), (0, 1), (1, 2)\}$; domain: $\{1, 0, -1\}$; range:
$\{-2, -1, 0, 1, 2\}$; not a function **b.** $\left\{(0, -2), \left(\frac{1}{2}, -1\right), (1, 0),\right.$
$\left.\left(-\frac{1}{2}, -1\right), (-1, 0)\right\}$; domain: $\left\{-1, -\frac{1}{2}, 0, \frac{1}{2}, 1\right\}$; range:
$\{-2, -1, 0\}$; a function **c.** $y = -2x + 4$; domain: $x \in R$;
range: $y \in R$; a function **d.** $x = 4 - y^2$; domain: $x \leq 4, x \in R$;
range: $y \in R$; not a function **e.** $y = 2x - 2$; domain: $x \in R$;
range: $y \in R$; a function **f.** $x = \sqrt{y + 1}$; domain: $x \geq 0, x \in R$;
range: $y \geq -1, y \in R$; a function **3. a. i)** domain:
$\{-2, 0, 2, 4\}$; range: $\{1, 2, 3\}$ **ii)** domain: $-2 \leq x \leq 1, x \in R$;
range: $-2 \leq y \leq 4, y \in R$ **iii)** domain: $-4 \leq x \leq 5, x \in R$;
range: $-1 \leq y \leq 2, y \in R$ **5. a.** 45 m **b.** 2 s **c.** height above the
ground from which the projectile was fired **d.** the time for the
projectile to "hit" the ground **6. a.** $f(-2) = 3, f(-1) = 0,$
$f(2) = 3$ **b.** domain: $x \in R$; range: $y \geq -1, y \in R$
7. a. $f(1) = 2, f(-2) = 1, f(-4)$ is not real.
b. domain: $x \geq -3, x \in R$; range: $y \geq 0, y \in R$
8. a. $f(-2) = -\frac{1}{2}, f(0) = -1, f\left(\frac{1}{2}\right) = -\frac{4}{3}$ **b.** domain: $x \neq 2$,
$x \in R$; range: $y \neq 0, y \in R$ **9. a.** $f(-2) = 1, f\left(\frac{1}{2}\right) = 1$,
$\frac{1}{f(2)} = 1$ **10. a.** 5 **b.** -2 **c.** 5 **d.** 2 **12.** $y = \frac{x + 4}{2}$
13. a. vertex: (2, 1); range: $y \leq 1, y \in R$ **b.** vertex: $(-4, 0)$;
range: $y \geq 0, y \in R$ **c.** vertex: (0, 16); range: $y \leq 16, y \in R$
14. a. $y = (x + 1)^2 + 1$ **b.** $y = -2(x - 2)^2$
c. $y = \frac{1}{2}(x + 2)^2 - 3$ **15. a.** domain: $x \in R$; range: $y \geq -3$,
$y \in R$ **b.** domain: $x \geq 1, x \in R$; range: $y \geq -3, y \in R$

c. domain: $x \neq 1, x \in R$; range: $y \neq -3, y \in R$
17. a. $x - 2 = 0, y = 1$ **b.** $x + 1 = 0, y = 3$

Chapter 2

Exercise 2.1

1. a. $-7x^3$ **b.** $120x^9$ **c.** $2x^2y$ **d.** $-6x^6y^3$ **e.** $2x$ **f.** $3 - 2x$
g. $x^2 - 3x + 3$ **h.** $3x^3y^2$ **i.** $2x^2y - 2xy^2 + 3xy$ **2. a.** $-4x^6$ **b.** $5x^3$
c. -4 **d.** $-24x^4$ **e.** $-8x^4$ **f.** $27x^3$ **g.** $6x^2 + 8x$ **h.** $2x^3 - 3x^2$
i. $3x^3y - 6x^2y^2 + 9xy^3$ **j.** $2x$ **k.** $x - 2$ **l.** $4x - 1$ **3. a.** $x^2 + 3x + 2$
b. $2x^2 - 11x + 12$ **c.** $3x^2 + 16x - 12$ **d.** $x^2 - 9$ **e.** $x^2 + 6x + 9$
f. $16 - 8x + x^2$ **g.** $25 - 4x^2$ **h.** $x^2 - y^2$ **i.** $x^2 + 6xy + 9y^2$
4. a. $-x^2, -36$ **b.** $x^2 - x, 30$ **c.** $3x, 18$ **5. a.** $x + 1$
b. $x^3 + x^2 - x - 1$ **c.** $6x^2 - 6$ **d.** $2x^2 + 14x$ **e.** $24x^2 - 58x + 35$
f. $2x^3 - x^2 - 2x + 1$ **g.** $x^3 - 8$ **h.** $9x^4 - 16$ **i.** $25x^2 - 60x + 36$
j. $24x^3 - 4x^2 - 4x$ **6. a.** $3x^2 + 4x - 8$ **b.** $5x^2 - x + 2$
c. $2x^2 - 6x + 9$ **d.** $2x^2 - 8x$ **e.** $8x^3 - 36x^2 + 46x - 15$
f. $8x^3 - 12x^2 + 6x - 1$ **g.** $x^4 - 4x^3 + 6x^2 - 4x + 1$
h. $20x^2 - 1$ **7. a.** $6x$ cm; $2x^2$ cm² **b.** $(4x + 8)$ cm;
$(x^2 + 4x + 4)$ cm² **c.** $3\pi x$ cm; $\frac{9\pi x^2}{4}$ cm² **8. a.** 17 **b.** 7 **c.** $\frac{85}{9}$
d. $a^2 + 7a + 17$ **e.** $a^4 + 3a^2 + 7$ **f.** $4a^2 - 10a + 11$
9. a. $6x + 10$ **b.** $16x^2 + 20x$ **c.** $10x + 20$ **d.** $68x^2 + 160x + 100$
10. a. $x^2 - 4x + 4$ **b.** $3x^2 - 12$ **c.** $x^2 - 4$ **d.** $9x^2 - 12x$
e. $x^2 - 6x + 5$ **f.** x^2 **g.** $-x^2 + 4x$ **h.** $x^2 + 4x$ **i.** $4x^2 - 8x - 4$
11. a. $x^2 + 2xy + y^2$ **b.** $x^3 + 3x^2y + 3xy^2 + y^3$
c. $x^4 + 4x^3y + 6x^2y^2 + 4xy^3 + y^4$
d. $x^5 + 5x^4y + 10x^3y^2 + 10x^2y^3 + 5xy^4 + y^5$ **12. a.** $4x + 1$
b. $4x - 2$ **13. a.** $4x^2 - 4x$ **b.** $2x^2 - 3$ **14. a.** x **b.** x **15.** f and g are
inverses of each other.

Exercise 2.2

1. a. 11 **b.** 25 **c.** $\frac{2}{3}$ **d.** 0.1 **e.** $\frac{3}{2}$ **f.** 3 **g.** 4 **h.** 17 **2. a.** $\sqrt{35}$ **b.** $6\sqrt{15}$
c. $4\sqrt{5}$ **d.** $-3\sqrt{7}$ **e.** $2\sqrt{3} + 6$ **f.** $\sqrt{2} + 2$ **g.** $3\sqrt{6} - 3\sqrt{10}$
h. $2\sqrt{3}$ **3. a.** $2\sqrt{3}$ **b.** $3\sqrt{2}$ **c.** $3\sqrt{5}$ **d.** $5\sqrt{3}$ **e.** $10\sqrt{2}$ **f.** $16\sqrt{2}$
g. $6\sqrt{6}$ **h.** $15\sqrt{7}$ **i.** $54\sqrt{2}$ **j.** $15\sqrt{5}$ **4. a.** $2\sqrt{5}$ **b.** $4\sqrt{3}$ **c.** $10\sqrt{3}$
d. $18\sqrt{2}$ **e.** 30 **f.** $30\sqrt{10}$ **g.** 108 **h.** $84\sqrt{3}$ **i.** $420\sqrt{6}$
5. a. $\sqrt{3} + 3\sqrt{5}$ **b.** $2\sqrt{7} - 14$ **c.** $3\sqrt{3}$ **d.** 0 **e.** $\sqrt{3}$ **f.** $6\sqrt{2}$ **g.** 0
h. $6 + 3\sqrt{2}$ **i.** $-2\sqrt{6}$ **6. a.** $5 - 5\sqrt{2}$ **b.** $6\sqrt{6} + 6$ **c.** $6\sqrt{3} + 18$
d. $9 + 5\sqrt{3}$ **e.** $-5\sqrt{6}$ **f.** 2

Exercise 2.3

1. a. $x < 5$ **b.** $x \leq 1$ **c.** $x \leq \frac{1}{2}$ **d.** $x < 2$ **e.** $x > 1$ **f.** $x \geq 9$
g. $x < \frac{1}{8}$ **h.** $x \leq 2$

2.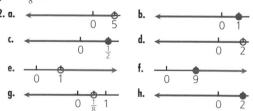

3. a. $x \leq 3$ **b.** $x > -3$ **c.** $-2 < x < 3$ **d.** $x < -2$ or $x > 2$
e. $-1 \leq x \leq 6$ **f.** $x < 0$ or $x \geq 4$

4.

c. $x > -\frac{3}{2}$

d. $x > -\frac{1}{2}$

e. $x \le -2$

f. $x \le 1$

g. $x \le 12$

h. $x \ge -5$

5. a. $x \le 2$

b. $x \le 5$

c. $x > 7$

d. $x < -2$

e. $x < -1$

f. $x > 0$

g. $x < 1$

h. $x < 2$

6. a. $2 < x < 6$

b. $x < 0$ or $x > 2$

c. $-2 \le x \le 3$

d. no values

e. $x \neq -1$

f. $x \le -4$ or $x \ge 4$

7. a. $x > 3$ or $x < -2$ b. $-7 \le x \le -4$ 8. -3 9. $\frac{-4}{3} < x < 1$ or $x > \frac{3}{2}$

Exercise 2.4

1. a. $x(x+1)$ b. $2(x-4)$ c. $b(a-c)$ d. $xy(w+z)$
e. $m^2(m^4+1)$ 2. a. $(y-1)(y+1)$ b. $(m-4)(m+4)$
c. $(2-a)(2+a)$ d. $(xy-3)(xy+3)$ e. $(5-t^2)(5+t^2)$
3. a. $(x-2)(x-1)$ b. $(y+5)(y+1)$ c. $(x-6)(x+1)$
d. $(t+4)(t+3)$ 4. a. $3a(3a^2-4)$ b. $ab(5a+b)$
c. $2ab(2a^2b^3-3ab+1)$ d. $\pi r(r+h)$ e. $(2x-1)(x+2)$
f. $x(x+3)$ 5. a. $(x+7)(x-7)$ b. $(10-y)(10+y)$
c. $(9x-2y)(9x+2y)$ d. $(5-xy)(5+xy)$ e. $(x-1)(x+7)$
f. $(x+5)(3x+1)$ g. $x(x-1)(x+1)$ h. $\pi(R-r)(R+r)$
i. $(x^2+y^2)(x+y)(x-y)$ 6. a. $(x-2)(x-9)$
b. $(x+14)(x-3)$ c. $(x+6)(x+9)$ d. $(x-3)(x-18)$
e. $(x-8)(x-8)$ f. $(x-20)(x+4)$ g. $(x^2+10)(x^2+5)$
h. $(x-3)^2(x+3)^2$ i. $(x+7y)(x-y)$ 7. a. $(3x+2)(2x-9)$
b. $(3x-2)(4x+1)$ c. $(4x+1)(x+4)$ d. $(3x-5)^2$
e. $(4x^2+1)(x+1)(x-1)$ f. $6(x-3)(x+1)$
g. $x(2x-1)(x-1)$ h. $(2x-3)(2x+3)(x+1)(x-1)$
i. $(8x^2+1)(x+2)(x-2)$ j. $(5x+2)(3x-7)$
k. $(7x-5)(3x-2)$ l. $(22x-1)(x+2)$ 8. a. $(x-4)(x^2+3)$
b. $(x-3)(2x^2-3)$ c. $(x+2)(2x+1)(2x-1)$
d. $2(x-3)(x^2+5)$ e. $(x+2)(x-2)(3x^2-1)$
f. $2x(x+2)(x-2)^2$ 9. a. $(x-2)(x-11)$ b. $3x(x-3)$
c. $(3x+7)(2x-1)$ d. $(5x-8)(2x-1)$ e. $(x-3)(x^2+4)$

f. $(5x-7y)(5x+7y)$ g. $(2x+5)(2x+5)$ h. $5(x-3)^2$
i. $(4x-1)(x+5)$ j. $4(5a-3b)(5a+3b)$ k. $(2x^2+7)^2$
l. $4(3x^2+2x+7)$ 10. a. 200 b. 200 c. 2320 d. 19 999
11. a. $x(x+1)$ b. $x(x-2)(x^2-2x+2)$
c. $(x-4-2y)(x-4+2y)$ 12. 30 13. a. 19×37
b. $3 \times 7 \times 19$ c. 73×89 d. 167×191
e. $3 \times 3 \times 11 \times 23 \times 23$

Exercise 2.5

1. a. $x \neq 0$ b. $x \neq -2$ c. $x \neq \frac{5}{2}$ d. $x \neq 0, -2$ e. $x \neq 0, -1$
f. $x \neq 0, 1, -1$ g. $x \neq y$ h. no restriction on x 2. a. $\frac{2}{x}, x \neq 0$
b. $\frac{1}{2}, x \neq y$ c. $\frac{x-1}{2(x+1)}, x \neq 0, -1$ d. $-1, x \neq 2$ 3. a. $\frac{x+2}{4}$
b. $\frac{x}{2(x-2)}$ c. $\frac{2}{3}, x \neq -4$ d. $\frac{1}{x}, x \neq -3$ e. $\frac{x-4}{x+4}, x \neq 0$
f. $\frac{x-1}{5}, x \neq 1$ g. $-2, x \neq 2$ h. $\frac{x}{2}$ 4. a. $\frac{1}{x-2}, x \neq -2$
b. $-3 - x, x \neq 3$ c. $\frac{2}{x+2}, x \neq 2$ d. $\frac{4}{x-4}, x \neq 0, -4$
e. $\frac{x+1}{x-1}, x \neq 1$ f. $\frac{x-2}{2}, x \neq 3$ g. $\frac{x+5}{x-1}, x \neq 2$
h. $\frac{3}{x-2}, x \neq 0, 2$ 5. a. $\frac{2(x-2)}{3(x+2)}, x \neq 5$
b. $\frac{x+4}{3x(x-3)}, x \neq 0, -3$ c. $\frac{x+3}{2x-3}, x \neq -\frac{3}{2}$
d. $(x+2)(x-1), x \neq 2, -1$ e. $\frac{2x-5}{2x+5}, x \neq -\frac{3}{2}$
f. $-\frac{4+3x}{x+1}, x \neq \frac{4}{3}$ 6. a. $-3, 1, 3, 7$ b. $-9, -6, -5, -4, -2,$
$-1, 0, 3$ 7. a. $(-4, 0), (-1, -1), (0, -2), (1, -5), (3, 7),$
$(4, 4), (5, 3), (8, 2)$ b. $(3, 13), (4, 8), (7, 5), (12, 4), (-8, 2),$
$(-3, 1), (1, -7), (0, -2)$ c. $(0, 3), (2, 1), (4, 7), (6, 5)$
d. $(1, -9), (5, 3)$

Exercise 2.6

1. a. $\frac{x}{2}, x \neq 0, 2$ b. $3(x+2), x \neq -2, 2$ c. $-2, x \neq -3, 1$
d. $\frac{-x^2}{3}, x \neq 0, \frac{3}{2}$ e. $\frac{x}{x+1}, x \neq -2, -1$ f. $\frac{x^2}{2}, x \neq -2, 2, 0$
2. a. $\frac{x(x+6)(x-1)}{(x-2)(x+3)(x+3)}, x \neq -2, 2, -3$ b. $\frac{x+1}{x}, x \neq 0, 1$
c. $1, x \neq -3, 1, 2, 3$ d. $\frac{2x-1}{2}, x \neq -\frac{1}{2}, \frac{1}{2}, 2$ e. $\frac{2x(x+1)}{3}, x \neq 1, 3$
f. $\frac{3x+4}{3x+1}, x \neq \frac{1}{3}, \frac{4}{3}, -\frac{1}{3}$ 3. a. $\frac{1}{2x}, x \neq -2, -1, 0$ b. $\frac{3x-1}{2x-1},$
$x \neq -\frac{2}{3}, -\frac{1}{2}, \frac{1}{2}$ c. $\frac{1}{2}, x \neq -4, 3, 4, 5$ d. $-\frac{x(x+2)}{2(x+3)}, x \neq -2, 0,$
$2, 3, -3$ 4. a. $x(x-2)(x+1), x \neq 0, 2, 3$ b. $\frac{2x-1}{2x-5}, x \neq -1,$
$-\frac{1}{2}, \frac{3}{7}, 7, \frac{5}{2}$ 5. a. 0 b. -1 c. $\frac{2}{3}$ 7. $-\frac{1}{2}, -5, 1$

Exercise 2.7

1. a. $2x$ b. $\frac{6}{x}$ c. $\frac{2}{x+1}$ d. $\frac{7m}{3x}$ e. $\frac{3y}{2x+1}$ f. $\frac{-2x}{x-3}$ 2. a. $\frac{11x}{12}$ b. $\frac{x}{24}$
c. $\frac{7}{2x}$ d. $\frac{13}{12x^2}$ e. $\frac{3x+2}{6}$ f. $\frac{6}{x-1}$ g. $\frac{4x-2}{x}$ h. $\frac{5x}{3}$ i. $\frac{1}{x+1}$
3. a. $\frac{3x-16}{4x^3}$ b. $\frac{x^2+x-4}{2x^2}$ c. $\frac{x}{21}$ d. $\frac{x+4}{x+1}$ e. $\frac{4x^2+x-2}{2x}$
f. $\frac{x^2-x-1}{x}$ 4. a. $\frac{3x^2+4x}{(x+1)(x+2)}$ b. $-\frac{1}{x(x+1)}$ c. $\frac{2x^2-4x-3}{(2x-3)(x-3)}$
d. $\frac{6x+4}{(x+2)(x-2)}$ e. $\frac{5x^2-2}{(x-1)(2x+1)}$ f. $\frac{x^2-x-4}{(x-2)^2}$
5. a. $\frac{1}{(x-1)(x+1)}$ b. $\frac{3x-2}{(x+2)(x-2)}, x \neq 0$ c. $\frac{x^2-2}{2x(x+2)}$
d. $\frac{x+8}{(x-2)(x-1)(x+2)}$ e. $\frac{-1}{(x-1)(x-2)}, x \neq -2$
f. $\frac{-x^2+5x-4}{x(x-3)(x+3)}$ 6. a. $\frac{-x^3-10x^2+34x-20}{x(x+4)(2x-1)}$ b. $\frac{3x^2+x-6}{x(x+1)(x-3)}$
7. $\frac{4x^2+15x}{(x+1)(x+2)(x+3)}$ 8. a. $\frac{3}{2}$ b. $\frac{3x^2-14}{(x+2)(5x-17)}$
10. a. $\frac{5}{x+2} - \frac{3}{x+1}$ b. $\frac{3}{x+3} + \frac{2}{x-2}$

2.8 Review Exercise

1. a. $4x^3 - 2x^2 - 4x + 3$ **b.** $120x^6$ **c.** $32x^6$ **d.** $6x^2$
e. $6x^4 - 4x^3 + 2x^2$ **f.** $5x^3$ **g.** $-21x^2$ **h.** x^4 **i.** x^3 **2. a.** $4x^2 - 7x + 3$
b. $2x^3 + x^2 + 3x - 2$ **c.** $6x^3 + 9x^2 - 6x$ **d.** $36x^2 - 48x + 16$
e. $27x^3 + 27x^2 + 9x + 1$ **f.** $2x$ **g.** $3x^2 - 3$ **h.** $1 - x + 2x^2 - x^3$
3. a. $4x - 1$ **b.** $12x^2 + 30x + 12$ **c.** $5x^2 + 8x + 5$
d. $x^2 - 2x + 1$ **4. a.** $2x^2 + x$ **b.** $8x^2 - 6x + 1$ **c.** $2x^2 + 3x + 1$
d. $4x^2 - 14x$ **5.** $7x + 2$, $2x^2 + 3x + 1$ **6. a.** $8\sqrt{2}$ **b.** $18\sqrt{2}$
c. $20\sqrt{2}$ **d.** $20\sqrt{2}$ **e.** $15\sqrt{7}$ **7. a.** $6\sqrt{35}$ **b.** $24\sqrt{2}$ **c.** $4\sqrt{5} - 6\sqrt{3}$
d. $6\sqrt{5} + 2\sqrt{10}$ **e.** $14\sqrt{3}$ **f.** $6 - 3\sqrt{6}$ **g.** $11 + 6\sqrt{2}$ **h.** 3 **i.** -62
8. a. $9x(x - 2)$ **b.** $(x - 19)(x + 1)$ **c.** $9(x - 1)(x - 1)$
d. $(x^2 + 1)(x^2 + 4)$ **e.** $(x - 3)(x + 3)(x - 1)(x + 1)$
f. $(4x - 1)(x + 4)$ **g.** $4x(x + 3)$ **h.** $(x + 1)(x - 2)$
i. $(2x - 1)(x - 2)(x + 2)$ **j.** $(6x - 5)(x + 2)$
k. $x(2x - 3)(2x - 3)$ **l.** $x(x - 2)(5x^2 - 3)$ **9. a.** $x > -1$ **b.** $x < 4$
c. $x \le -4$ **d.** $x < \frac{3}{2}$ **e.** $x \ge 0$ **f.** $x > -1$ **10. a.** $-5 \le x \le 1$
b. $x < 2$ or $x > 4$ **11. a.** no restrictions **b.** $x \ne 2$ **c.** $x \ne \frac{1}{2}$
d. $x \ne 2, -2$ **e.** $x \ne -1, 0$ **f.** $x \ne 1$ **g.** $x \ne -4, 1$ **h.** $x \ne 0, 2, -1$
i. $x \ne 3, 5$ **12. a.** $\frac{1}{2x - 3}$ **b.** $\frac{2}{x + 1}$, $x \ne 0, -1$ **c.** $\frac{1}{x - 1}$,
$x \ne -1$ **d.** $\frac{2}{3}$, $x \ne -2$ **e.** $\frac{x + 2}{2}$, $x \ne 2$ **f.** $\frac{3}{x}$, $x \ne 1, -1, 0$
g. $\frac{x + 3}{x - 1}$, $x \ne 1$ **h.** $\frac{x + 2}{x - 2}$, $x \ne 2$ **i.** $\frac{2x - 3}{2x + 3}$, $x \ne -4$ **13. a.** $\frac{x^2}{8}$
b. $\frac{2}{5}$, $x \ne 2$ **c.** $\frac{x - 2}{x}$, $x \ne -2, 2$ **d.** $-\frac{2}{x(x + 3)}$, $x \ne 0, 3$
e. $\frac{1}{6}$, $x \ne -3, -2, 3$ **f.** -2, $x \ne -1, 0, 1, 3$ **14. a.** $\frac{5x - 3}{x(x - 1)}$
b. $\frac{x + 5}{(x - 1)(x + 2)}$ **c.** $\frac{2x + 1}{x(x + 1)}$ **d.** $\frac{3x}{x - 1}$ **e.** $\frac{4x + 9}{(x - 1)(x + 1)}$
f. $\frac{2x^2 + 2}{x(x - 1)(x + 1)}$ **15. a.** 1, $x \ne -1, 0, 1, 2$
b. $\frac{x + 8}{(x - 2)(x + 2)}$, $x \ne 0$

Chapter 3

Exercise 3.1

1. a. 0, 4 **b.** $-2, 2$ **c.** -1 **d.** no real roots **2. a.** ± 9 **b.** ± 20
c. $\pm\sqrt{10}$ **d.** ± 6 **e.** $\pm\frac{5}{2}$ **f.** $\pm\sqrt{3}$ **g.** $\pm\frac{3\sqrt{5}}{2}$ **h.** -11 or 1 **i.** 10 or -6
j. $\frac{7}{2}$ or $\frac{-3}{2}$ **k.** $\pm 5i$ **l.** $\pm 3\sqrt{2}i$ **3. a.** 0 or -5 **b.** -4 or -1 **c.** -8 or 3
d. 6 or -1 **e.** 0 or 4 **f.** 2 **g.** -9 or 1 **h.** 0 or 6 **4. a.** $-1 \pm \sqrt{2}$
b. $\frac{3 \pm \sqrt{41}}{4}$ **c.** $\frac{1}{3}$ or -1 **d.** $\frac{1 \pm \sqrt{13}}{2}$ **e.** $\frac{3}{2}$ or -7 **f.** $5 \pm \sqrt{35}$
g. $\frac{-1}{2}$ or 5 **h.** $\frac{1 \pm \sqrt{5}}{2}$ **i.** $1 \pm i$ **j.** $\frac{-3 \pm \sqrt{41}}{4}$ **k.** $\frac{3 \pm \sqrt{23}i}{4}$ **l.** 1
5. a. 3 or -1 **b.** $\frac{13 \pm 3\sqrt{21}}{10}$ **c.** 1 or $\frac{3}{2}$ **d.** $-1 \pm \sqrt{3}$ **e.** $\frac{1 \pm \sqrt{37}}{6}$
6. a. 0, 2 **b.** 1, 6 **c.** $\frac{\sqrt{10}}{2}$, $\frac{-\sqrt{10}}{2}$ **d.** 1, $\frac{-2}{3}$ **e.** $-1 + \sqrt{5}$,
$-1 - \sqrt{5}$ **7. a.** $x \le -2$ or $x \ge 2$ **b.** $0 < x < 8$ **c.** $-2 < x < 6$
d. $x < -3$ or $x > 1$ **e.** no values **f.** $-2 < x < 1$ **8.** 10 m/s
9. a. 4 s **b.** 1 s or 3 s **10.** $t > 6$ **11. a.** $-1 \pm \sqrt{1 - k}$
b. $\frac{2k \pm \sqrt{k^2 - k}}{k}$ **c.** 1 or $-k$ **12. a.** $k \le 1$ **b.** $k < 0$ or $k \ge 1$
c. no restrictions on k **13. a.** $x = 3y$ or $x = y$ **b.** $x = 1$
c. $x = \frac{(3 + \sqrt{3})}{2}y$ or $x = \frac{3 - \sqrt{3}}{2}y$ **14. a.** $\pm 3, \pm 1$ **b.** $\pm 3, \pm\sqrt{5}$
c. $\pm\sqrt{10}, \pm i$ **15. a.** $-3, 1$ **b.** $1 \pm \sqrt{3}$ **c.** $\frac{2}{3}$
16. a. $p + q = \frac{-b}{a}$, $pq = \frac{c}{a}$ **c.** $x^2 - 9x + 4 = 0$

Exercise 3.2

1. 11, 13 or -11, -13 **2.** 24, 72 **3.** 3, 5, 7 or 1, 3, 5 **4.** 6.25 or
0.16 **5.** 3.5 m by 6.0 m **6.** 120 cm and 50 cm **7.** 4000 cm³
8. 108 m **9.** $4\sqrt{5}$ cm **10.** 5 cm **11.** 50 cm by 30 cm

12. 240 km/h **13.** 64 km/h **14.** 32 **16.** 6 km/h **17.** $y = 2x - 3$
or $y = 2x + 2$

Exercise 3.3

1. a. 12 **b.** 9 **c.** 25 **d.** 0 **e.** 40 **f.** 25 **g.** 0 **h.** 9 **i.** 41 **2. a.** $D > 0$
b. $D > 0$ **c.** $D < 0$ **d.** $D = 0$ **3. a.** real, unequal **b.** non-real
c. real, unequal **d.** real, equal **e.** real, unequal **f.** non-real **g.** real,
unequal **h.** real, unequal **i.** real, unequal **j.** real, equal **k.** real,
unequal **l.** real, unequal **4. a.** $k = 9$ **b.** $k \in R$ **c.** $k > 1$ **d.** $k = 0$ or
$k = 8$ **e.** $\frac{-3}{4} < k < \frac{3}{4}$ **5.** $0 < k < 4$ **6.** $b^2 - 4ac$ is a square
7. on the x-axis **9. a.** $m = 1$ **b.** $m > -3$, $m \ne -1$
10. $k \le \frac{25}{48}$ **11.** $a > 3 + 2\sqrt{2}$ or $a < 3 - 2\sqrt{2}$, $a \ne 0$

Exercise 3.4

1. a. $(0, 9)$ **b.** $(2, 4)$ **c.** $(2, 0)$ **d.** $(3, -4)$ **e.** $(0, 4)$ **f.** $(-2, -9)$
g. $(2, -21)$ **h.** $(1, 7)$ **2. a.** minimum -5, $x = 5$ **b.** minimum -18,
$x = -3$ **c.** maximum 105, $x = 10$ **d.** minimum -60, $x = -4$
e. maximum 21, $x = -1$ **f.** maximum 9, $x = 0$
g. minimum $\frac{-71}{4}$, $x = \frac{5}{2}$ **h.** maximum $\frac{23}{4}$, $x = \frac{1}{2}$ **3.** 6 and -6
4. 200 **5.** $\frac{-1}{4}$ **6.** $\frac{1}{2}$ **7.** 22 500 m², 150 m by 150 m **8.** 15 000 m²,
100 m by 150 m **9.** 45 000 m², 150 m by 300 m **10.** 320 m
11. 21.2 m, 4 s **12.** $1000 **13.** $8.50, $162 **14.** 130 **15.** 7.5 cm
16. 5 m × 10 m × 10 m × 5m **18.** 1500 cm² **19.** $\frac{(a - b)^2}{2}$ **20.** $\frac{1}{2}$

Exercise 3.5

1. a. $3 - 7i$ **b.** 7 **c.** $-1 - 6i$ **d.** $-1 + i$ **e.** 4 **f.** $-3 + i$ **2. a.** 6
b. -1 **c.** $4 + 4i$ **d.** $-3 + 9i$ **3. a.** $-4 + 7i$ **b.** $7 - i$ **c.** $-10 + 10i$
d. 41 **e.** 3 **f.** $5 - 12i$ **4. a.** $10 - 4i$ **b.** $-4 + 4i$ **c.** $-4 + 2i$
d. $3 - 2i$ **e.** $10i$ **f.** $2 - 11i$ **5. a.** $3 + 4i$ **b.** 6 **c.** $-8i$ **d.** 25
6. a. $\frac{9}{5} - \frac{3}{5}i$ **b.** $\frac{6}{5} + \frac{3}{5}i$ **c.** $\frac{1}{2} - 2i$ **d.** $\frac{3}{5} - \frac{4}{5}i$ **e.** 3 **f.** $\frac{4}{7} + \frac{2\sqrt{3}}{7}i$
g. $\frac{-5}{2} + \frac{1}{2}i$ **h.** $2 + i$ **7. a.** $7 - 2i$ **b.** $-1 - 9i$ **c.** $15 - 5i$
d. $\frac{3}{50} + \frac{1}{50}i$ **8. a.** $-4 + 3i$ **b.** $1 - i$ **c.** $2 - i$ **d.** $-1 - \frac{3}{2}i$
e. $\frac{3}{2} - \frac{1}{2}i$ **f.** $1 + i$ or $1 - i$ **9. a.** $2i, -4, -2i, -4$
b. $-2i, -4, 2i, -4$ **c.** $1 + i - 1, -i, -1 + i, 1 - i$

3.6 Review Exercise

1. a. $\pm\sqrt{5}$ **b.** 0, 5 **c.** $\frac{5 \pm \sqrt{5}}{2}$ **d.** $\frac{5}{2}$, 1 **e.** $\pm 2i$ **f.** -3, 1
g. $-1 \pm \sqrt{5}$ **h.** $\frac{-1 \pm \sqrt{7}i}{2}$ **i.** $\frac{3}{4}$, -2 **2. a.** $-10 + 6\sqrt{3}$
b. $3 \pm 2\sqrt{3}$ **c.** 1, 0 **3. a.** $x \le -2$ or $x \ge 2$ **b.** $0 < x < 3$ **c.** $x \in R$
d. $x \le -3$ or $x \ge 3$ **e.** $-1 < x < 5$ **f.** $-2 < x < 1$
4. a. x-intercepts: 0, 2; vertex: $(1, -2)$
b. x-intercepts: $\frac{2 + \sqrt{2}}{2}$, $\frac{2 - \sqrt{2}}{2}$; vertex: $(1, -1)$
c. x-intercept: 1; vertex: $(1, 0)$ **d.** x-intercept: $2\sqrt{2}$, $-2\sqrt{2}$;
vertex: $(0, 8)$ **e.** no x-intercepts; vertex: $\left(\frac{-3}{2}, \frac{1}{2}\right)$
f. x-intercepts: -1, $\frac{1}{4}$; vertex: $\left(\frac{-3}{8}, \frac{25}{16}\right)$ **5.** Perimeter is $6x$,
area is $(x^2 + x)$. **6.** 15, 17 and -15, -17 **7.** 63 cm **8.** 51
9. 12 **10. a.** real, unequal **b.** non-real **c.** non-real **11. a.** ± 6 **b.** 2
12. a. $m < 2$ **b.** $m < 0$ or $m > 32$ **13.** $-6 < b < 6$
14. a. minimum -10 at $x = -4$ **b.** minimum $\frac{-9}{4}$ at $x = \frac{3}{4}$
c. maximum 5 at $x = -1$ **d.** minimum 0 at $x = \frac{-5}{2}$
e. maximum $\frac{-1}{2}$ at $x = \frac{3}{2}$ **f.** maximum 9 at $x = 0$ **15. a.** 4 m
b. 84 m **c.** 8.2 s **16.** 6 s **17.** 15 cm by 30 cm **18.** $6.20
19. a. $-2 - 4i$ **b.** $-1 + 2i$ **c.** $-3 + 4i$ **d.** -2 **e.** 5

Cumulative Review 1–3

1. a. $-4x^2 + 6x - 8$ b. $6x^2 - 8$ c. $-x^3 + 3x^2 - x$ 2. a. 3
b. -45 c. $\frac{7}{4}$ d. $8x - 4x^2$ e. $-x^2 + 8x - 12$ 3. a. $6\sqrt{30}$ b. 6
c. $\sqrt{2}$ d. 6 e. 10 f. $12 + 7\sqrt{2}$ 4. a. $4x(x-4)$ b. $4(x-5)(x+1)$
c. $(x+2)^2(x-2)^2$ d. $3x(x-2)^2(x+2)$ e. $(4x+1)(x-4)$
f. $(4x-3y)(3x+2y)$ 5. a. $\frac{x}{x-2}$ b. $\frac{x+2}{2}$, $x \neq 2$ c. $\frac{x-5}{1-x}$,
$x \neq -1$ d. $\frac{x-3}{2x-3}$, $x \neq \frac{1}{2}$ 6. a. $\frac{x}{3(x+3)}$, $x \neq 0, 3$ b. $\frac{x^2}{2(x-2)}$,
$x \neq 1, 0$ c. $\frac{2}{1-x}$, $x \neq 1, -1, 2$ d. $\frac{x^2+2x-2}{x(x-1)}$ e. $\frac{2}{x^2-1}$
f. $\frac{-(x^2+4)}{2x(x+2)}$ 7. a. -2 b. $4, -1$ c. $0, 4$ d. $-3 \pm \sqrt{17}$ e. $6, -2$
f. $\frac{2}{3}, -\frac{3}{2}$ 8. a. $x \leq \frac{5}{2}$ b. $-4 \leq x \leq 2$ c. $x > 5$ or $x < -2$
9. a. $x \in R, y \geq 1, y \in R$ b. $x \geq 4, x \in R, y \geq 0, y \in R$
c. $x \in R, y \leq 4, y \in R$

10.	x-intercepts	y-intercepts
a.	$\pm\sqrt{2}$	8
b.	12, −3	−36
c.	2, $\frac{3}{2}$	6
d.	4	2

11. a. vertex (3, −36); min. value b. vertex (−1, 7); max. value
c. vertex (−1, 5); min. value 12. $0 < k < 16$
13. a. $-3 \leq x \leq 4, y \in R$ b. no d. yes 15. Reflect $\frac{2}{x-3}$ in the
x-axis and translate up one unit. 16. a. $3 + 6i$ b. $13 - i$
c. $15 - 8i$ d. $\frac{6}{5} + \frac{3}{5}i$ 17. 10.5 m 18. a. 500 m b. 20 s 19. 150
20. 300 m by 600 m

Chapter 4

Exercise 4.1

1. a. 81 b. 3 c. 16 d. $\frac{1}{8}$ e. $\frac{-8}{27}$ f. -16 g. $\frac{4}{25}$ h. $\frac{-27}{8}$ i. $\frac{-4}{3}$
2. a. 5^{11} b. a^8 c. r^{10} d. $\left(\frac{5}{7}\right)^{15}$ e. $-m^5$ f. $(2ab)^{13}$ 3. a. 6^6 b. t^6
c. $-128m^7$ d. $2q^2$ e. $5k^8$ f. $4b^2$ 4. a. m^2 b. -3^{17} c. a^{10} d. $64c^6$
e. 2^6 f. $\frac{81}{256}$ 5. a. 4^{18} b. $\left(\frac{3}{2}\right)^{15}$ c. m^{18} d. 3^{15} e. k^5 f. h^{13} 6. a. $6m^{13}$
b. $2f^5$ c. $5y^4$ d. $2p$ e. 6 f. $2x^2$ 7. a. $8x^{21}$ b. $-128m^{14}$ c. $16a$
d. $48e^{18}$ e. $25m^4$ f. $-72p^8$ 8. a. not b. yes c. not d. not
9. a. 11 664 b. $\frac{-9}{8}$ c. 12 d. $\frac{729}{4}$ 10. a. x^2yw^3 b. $3a^2b^5$ c. 432
d. 288 11. a. 10^3 b. 4^4 c. 4^3 d. 2^9 12. 5^{24} 13. a. 27 b. 8 c. 32 d. 25

14. a.

a	1	2	3	6
b	6	3	2	1

b.

a	1	2	3	6	9	18
b	18	9	6	3	2	1

c.

a	1	3	9
b	9	3	1

d.

a	1	2	4	8
b	8	4	2	1

15. a. x^{2p} b. x^{5p-q} c. $y^{a^2-b^2-a+b}$ d. $m^{p^3-q^3}$ e. $x^{a+b} \cdot y^{-a+b}$ f. r^{20x+4y}
16. a. 10 125 b. 7

Exercise 4.2

1. a. $\frac{1}{5^3}$ b. 4^3 c. x^3y^2 d. $\left(\frac{5}{3}\right)^4$ e. $\frac{2y^3w^4}{3x^2}$ f. $\frac{b^2}{a^3}$ 2. a. $\frac{1}{8}$ b. $\frac{1}{25}$ c. $\frac{27}{125}$ d. 1
e. 18 f. $\frac{1}{40}$ g. $\frac{25}{729}$ h. $\frac{16}{9}$ i. 1 3. a. m^3 b. a^2 c. $\frac{y}{x^2}$ d. b^3 e. x^3 f. $\frac{30}{r^7}$
g. $4a^7$ h. $\frac{6}{a}$ 4. a. $\frac{3}{4}$ b. 36 c. $\frac{24}{25}$ d. -14 e. -4 f. $\frac{5}{2}$ g. 2 h. $\frac{1}{9}$
5. a. $\frac{1}{27}$ b. $\frac{1}{2}$ c. $\frac{1}{32}$ d. 625 6. a. 2^{4a+4} b. 3^{-2m-14} 7. a. $\frac{-5}{6}$ b. 4

Exercise 4.3

1. a. $14^{\frac{1}{3}}$ b. $5^{\frac{1}{3}}$ c. $9^{\frac{1}{4}}$ d. $(-8)^{\frac{1}{5}}$ e. $35^{\frac{1}{2}}$ f. $-4^{\frac{2}{3}}$ g. $5^{\frac{1}{3}}$ h. $(-6)^{\frac{3}{5}}$ i. $3^{\frac{3}{4}}$
2. a. $\sqrt{12}$ b. $\sqrt[3]{5}$ c. $\sqrt[5]{7}$ d. $-\sqrt[3]{6}$ e. $-\sqrt[4]{9}$ f. $\sqrt{7^3}$ g. $\sqrt[5]{8^2}$ h. $\sqrt{19^5}$

i. $\sqrt[3]{13^2}$ 3. a. 28.730 b. 9.699 c. 0.004 4. a. 2 b. 3 c. 5 d. 4 e. $\frac{1}{7}$
f. -16 g. $\frac{1}{8}$ h. $\frac{1}{32}$ i. $\frac{1}{9}$ 5. a. $\frac{1}{3}$ b. 2 c. $\frac{1}{2}$ 36 e. $\frac{5}{6}$ f. 1 g. $\frac{3}{4}$ h. $\frac{7}{6}$
6. a. -8 b. $\frac{-1}{16}$ c. 12 d. 2 e. 3 f. $\frac{11}{18}$ 7. a. $\frac{1}{5}$ b. 6 c. -128 d. $\frac{5}{9}$
e. $\frac{125}{8}$ f. 1 g. $\frac{7}{12}$ h. $\frac{27}{8}$ 8. a. $\frac{3}{5}$ b. $\frac{1}{32}$ c. $\frac{9}{4}$ d. $\frac{-1}{3}$ e. $\frac{4}{27}$ 9. a. $a^{\frac{2}{3}}$
b. $m^{\frac{-3}{20}}$ c. $p^{\frac{-28}{15}}$ d. $\frac{1}{2}x^{\frac{1}{2}}$ e. $x^{\frac{3}{4}}$ f. $x^{\frac{3}{5}}$

Exercise 4.4

4. a. 4, 20, 100, 500, 2500 b. 5 c. same as the base 5. a. $\frac{-2}{3}$,
$\frac{-2}{9}, \frac{-2}{27}, \frac{-2}{81}, \frac{-2}{243}$ b. $\frac{1}{3}$ c. same as the base 6. $y = 7^x$: a. 6, 42,
294, 2058, 14 406 b. 7 c. same as the base; $y = \left(\frac{1}{4}\right)^x$: a. $\frac{-3}{4}$,
$\frac{-3}{16}, \frac{-3}{64}, \frac{-3}{256}, \frac{-3}{1024}$ b. $\frac{1}{4}$ c. same as the base; $y = 6^x$: a. 5, 30,
180, 1080, 6480 b. 6 c. same as the base; $y = \left(\frac{1}{5}\right)^x$: a. $\frac{-4}{5}, \frac{-4}{25}$,
$\frac{-4}{125}, \frac{-4}{625}, \frac{-4}{3125}$ b. $\frac{1}{5}$ c. same as the base

7.

	a.	b.	c.
x	y	First Difference	Second Difference
0	1	1	1
1	2	2	2
2	4	4	4
3	8	8	8
4	16	16	16
5	32	32	32
6	64	64	
7	128		

d. 1, 2, 4, 8, 16 8. $y = 2^{-x}, y = 5^{-x}, y = 10^{-x}, y = 4^{-x}, y = 12^{-x}$
9. $f(n) = 5^n, f(n+1) = 5^{n+1}, f(n+1) - f(n) = 4 \cdot 5^n$
10. $g(n) = b^n, g(n+1) = b^{n+1}, g(n+1) - g(n) = (b-1)b^n$
12. b. (0, 1), (1, 2), (4.2575, 19.126) c. i) $x < 0, 1 < x < 4.2575$
ii) $0 < x < 1, x > 4.2575$

Exercise 4.5

1. i) a. 1 b. 2 c. 2 d. $y = 2^x$ e. 4, 8, $\frac{1}{2}, \frac{1}{8}$ ii) a. 1 b. 4 c. 4 d. $y = 4^x$
e. 16, 64, $\frac{1}{4}, \frac{1}{64}$ iii) a. 1 b. 8 c. 8 d. $y = 8^x$ e. 64, 512, $\frac{1}{8}, \frac{1}{512}$
2. i) a. 1 b. 2 c. $\frac{1}{2}$ d. $y = \left(\frac{1}{2}\right)^x$ e. $\frac{1}{4}, \frac{1}{8}, \frac{1}{2}, 8$ ii) a. 1 b. 4 c. $\frac{1}{4}$
d. $y = \left(\frac{1}{4}\right)^x$ e. $\frac{1}{16}, \frac{1}{64}, \frac{1}{4}, 64$ iii) a. 1 b. 8 c. $\frac{1}{8}$ d. $y = \left(\frac{1}{8}\right)^x$
e. $\frac{1}{64}, \frac{1}{512}, \frac{1}{8}, 512$

Exercise 4.6

1. a. 6 b. $\frac{11}{3}$ c. $\frac{-3}{2}$ d. -2 e. 2 or 1 f. -1 g. $\frac{-6}{7}$ h. -2 or -3
2. a. 4 b. $\frac{-3}{2}$ c. 1 d. -2 or -1 3. a. 9 b. 9 or -1 c. 0 or -2
d. $\frac{-1}{2}$ 4. a. 2 or 1 b. 3 or 1 c. 2 or 1 d. 1 or 2 e. 3 or 2 f. 1 or -1
g. 2 h. 1 5. a. 2 or 1 b. -6 or 1 c. 2 d. 1 e. 9 f. 2 g. $\frac{-8}{3}$ h. $\frac{2}{3}$
6. a. 4.209 b. 9.850 c. 6.034 d. 7.555 e. 3.253 f. 3.548
7. a. -1 or 1 b. 2 c. 3 or 2 d. 0 e. 1 8. a. 2.756, -3.936
b. $-2.978, 1.334$ c. $-1.979, 1.142$ 9. a. $x = 2, y = 1$
b. $x = 3, y = -2$

Exercise 4.7

1. a. $\log_3 243 = 5$ b. $\log_2 128 = 7$ c. $\log_7 343 = 3$
d. $\log_{11} 14\,641 = 4$ 2. a. $4^5 = 1024$ b. $7^1 = 7$ c. $6^5 = 7776$
d. $a^r = p$ 3. a. 2 b. 3 c. 0 d. 5 4. a. 2.51 b. 2.33 c. -3 d. -1.41
5. a. 1.56 b. 1.49 c. 2.01 d. 2.03 6. a. 3.15 b. 2.60 c. 2.20
d. 17.75 e. 4.25 f. 65.38 g. 19.58 h. 23.95 i. 68.06

4.8 Review Exercise

1. a. x^8 b. $-32a^5b^2$ c. $9ab^2$ d. 1 e. $27b^9$ f. $2a^2b$ 2. a. $\frac{1}{9}$ b. 8 c. $\frac{9}{16}$
d. 36 e. $\frac{16}{3}$ f. $\frac{1}{3}$ 3. a. $22^{\frac{1}{3}}$ b. $2^{\frac{4}{5}}$ c. $12^{\frac{1}{2}}$ 4. a. $\sqrt{3}$ b. $\sqrt[5]{17^2}$ c. $-\sqrt[4]{4}$

5. a. $\frac{6}{5}$ **b.** $\frac{1}{6}$ **c.** 16 **d.** $-\frac{4}{3}$ **6. a.** 16 **b.** 3 **7. a.** $\frac{1}{8}$ **b.** $\frac{1}{3}$ **c.** $\frac{2}{5}$ **d.** $\frac{25}{36}$
8. a. 3 **b.** $\frac{-3}{2}$ **c.** 1 or -2 **d.** 2 or 0 **9. a.** 5.79 **b.** 4.59 **c.** 23.45
d. 3.32 **10. a.** x^{a+4b} **b.** $y^{a^2-b^2}$ **c.** $5^{-4m+6n-3}$ **d.** $x^{\frac{1}{4}}$

Chapter 5

Exercise 5.1

1. a. 81, 243 **b.** 16, 20 **c.** $\frac{1}{3}$, $\frac{1}{9}$ **d.** 85, 80 **e.** $\frac{1}{16}$, $\frac{1}{32}$ **f.** 5000,
50 000 **g.** -8, -10 **2. a.** 3, 6, 9, 12 **b.** -1, 0, 1, 2 **c.** 1, $\frac{1}{2}$, $\frac{1}{3}$, $\frac{1}{4}$
d. 0, 3, 8, 15 **e.** 2, 1, 0, -1 **f.** 1, $\frac{4}{3}$, $\frac{3}{2}$, $\frac{8}{5}$ **g.** 3, 12, 27, 48 **h.** 0,
-7, -26, -63 **i.** 5, 7, 9, 11 **j.** $\frac{1}{3}$, $\frac{1}{9}$, $\frac{1}{27}$, $\frac{1}{81}$ **3. a.** 5, 7, 9, 11
b. -5, -10, -20, -40 **c.** 3, 3, 3, 3 **d.** -1, 1, 1, 1 **4. a.** add 5
b. multiply by 6 **c.** add 9 **d.** multiply by 2 **5. a.** $2n$
b. $(-1)^{n+1}$ **c.** 2^n **d.** 2^{2n-1} **e.** $-2(3)^{n-1}$ **f.** $\frac{1}{n^2}$ **6. a.** 1, 1, 2, 3, 5, 8
b. 3, -2, 7, 1, 22, 25

Exercise 5.2

1. a. 2 **c.** 5 **d.** -3 **f.** $-\frac{1}{2}$ **g.** 2.5 **h.** $2x$ **j.** $-b$ **2. a.** 3, 9, 15, 21
b. 6, 4, 2, 0 **c.** -1, 6, 13, 20 **d.** 10, 5.5, 1, -3.5 **e.** $2x$, $5x$, $8x$,
$11x$ **f.** 3, $3\frac{1}{4}$, $3\frac{1}{2}$, $3\frac{3}{4}$ **g.** $\frac{3}{2}$, $\frac{13}{6}$, $\frac{17}{6}$, $\frac{21}{6}$ **h.** $x + 1$, $2x + 3$, $3x + 5$,
$4x + 7$ **i.** 7, 11, 15, 19 **j.** 7, 3, -1, -5 **3.** 5, 7, 9, 11, 13
4. a. $t_n = 2n - 1$, 62 **b.** $t_n = 80 - 5n$, 21 **c.** $t_n = \frac{2n+3}{4}$, 19
d. $t_n = 0.5n + 3.5$, 41 **e.** $t_n = 3n - 48$, 21
f. $t_n = 4\sqrt{2}n - 3\sqrt{2}$, 14 **5. a.** $3n - 2$ **b.** $2n$ **c.** $7n - 24$
d. $30 - 5n$ **e.** $24 - 4n$ **f.** $15 - 2n$ **6.** 37 **7.** 165 **8.** 53 **9.** 17
10. 37 **11.** 7 **12.** 59 **13.** $10 - \frac{1}{2}n$ **15.** $4n - 2$ **16.** 8 **17.** 2

Exercise 5.3

1. a, c, and g are arithmetic; b, e, and f are geometric; d and h
are neither **2. a.** 6, -12, 24 **b.** 1, 5, 25 **c.** -2, 6, -18 **d.** 5, $\frac{5}{2}$, $\frac{5}{4}$
e. -3, -6, -12 **f.** p, $2pq$, $4pq^2$ **3. a.** -2^n **b.** $6 \cdot 3^{n-1}$ **c.** $\frac{(-5)^{n-1}}{2}$
d. $\sqrt{3} \cdot \sqrt{2}^{n-1}$ **e.** $\frac{-2 \cdot 5^{n-1}}{3^n}$ **f.** $\frac{m^{2n}}{(3p)^{n-1}}$ **4. a.** 3^{n-1}, 729 **b.** -2^{5-n},
$\frac{-1}{8}$ **c.** $-(-10)^{n-3}$, 1000 **d.** $(-1)^{n-1}$, -1 **e.** $\sqrt{3}^n$, 81
f. $\frac{a^n}{2^{n-1}b^{n+1}}$, $\frac{a^9}{256b^{10}}$ **5. a.** 10 **b.** 9 **c.** 11 **d.** 7 **e.** 8 **f.** 6 **6. a.** 162 or
-162 **b.** 3 or -3 **c.** 1280 **d.** 4 **e.** $\frac{3}{4}$ or $\frac{-3}{4}$ **f.** 36 **7. a.** 58 **b.** 9
c. $4 \cdot 3^{n-1}$ **d.** 30 **8.** -3 **9. a.** -3 **b.** 6 **10.** 512 **11.** $288.12
12. 1 600 000 **13.** 4, -12, 36, -108; $4(-3)^{n-1}$
14. a. arithmetic; $8 - 3n$ **b.** geometric; $5(-3)^{n-1}$ **15.** 14, 25, 36
and -26, -15, -4 **16.** $\frac{-1}{3}$, 9 **17.** -6 **18. a.** 1, $\frac{4}{3}$, $\frac{5}{3}$, 2, ...
b. 106

Exercise 5.4

1. a. 105 **b.** -120 **c.** 105 **d.** 1120 **e.** -150 **f.** 0 **2. a.** 2, 8, 18, 32,
50 **b.** 2, 6, 10, 14, 18 **c.** 78 **3. a.** 60 300 **b.** $-180\ 000$ **c.** 2190
d. 157 200 **e.** 5025 **f.** 40 000 **4. a.** 250 500 **b.** $-125\ 250$ **c.** 42
d. 210.9 **5.** 3, 7, 11, 15; $4n - 1$ **6.** 210 **7.** $535.50 **8.** 469 **9.** 945
10. $\frac{15n(n+1)}{2}$ **12.** $\frac{n(3n+1)}{2}$ **13. a.** 1275 **b.** 5050 **c.** 3775 **14. a.** n^2
b. $n(n + 1)$ **c.** sum of the first $2n$ natural numbers
15. a. $4n - 5$ **b.** $6n + 1$ **16.** $\frac{n}{2}[2p(n - 3) + q(n + 5)]$ **17.** $\frac{-1}{2}$
18. 1991 **19.** 2700

Exercise 5.5

1. a. 242 **b.** 0 **c.** 31 100 **d.** 215 **2. a.** 1275 **b.** 3280 **c.** -85
d. -8500 **3. a.** 88 572 **b.** 6560 **c.** -1456 **d.** 855 **4.** 5 **5. a.** $\frac{255}{8}$
b. $\frac{55}{8}$ **c.** 58 824 **d.** 2044 **6.** 121 **7. a.** $638 140.78
b. $3 400 956.41 **8.** $12 750 **9. a.** 124 **b.** 496 **10. a.** 260.4 mg

b. 260.417 mg **c.** $\frac{250(1 - 0.4^n)}{0.96}$ **12. a.** 29 m **b.** 52.4 m
13. b. i) 8 **ii)** $\frac{8}{3}$ **iii)** $\frac{15}{2}$ **iv)** 18 **14. a.** 1, 1, 1, ... and 4, -2, 1, $-\frac{1}{2}$,
$\frac{1}{4}$, ... **b.** $-\frac{7}{3}$ **15. a.** $\frac{9}{4}$, $\frac{16}{7}$, $\frac{23}{10}$ **b.** $\frac{9}{4}$, $\frac{1}{28}$, $\frac{1}{70}$
c. $a_n = \frac{1}{(3n + 1)(3n - 2)}$ **d.** $\frac{7}{3}$

5.6 Review Exercise

1. a. 4, 7, 10, 13 **b.** -2, 1, 6, 13 **c.** 3, 5, 9, 17 **d.** 1, $\frac{4}{3}$, $\frac{3}{2}$, $\frac{8}{5}$
e. -1, 6, 25, 62 **f.** 1, 3, 6, 10 **2.** 3, -3, -15 **3. a.** $4n - 2$
b. $2 \cdot 3^{n-1}$ **c.** $18 - 5n$ **d.** $\frac{n-1}{n^2}$ **4. a.** arithmetic, $3n + 2$, 62
b. geometric, 3^n, 6561 **c.** arithmetic, $60 - 6n$, -30
d. geometric, $\frac{(-1)^{n-1}}{3 \cdot 2^{n-1}}$, $\frac{-1}{384}$ **e.** arithmetic, $-2 - 3n$, -56
5. a. arithmetic, 32 **b.** geometric, 7 **c.** geometric, 10
d. arithmetic, 25 **6.** 126 **7. a.** 48 m **b.** 15.7 m **8. a.** arithmetic,
13 050 **b.** geometric, -341 **c.** geometric, $\frac{2059}{2187}$
d. arithmetic, 5777 **9.** $x = -4$; -3, 4, 11 **10.** 12 **11.** 330
12. 3995 **13.** $x = 18$, $y = 12$ **14.** $n = 20$

Chapter 6

Exercise 6.1

1. a. $1158.92 **b.** $1191.12 **c.** $1219.64 **d.** $1225.35
2. a. 18 years **b.** 8 years **c.** 13 years **d.** 11 years **3.** $231.70
4. the bank offering $5\frac{1}{4}$% **5.** 11 years **6.** $6584.05 **7.** 6%
8. $4500 **9. a.** $1749.19 **b.** $750.81 **10.** 8% **11.** $39 507.13
12. 7.314×10^{24} dollars **13.** Invest at $5\frac{1}{2}$% **14.** $5235.04
15. a. $8512.17 **b.** $3536.98 **16. a.** 6.09% **b.** 8.24%

Exercise 6.2

1. $911.32 **2. a.** $31 834.25 **b.** $16 229.35 **c.** $16 518.95
3. a. $416.45 **b.** $219.99 **4.** $1017.28 **5.** $5816.37
7. $101 281.17 **8.** $22 860.49 **9.** $816.53 **10.** $899.65
11. a. $11 765.60 **b.** $965.60

Exercise 6.3

1. a. $1731.79 **b.** $7897.00 **c.** $4564.61 **2.** $17 525.75
3. $1791.40, $350.60 **4.** $20 413.81 **5.** $109.71 **6.** $19 131.26
7. $17 021.28 **8.** $3260.67 **9. a.** "forever" **b.** 10 years
10. a. $444 274.26 **b.** $63 254.66

Exercise 6.4

1. $7514.40 **2.** $29 651.04 **3.** $676.58 **4.** $1200.57
5. b. $53 138.37 **6. a.** $148 394.00 **c.** $82 424.77 **d.** $73 177.55
e. 21 years **f.** $27 148.08 **7. b.** $20 666.36 **8. a. i)** $758.52,
$136 533.60 **ii)** $610.57, $183 171.00 **iii)** $579.77, $208 717.20
b. $147.95, $46 637.40 **c.** $30.80, $25 546.20 **9. a. i)** $383.88,
$115 164.00 **ii)** $457.93, $137 379.00 **iii)** $536.69, $161 007.00
10. a. i) $828.36, $248 806.40 **ii)** $662.69, $159 044.64
iii) $414.18, $99 403.20 **12. a.** $1861.05 **b.** $51 085.43
c. $4450.96

6.5 Review Exercise

1. a. $2012.76 **b.** $2088.13 **2. a.** $10 079.37 **b.** $2579.37
3. a. $6805.83 **b.** $3820.74 **4.** $1097.44 **5. a.** $19 668.05
b. $22 327.49 **6.** $6027.71 **7.** $1802.95 **8.** $149 232.95
9. $66 619.25 **10.** $753.39 **11. b. i)** $76 316.53 **ii)** $57 927.08
12. a. $1041.85 **b.** $302 575.98 **c.** $1070.32 **d.** $52 259.36

Cumulative Review 1–6

1. a. $\frac{-27}{125}$ **b.** $\frac{16}{9}$ **c.** $\frac{1}{4}$ **d.** 8 **e.** $\frac{-8}{3}$ **f.** $\frac{1}{125}$ **2. a.** $\frac{27m^5}{4}$ **b.** $\frac{3b^4}{2a}$ **c.** $3k^4$
d. $\frac{1}{3}$ **e.** x^{p-g} **f.** $x^{\frac{13}{18}}$ **3. a.** -1 **b.** -7 **c.** 3, -2 **d.** 0 **e.** 2, 3 **f.** 3.2211

4. a. 2, 7, 12, 17 **b.** 1, 3, 7, 15 **c.** 2, 9, 28, 65 **d.** $-2, -1, 1, 5$

5. a. arithmetic, $3n + 1$, 37 **b.** geometric, $3\left(\dfrac{-1}{3}\right)^{n-1}$, $\dfrac{-1}{729}$

c. arithmetic, $17 - 3n$, -43 **d.** geometric, $4\left(\dfrac{1}{10}\right)^{n-1}$, $\dfrac{1}{25\,000\,000}$

6. a. geometric, $\dfrac{1365}{128}$ **b.** arithmetic, 4810 **c.** arithmetic, -8757

d. geometric, 65 520 **7.** 24 576, $-24\,576$ **8.** 37 **9. a.** $\dfrac{-16}{5}$

b. sequence can never be arithmetic **10. a.** $13 146.74

b. $7146.74 **11.** $7083.88 **12.** 8% **13.** $6563.16 **14.** $3400.17,

$1394.07 **15. a.** $671.80 **b.** $96 540.00 **16. a.** domain:

$\{4, -3, 6, 4\}$; range: $\{0, 5, 2, -2\}$ **b.** domain: $x \in R$;

range: $y \geq 3$, $y \in R$ **c.** domain: $x \in R$, $x \neq -1$;

range: $y \in R$, $y \neq -2$ **d.** domain: $x \geq -3$, $x \in R$;

range: $y \geq -1$, $y \in R$ **17.** $f^{-1}(x) = \dfrac{-x+3}{2}$ **18. a.** -50

b. $-\dfrac{121}{8}$ **c.** $-18(x - 1)^2$ **19. a.** $6x^3 - 12x^5$ **b.** $3x^3 + 6x^2 - x - 2$

c. $2x^3 + 5x^2 - 11x + 4$ **d.** $-x^3 + 2x^2 - x + 2$

20. a. $3(x - 5)(x + 5)$ **b.** $(2x + 3)(x - 1)$

c. $(x^2 + 9)(x - 1)(x + 1)$ **d.** $(x^2 - 3)(x^2 - 7)$ **21. a.** $\dfrac{x + 2}{2}$,

$x \neq 2$ **b.** $2x^2 - x - 15$, $x \neq \dfrac{5}{2}$ **c.** $\dfrac{-x^2 - 4}{2x^2 + 4x}$ **22. a.** $\dfrac{5 \pm \sqrt{47}}{2}$

b. $x \geq 5$ or $x \leq -2$, $x \in R$ **c.** $\dfrac{-2 \pm 3i}{2}$ **23.** 21, -18

24. 20, 22, 24 **25.** 2.5 m **26. a.** 2 real, unequal roots

b. no real roots **27. a.** $(4, 5)$, minimum **b.** $(3, -2)$, maximum

28. 30, 30 **29.** 200 m by 300 m

Chapter 7

Exercise 7.1

1. a. $\dfrac{a}{r} = \dfrac{b}{s} = \dfrac{c}{t}$ **b.** $\dfrac{a}{x} = \dfrac{b}{b + 10} = \dfrac{c}{w}$ **c.** $\dfrac{m}{h} = \dfrac{h}{p}$ **2. a.** 17 **b.** 21

c. 105 **3. a.** $p = 13$, $q = 66$ **b.** $h = 56$, $d = \dfrac{169}{7}$ **4. a.** $\sin \theta = \dfrac{3}{5}$,

$\cos \theta = \dfrac{4}{5}$, $\tan \theta = \dfrac{3}{4}$ **b.** $\sin \theta = \dfrac{5}{13}$, $\cos \theta = \dfrac{12}{13}$, $\tan \theta = \dfrac{5}{12}$

c. $\sin \theta = \dfrac{8}{17}$, $\cos \theta = \dfrac{15}{17}$, $\tan \theta = \dfrac{8}{15}$ **5. a.** 0.8387, 0.5446

b. 0.8434, 1.5697 **c.** 0.2974, 0.2974 **d.** 0.7524, 0.7524

6. a. 33.0° **b.** 57.5° **c.** 17.3° **d.** 41.2° **7. a.** 226.4 **b.** 4.7 **c.** 74.0

d. 35.5° **8. a.** $\dfrac{14}{5}$ **b.** $a = 8$, $b = \dfrac{136}{15}$ **c.** $a = 16$, $b = 17\sqrt{2}$

d. $a = 3$, $b = 12$ **9.** 200 m **10.** no **11. a.** $\sin \alpha = \dfrac{5}{13}$,

$\sin \beta = \dfrac{12}{13}$, $\cos \alpha = \dfrac{12}{13}$, $\cos \beta = \dfrac{5}{13}$, $\tan \alpha = \dfrac{5}{12}$, $\tan \beta = \dfrac{12}{5}$

b. $\sin \alpha = \dfrac{35}{37}$, $\sin \beta = \dfrac{12}{37}$, $\cos \alpha = \dfrac{12}{37}$, $\cos \beta = \dfrac{35}{37}$,

$\tan \alpha = \dfrac{35}{12}$, $\tan \beta = \dfrac{12}{35}$ **c.** $\sin \alpha = \dfrac{3}{\sqrt{10}}$, $\sin \beta = \dfrac{1}{\sqrt{10}}$,

$\cos \alpha = \dfrac{1}{\sqrt{10}}$, $\cos \beta = \dfrac{3}{\sqrt{10}}$, $\tan \alpha = 3$, $\tan \beta = \dfrac{1}{3}$

d. $\sin \alpha = \dfrac{4}{\sqrt{17}}$, $\sin \beta = \dfrac{1}{\sqrt{17}}$, $\cos \alpha = \dfrac{1}{\sqrt{17}}$,

$\cos \beta = \dfrac{4}{\sqrt{17}}$, $\tan \alpha = 4$, $\tan \beta = \dfrac{1}{4}$ **12.** 15.1 m **13.** 15.5 m

14. 29.4° **15.** 60° **16.** 80 m **17.** 128 m **18.** 2279 m; 22 m

19. a. 1 **b.** 5 **c.** $\dfrac{\sqrt{6}}{4}$ **d.** 0 **e.** $\dfrac{3\sqrt{2}}{2}$ **f.** 0 **g.** $\dfrac{1 + \sqrt{3}}{2}$ **h.** $\dfrac{3}{2}$

20. a. 45.9 cm; 2638.5 cm² **21. a.** 30.5 cm; 365.6 cm² **23.** 25.7°

24. 77 m

Exercise 7.2

1. a. 390°; $-330°$ **b.** 680°; $-40°$ **c.** 210°; $-510°$ **d.** 135°;

$-585°$ **2. a.** coterminal **b.** coterminal **c.** coterminal **d.** coterminal **e.** not coterminal **f.** coterminal **3. a.** $\alpha = 180°$, $\theta = -180°$

b. $\alpha = 90°$, $\theta = -270°$ **c.** $\alpha = 135°$, $\theta = -225°$ **4. a.** 260°

b. 75° **c.** 199° **d.** 1° **e.** 243° **f.** 130° **g.** 70° **h.** 159° **5. a.** 25°

b. 150° **c.** $\alpha = 22.5°$, $\theta = 135°$ **6. a.** $-135°$ **b.** $-(135 + 360k)°$,

$k \in N_0$ **c.** 225° **d.** $(360k + 225)°$, $k \in N_0$ **e.** $P(-2\sqrt{2}, 2\sqrt{2})$,

$Q(-2\sqrt{2}, -2\sqrt{2})$

Exercise 7.3

1. a. $\sin \theta = \dfrac{4}{5}$, $\cos \theta = \dfrac{3}{5}$, $\tan \theta = \dfrac{4}{3}$ **b.** $\sin \theta = \dfrac{15}{17}$,

$\cos \theta = \dfrac{-8}{17}$, $\tan \theta = \dfrac{-15}{8}$ **c.** $\sin \theta = \dfrac{-3}{\sqrt{34}}$, $\cos \theta = \dfrac{-5}{\sqrt{34}}$,

$\tan \theta = \dfrac{3}{5}$ **d.** $\sin \theta = \dfrac{-5}{13}$, $\cos \theta = \dfrac{12}{13}$, $\tan \theta = \dfrac{-5}{12}$ **2. a.** 0.9848

b. -0.9063 **c.** -0.9659 **d.** -0.3420 **3. a.** -5.6713 **b.** 2.1445

c. -3.7321 **d.** 0.3640 **4. a.** $\dfrac{12}{13}, \dfrac{5}{13}, \dfrac{12}{5}$ **b.** $\dfrac{-21}{29}, \dfrac{20}{29}, \dfrac{-21}{20}$ **c.** $\dfrac{-8}{17}$,

$\dfrac{-15}{17}, \dfrac{8}{15}$ **d.** $\dfrac{12}{13}, \dfrac{-5}{13}, \dfrac{-12}{5}$ **5. a.** $\dfrac{-5}{12}$ **b.** $\dfrac{-15}{17}$ **c.** $\dfrac{4}{3}$ or $\dfrac{-4}{3}$ **d.** $\dfrac{\sqrt{21}}{5}$ or

$\dfrac{-\sqrt{21}}{5}$ **6. a.** $(337 + 360k)°$, $k \in N$; $\dfrac{-5}{13}$ **b.** $-118°$; $\dfrac{-8}{17}$ **7. a.** $\dfrac{5}{13}$,

$\dfrac{5}{12}, \dfrac{5}{12}$ **b.** $\dfrac{8}{17}, \dfrac{-15}{17}, \dfrac{-8}{15}$ **c.** $\dfrac{-15}{17}, \dfrac{15}{8}, \dfrac{15}{8}$ **d.** $\dfrac{21}{29}, \dfrac{-20}{21}, \dfrac{-20}{21}$

8. a. 74°, 286° **b.** 9°, 171° **c.** 121°, 239° **d.** 79°, 259° **e.** 202°, 338°

f. 148°, 328° **9. a.** $\sin 300° = \dfrac{-\sqrt{3}}{2}$, $\cos 300° = \dfrac{1}{2}$,

$\tan 300° = -\sqrt{3}$ **b.** $\sin 225° = -\dfrac{1}{\sqrt{2}}$, $\cos 225° = -\dfrac{1}{\sqrt{2}}$,

$\tan 225° = 1$ **10. a.** $(-16.4, -11.5)$ **b.** $(80.3, -95.8)$

c. $(-8.2, 2.2)$ **d.** $(\cos \alpha, \sin \alpha)$ **11. a.** $\sin \beta = \dfrac{15}{17}$, $\cos \beta = \dfrac{8}{17}$,

$\tan \beta = \dfrac{15}{8}$ **b.** $\dfrac{-15}{8}$; $\dfrac{15}{17}$ **c.** $(-8, -15)$; $(8, -15)$ **d.** $\dfrac{-8}{17}$; $\dfrac{-15}{17}$

12. a. $\sin \theta$ **b.** $\cos \theta$ **c.** $-\tan \theta$ **d.** $-\sin \theta$ **e.** $-\cos \theta$ **f.** $\tan \theta$

g. $-\sin \theta$ **h.** $\cos \theta$ **i.** $-\tan \theta$ **j.** $-\sin \theta$ **k.** $\cos \theta$ **l.** $-\tan \theta$ **14.** 8

Exercise 7.4

1. a. 29.9 **b.** 6.6 **c.** 72° or 108° **2. a.** 5.4 **b.** $a = 200.5$, $b = 63.3$

3. a. 35.26° **b.** $\theta = 47.36°$, $\beta = 59.39°$ **4.** Parts a, b, and d will

solve to a unique solution. **5. a.** 10.4 **b.** 43.6° **c.** 53.9° **6. a.** 35.8

b. 72 880.6 **c.** 691.6 **7.** 1652 m **8.** 68.3° or 111.7°

9. a. $A = 79°$, $b = 13.6$, $c = 17.5$ **b.** $L = 37.9°$, $M = 37.1°$,

$m = 68.7$ **c.** $R = 68.3°$, $t = 14.5$, and $T = 63.7°$ or

$R = 111.7°$, $T = 20.3°$, and $t = 5.6$ **10.** 9200 m **11. a.** 7.32 m

b. 4.65 m **c.** 2.29 m **12. a.** 106.5 cm **b.** 513 cm² **13.** 3438 m

15. a. 5.3 cm **b.** 31.9 cm²

Exercise 7.5

1. a. ± 5.39 **b.** 5.80, -13.80 **2. a.** 29.54° **b.** 102.91° **3.** Unique

solutions possible for parts a, b, and d. **4. a.** 56.3° **b.** 782.2

c. 23.0; 7.6 **5. a.** 46.5° **b.** 11.5 **c.** 65.2° **6.** 68.8; 13.1

7. a. $A = 108.0°$, $C = 30.8°$, $B = 41.2°$ **b.** $q = 4.4$, $R = 41.7°$,

$P = 91.3°$ **c.** $X = 25.2°$, $Z = 24.8°$, $z = 9.9$ **8. a.** 9.24 m **b.** 41°;

41° **9.** 1293 sq. units **10. a.** 2 cm **b.** $10\sqrt{3}$ cm² **11.** 70.7

13. 3.489° **14.** 30.3 **15.** $\dfrac{b^2 + c^2 - (a^2 + d^2)}{2(ad + bc)}$

Exercise 7.6

1. 12.8 km **2.** 450 m **3.** 11° **4. a.** 10.5 **b.** 18.4 **c.** 70.9; 238.3

5. 66.6 cm² **6. a.** 73 km **b.** $S12°W$ **7.** 86.9 **8.** 9820 m

9. 4 h 7.8 min **10.** 9590 m; 5320 m **11.** 5400 m **12.** 68.7 m

13. 84 km **14.** 1.52 au

7.7 Review Exercise

1. a. 10 **b.** 3 **c.** 16 **2. a.** $\dfrac{8}{17}, \dfrac{15}{17}, \dfrac{8}{15}$ **b.** $\dfrac{-9}{41}, \dfrac{40}{41}, \dfrac{-9}{40}$ **3. a.** 67.97

b. 98.01 **c.** 65.27 **d.** 27.47 **e.** $e = -19.15$, $k = 16.07$ **f.** 240.95°

4. 190 m **5. a.** 29°, 331° **b.** 15°, 165° **c.** 54°, 234° **d.** 198°, 342°

e. 161°, 199° **f.** 157°, 337° **6. a.** $\dfrac{-4}{5}$ **b.** $\dfrac{-40}{9}$ **c.** $\dfrac{-5}{\sqrt{74}}$ **7. a.** 9.4

b. 47.8° or 132.2° **c.** 7.6 **d.** 134.4° **8. a.** $a = 17.4$, $b = 9.6$, $\theta = 25°$ **b.** $\theta = 78.5°$ **c.** $a = 7.9$ **d.** $a = 63.4$, $\theta = 97.0°$

9. a. 51° **b.** 51.2 m **10. a.** $\frac{-9}{41}$, 1 **b.** $\frac{-\sqrt{5}}{3}$, 1 **11. a.** $A = 66°$, $c = 33.5$, $b = 23.0$ **b.** $T = 26.5°$, $W = 134.5°$, $z = 36.5$

12. a. 2599 m **b.** 39 s **13.** 13 m **14.** 606 cm **15.** 180.0

16. a. 1121 m **b.** 1509 m

Chapter 8

Exercise 8.1

1. a. period $\frac{1}{60}$ **c.** period 3 **d.** period 2 **f.** period 10

2. b. 12 months **c.** 6.4 h **3. a.** $P_1(5\sqrt{3}, 5)$, $P_2(5, 5\sqrt{3})$,

b. $P_3(0, 10)$, $P_4(-5, 5\sqrt{3})$, $P_5(-5\sqrt{3}, 5)$, $P_6(-10, 0)$,

$P_7(-5\sqrt{3}, -5)$, $P_8(-5, -5\sqrt{3})$, $P_9(0, -10)$, $P_{10}(5, -5\sqrt{3})$,

$P_{11}(5\sqrt{3}, -5)$, $P_{12}(10, 0)$

c.

Point	P_1	P_9	P_{10}	P_{11}
Angle	30°	270°	300°	330°
Y-coord.	5	-10	-8.6	-5

d.

P_{13}	P_{14}	P_{15}	P_{16}	P_{17}	P_{18}
390°	420°	450°	480°	510°	540°
5	8.6	10	8.6	5	0

P_{19}	P_{20}	P_{21}	P_{22}	P_{23}	P_{24}
570°	600°	630°	660°	690°	720°
-5	-8.6	-10	-8.6	-5	0

f. yes, 360° **g.** y-coordinate multiplied by 3 indicating a vertical stretch by a factor of 3

h.

Point	P_0	P_1	P_2	P_3	P_4	P_5	P_6	P_7	P_8
Angle	0	30°	60°	90°	120°	150°	180°	210°	240°
X-coord.	10	8.6	5	0	-5	-8.6	-10	-8.6	-5

P_9	P_{10}	P_{11}	P_{12}	P_{13}	P_{14}	P_{15}	P_{16}	P_{17}
270°	300°	330°	360°	390°	420°	450°	480°	510°
0	5	8.6	10	8.6	5	0	-5	-8.6

P_{18}	P_{19}	P_{20}	P_{21}	P_{22}	P_{23}	P_{24}
540°	570°	600°	630°	660°	690°	720°
-10	-8.6	-5	0	5	8.6	10

period: 360°; x-coordinate multiplied by 3 indicating a vertical stretch by a factor of 3 **4. a.** Input time as a decimal, e.g., input 07:44 as 7.73. **b.** period 12 months **c.** 3 h 7 min

5. a. period 12 since $f(x) = f(x + 12)$ **b.** 27 m **c.** 12 h

6. a. period 4 min **b.** 14 m **c.** 2 m **d.** 8 m **7. b.** yes, period is 10

Exercise 8.2

1. $\left(60°, \frac{1}{2}\right)$, $\left(300°, \frac{1}{2}\right)$, $\left(420°, \frac{1}{2}\right)$; $\theta = (60 + 360k)°$, $\theta = (300 + 360k)°$, $k \in I$ **2. a.** $(0°, 0)$, $(180°, 0)$, $(360°, 0)$; $\theta = (180k)°$, $k \in I$ **b.** $\left(-30°, -\frac{1}{2}\right)$, $\left(210°, -\frac{1}{2}\right)$, $\left(330°, -\frac{1}{2}\right)$; $\theta = (210 + 360k)°$, $\theta = (330 + 360k)°$, $k \in I$ **c.** $(-90°, -1)$, $(270°, -1)$, $(630°, -1)$; $\theta = (270 + 360k)°$, $k \in I$

d. $\left(\theta, -\frac{\sqrt{3}}{2}\right)$, $\left(240°, -\frac{\sqrt{3}}{2}\right)$, $\left(-60°, -\frac{\sqrt{3}}{2}\right)$, $\left(300°, -\frac{\sqrt{3}}{2}\right)$; $\theta = (240 + 360k)°$, $\theta = (300 + 360k)°$, $k \in I$ **3. a.** $(90°, 0)$, $(270°, 0)$, $(-90°, 0)$; $\theta = (2k + 1)90°$, $k \in I$ **b.** $(0, 1)$, $(360°, 1)$, $(-360°, 1)$; $\theta = 360k°$, $k \in I$ **c.** $\left(150°, -\frac{\sqrt{3}}{2}\right)$, $\left(210°, -\frac{\sqrt{3}}{2}\right)$, $\left(-210°, -\frac{\sqrt{3}}{2}\right)$; $\theta = (150 + 360k)°$, $\theta = (210 + 360k)°$, $k \in I$ **d.** $\left(30°, \frac{\sqrt{3}}{2}\right)$, $\left(330°, \frac{\sqrt{3}}{2}\right)$,

$\left(-30°, \frac{\sqrt{3}}{2}\right)$; $\theta = (30 + 360k)°$, $\theta = (330 + 360k)°$, $k \in I$

4. a. $(0°, 0)$, $(180°, 0)$, $(-180°, 0)$; $\theta = 180k°$, $k \in I$

b. $(45°, 1)$, $(225°, 1)$, $(-135°, 1)$; $\theta = (45 + 180k)°$, $k \in I$

c. $(60°, \sqrt{3})$, $(240°, \sqrt{3})$, $(-120°, \sqrt{3})$; $\theta = (60 + 180k)°$,

$k \in I$ **d.** $\left(150°, -\frac{1}{\sqrt{3}}\right)$, $\left(330°, -\frac{1}{\sqrt{3}}\right)$, $\left(-30°, -\frac{1}{\sqrt{3}}\right)$;

$\theta = (150 + 180k)°$, $k \in I$

Exercise 8.3

1. a. $\frac{\pi}{3}$ **b.** $\frac{5\pi}{6}$ **c.** $\frac{5\pi}{4}$ **d.** $\frac{35\pi}{18}$ **e.** $\frac{5\pi}{3}$ **f.** $\frac{11\pi}{4}$ **g.** $\frac{-2\pi}{3}$ **h.** $\frac{-19\pi}{4}$ **i.** $\frac{\pi}{2}$

j. $\frac{3\pi}{2}$ **2. a.** 60° **b.** 45° **c.** 36° **d.** 210° **e.** 166° **f.** 246° **3. a.** 0.8090

b. 0.3249 **c.** 0.8660 **d.** 0.9816 **e.** 0.5446 **f.** -2.7475 **g.** 0.4685

h. -1.8114 **i.** -0.0818 **5. a.** $\frac{1}{2}$ **b.** $\frac{1}{\sqrt{2}}$ **c.** $\sqrt{3}$ **d.** $-\frac{1}{2}$ **e.** $\frac{1}{2}$ **f.** $-\frac{1}{2}$

g. -1 **h.** $\frac{1}{\sqrt{2}}$ **i.** 1 **j.** $\frac{5}{4}$ **k.** $\frac{1}{2}$ **l.** 1 **8. a.** 2.183 **b.** 2.040 **c.** 3.265

d. 4.378 **e.** 4.871 **f.** 5.833 **9.** $\frac{3\pi}{20}$ rad or 27° **10.** 28.65 cm

11. $0 \le \theta \le \frac{\pi}{4}, \frac{5\pi}{4} \le \theta \le \frac{9\pi}{4}, \frac{13\pi}{4} \le \theta \le 4\pi$

12. $Q(3.80, 3.60)$

Exercise 8.4

1. a. vertical translation of +3 **b.** horizontal translation of $+\frac{3\pi}{4}$

c. vertical translation of -3 and horizontal translation of $+\frac{\pi}{3}$

d. vertical translation of +2 and horizontal translation of $-\frac{\pi}{2}$

2. a. vertical translation of -5 **b.** horizontal translation of $\frac{\pi}{6}$

c. vertical translation of -3 and horizontal translation of $-\frac{3\pi}{4}$

d. vertical translation of +2 and horizontal translation of $\frac{2\pi}{3}$

3. a. $y = \sin x - 3$ **b.** $y = \sin\left(x - \frac{\pi}{4}\right)$ **c.** $y = \sin\left(x + \frac{2\pi}{3}\right) + 2$

d. $y = \cos\left(x + \frac{\pi}{6}\right) - 2$ **4.** 3b: $y = \cos\left(x - \frac{3\pi}{4}\right)$;

3c: $y = \cos\left(x + \frac{\pi}{6}\right) + 2$ **6. a.** 8, 6 **b.** -2, -4 **7. a.** horizontal of

$\frac{\pi}{2}$ **b.** horizontal of $-\frac{4\pi}{3}$ and vertical of 5 **8. a.** $y = \sin \theta + 11$

b. $y = \sin \theta - 6$ **c.** $y = \sin \theta + \frac{9}{2}$ **10. a.** horizontal of $-\frac{\pi}{3}$ or

$+\frac{5\pi}{3}$ **b.** horizontal of $\frac{\pi}{6}$ or $\frac{13\pi}{6}$ **11.** $\frac{17\pi}{4}, \frac{19\pi}{4}$

12. $y = \cos\left(x - \frac{\pi}{6}\right) - 10$ **13.** $y = \sin\left(x - \frac{\pi}{3}\right) + 5$

14. $-\frac{\pi}{3} \le t \le 0, \frac{2\pi}{3} \le t \le 2\pi$

Exercise 8.5

1.

	Amplitude	Period
a.	3	2π
b.	1	3π
c.	5	$\frac{\pi}{3}$
d.	$\frac{5}{3}$	6π
e.	7	$\frac{2\pi}{5}$
f.	6	$\frac{10\pi}{3}$
g.	4	1
h.	4	3

2. a. $y = 2 \sin x$ **b.** $y = \sin 2x$ **c.** $y = \frac{1}{3} \sin 3x$ **d.** $y = 7 \sin \frac{1}{2}x$

e. $y = 3 \sin \frac{2}{3}x$ **f.** $y = \frac{3}{5} \sin \frac{2\pi}{3}x$ **3.** $k > 1$: horizontal

contraction factor $\frac{1}{k}$; $0 < k < 1$: horizontal stretch by a factor

$\frac{1}{k}$ **4. a.** $y = 2 \cos 2x$ **b.** $y = \sin \frac{1}{3}x$ **c.** $y = 1.5 \cos \frac{\pi}{4}x$

5. b. i) maximum points: $(0, 2)$, $(2\pi, 2)$; minimum point: $(\pi, -2)$
ii) maximum point: $\left(\frac{\pi}{2}, 1.5\right)$; minimum point: $\left(\frac{3\pi}{2}, -1.5\right)$
iii) maximum points: $(0, 3)$, $(\pi, 3)$; minimum points: $\left(\frac{\pi}{2}, -3\right)$
iv) maximum point: $(\pi, 2)$; minimum point: $(3\pi, -2)$
v) maximum point: $\left(\frac{\pi}{8}, 3\right)$; minimum point: $\left(\frac{3\pi}{8}, -3\right)$
vi) maximum points: $(0, 1)$, $\left(\frac{2\pi}{3}, 1\right)$; minimum point: $\left(\frac{\pi}{3}, -1\right)$
vii) maximum point: $(\pi, 1)$; minimum point: $\left(\frac{\pi}{3}, -1\right)$
viii) maximum points: $(0, 1)$, $(3\pi, 1)$; minimum point: $\left(\frac{3\pi}{2}, -1\right)$
6. $k = 2 + 8p$, $p \in I$ **7.** $k = \frac{1}{2} + p$, $p \in I$ **8. c.** 3
9. b. $y = \sqrt{2}\sin\left(x + \frac{\pi}{4}\right)$, $y = \sqrt{2}\cos\left(x - \frac{\pi}{4}\right)$
10. b. $y = \sqrt{13}\sin(x + 0.983)$, $y = \sqrt{13}\cos(x - 0.588)$
c. $a^2 = 2^2 + 3^2$ **d.** $a^2 = r^2 + s^2$

Exercise 8.6

1.	Amplitude	Period
a.	7	$\frac{\pi}{5}$
b.	220	$\frac{1}{50}$
c.	3	12
d.	10	$\frac{16\pi}{5}$

4. a. period $\frac{2}{5}$, amplitude 25 **b.** period $\frac{\pi}{10}$, amplitude 0.25
c. period 20, amplitude 150 **d.** period $\frac{7\pi}{2}$, amplitude $\frac{3}{5}$

Exercise 8.7

2.	Amplitude	Period	Phase Shift	Vertical Translation
a.	20	$\frac{2\pi}{3}$	0	10
b.	$\frac{1}{5}$	$\frac{\pi}{2}$	$-\frac{\pi}{2}$	-3
c.	1	π	$-\frac{\pi}{16}$	0
d.	5	π	π	-3
e.	$\frac{1}{3}$	$\frac{2\pi}{3}$	$-\frac{\pi}{12}$	10

3. a. i) 2 **ii)** 2π **iii)** $+\frac{2\pi}{3}$ **iv)** -1 **v)** $y = -1$ **vi)** -3, $x = \frac{\pi}{6}$
vii) 1, $x = \frac{7\pi}{6}$ **viii)** $y = 2\sin\left(x - \frac{2\pi}{3}\right) - 1$ **b. i)** 20 **ii)** 4π **iii)** 0
iv) $+10$ **v)** $y = 10$ **vi)** -10, $\theta = 3\pi$ **vii)** 30, $\theta = \pi$
viii) $y = 20\sin\frac{1}{2}\theta + 10$ **c. i)** 1 **ii)** π **iii)** $\frac{\pi}{3}$ **iv)** $+2$ **v)** $y = 2$
vi) 1, $x = \frac{13\pi}{12}$ **vii)** 3, $x = \frac{7\pi}{12}$ **viii)** $y = \sin 2\left(x - \frac{\pi}{3}\right) + 2$ **d. i)** 3
ii) 2π **iii)** $\frac{\pi}{6}$ **iv)** -1 **v)** $y = -1$ **vi)** -4, $x = \frac{7\pi}{6}$ **vii)** 2, $x = \frac{\pi}{6}$
viii) $y = 3\sin\left(x + \frac{\pi}{3}\right) - 1$ **4. a.** $y = 2\cos\left(x + \frac{\pi}{3}\right) - 1$
b. $y = 20\cos\frac{1}{2}(x - \pi) + 10$ **c.** $y = \cos 2\left(x - \frac{7\pi}{12}\right) + 2$
d. $y = 3\cos\left(x - \frac{\pi}{6}\right) - 1$ **5. a. i)** amplitude 5, phase shift 0,
period π, vertical translation 3 **iii)** $x \in R$, $-2 \le y \le 8$, $y \in R$
b. i) amplitude 2, phase shift $\frac{\pi}{3}$, period $\frac{2\pi}{3}$,
vertical translation 0 **iii)** $x \in R$, $-2 \le y \le 2$, $y \in R$
c. i) amplitude 3, phase shift $\frac{2\pi}{3}$, period π,
vertical translation -5 **iii)** $x \in R$, $-8 \le y \le -2$, $y \in R$
d. i) amplitude 1, phase shift $\frac{\pi}{4}$, period 4π,
vertical translation $+2$ **iii)** $x \in R$, $1 \le y \le 3$, $y \in R$
6. a. ii) minimum points: $\left(\frac{2\pi}{3}, 3\right)$, $(0, 3)$;
maximum points: $\left(\frac{\pi}{3}, 13\right)$, $(\pi, 13)$
b. ii) minimum points: $\left(\frac{\pi}{4}, -16\right)$, $\left(\frac{3\pi}{4}, -16\right)$;
maximum points: $(0, 4)$, $\left(\frac{\pi}{2}, 4\right)$, $(\pi, 4)$

c. ii) minimum points: $\left(\frac{7\pi}{3}, 1\right)$, $\left(\frac{19\pi}{3}, 1\right)$;
maximum points: $\left(\frac{\pi}{3}, 7\right)$, $\left(\frac{13\pi}{3}, 7\right)$
d. ii) minimum points: $\left(\frac{7\pi}{12}, -33\right)$, $\left(\frac{19\pi}{12}, -33\right)$;
maximum points: $\left(\frac{\pi}{12}, 27\right)$, $\left(\frac{13\pi}{12}, 27\right)$
e. ii) minimum points: $(0, -4)$, $(\pi, -4)$, $(25, -4)$;
maximum points: $\left(\frac{\pi}{2}, 0\right)$, $\left(\frac{3\pi}{2}, 0\right)$ **7. a.** 1 **9. b.** π **10. b.** 2π
11. b. π **12.** 2π **13.** $\frac{2\pi}{3}$

Exercise 8.8

1. b. 12.4, 3.5 **c.** 7.27 m, 4.34 m **d.** 4.30 m **e.** 00:24, 12:24
f. from 01:54 to 10:13 and 14:18 to 22:43
2. $y = 7\cos 0.52(x - 9.06) + 8.5$ **3. a.** 25, 4 **b.** Amplitude is
the radius of the tire, the period is the time for the tire to make
1 revolution. **c.** $y = -25\sin\frac{\pi}{2}t + 25$ **d.** 42.7 cm
4. a. $y = 9\cos 5\pi t + 11$ **b.** 20 cm **5. a.** 2, 12 **b.** 11 **c.** $y = 9$
d. Point B: closest to the CBR; Point D: in a vertical
position **e.** $y = -2\cos\frac{\pi}{6}t + 9$ **f. 9 6.** $y = 2400\cos\frac{4\pi}{3}t$
7. a. $y = -207\cos\frac{2\pi}{365}t + 407$ **b.** 03:23 **c.** August 29 and
April 13 **d.** 102 **e.** No. The domain is the set of integers from
0 to 365. **8. a.** $y = 30\sin 2\pi t$ **b.** 28.5 cm **c.** 0.065 s, 0.435 s,
0.565 s, 0.935 s **9. a.** $y = 1200\sin 600\pi t$ **b.** $y = 600\sin 480\pi t$
10. $x = 5\cos 1000\pi t$ (t is in minutes) **11.** $0 \le t \le \frac{2\pi}{3}$ or
$\frac{4\pi}{3} \le t \le \frac{8\pi}{3}$ **12.** Each unit along the vertical becomes 5 units
and the units along the horizontal are halved.
13. a. $y - 20\sin 10\left(t + \frac{\pi}{30}\right)$ **b.** 17.8 cm **c.** $0.145 + \frac{\pi}{5}k$,
$0.273 + \frac{\pi}{5}k$, $0.459 + \frac{\pi}{5}k$, $0.588 + \frac{\pi}{5}k$, $k \in N_0$
15. a. $\theta \ge 69.4°$ **b.** 1.25 m **c.** $y = \frac{8 - 3\tan\theta}{4\tan\theta}$, $5° \le \theta \le 65°$

8.9 Review Exercise

1. a. $\frac{\pi}{4}$ **b.** $\frac{2\pi}{3}$ **c.** $\frac{7\pi}{6}$ **d.** $\frac{14\pi}{3}$ **e.** $-\frac{5\pi}{6}$ **f.** $-\frac{35\pi}{9}$ **2. a.** 30° **b.** 202.5°
c. $-315°$ **d.** $-780°$ **e.** 1140° **f.** $-297.9°$ **3. a.** 3, $\frac{2\pi}{5}$ **b.** $\frac{2}{3}$, $\frac{\pi}{6}$
c. 4, $\frac{1}{2}$ **d.** $\frac{2}{7}$, $\frac{\pi}{6}$ **e.** 8, $\frac{10}{3}$ **f.** 7, $\frac{1}{30}$ **4. a.** 3, $\frac{\pi}{6}$ **b.** -4, π **c.** 1, $\frac{\pi}{6}$
d. 2, $\frac{\pi}{4}$ **e.** -5, $\frac{\pi}{6}$ **f.** 4, $\frac{\pi}{5}$ **5. a. i)** 2π, 4, 4, $y = 4$
ii) $y = 4\sin(x - \pi) + 4$ **iii)** $y = 4\cos\left(x + \frac{\pi}{2}\right) + 4$ **b. i)** $\frac{2\pi}{3}$, 4,
-2, $y = -2$ **ii)** $y = 4\sin 3x - 2$ **iii)** $y = 4\cos 3\left(x - \frac{\pi}{6}\right) - 2$
c. i) 4, 30, 10, $y = 10$ **ii)** $y = 30\sin\frac{\pi}{2}(t + 1) + 10$
iii) $y = 30\cos\frac{\pi}{2}t + 10$ **d. i)** 16, 8, $y = -2$
ii) $y = 8\sin\frac{\pi}{8}(x + 2) - 2$ **iii)** $y = 8\cos\frac{\pi}{8}(x - 2) - 2$
7. a. 2 **b.** -4 **8.** $\frac{\pi}{12}$ or $\frac{5\pi}{12}$ **9.** $T = 10$, $y = 2.9$ when $t = 4$ s
10. $p = -\frac{\pi}{3}$, $d = -3$ **11. a.** $h = 10\cos\frac{\pi}{24}(t + 24) + 12$
b. 5 m **c.** $(9.68 + 48k)$ s and $(38.32 + 48k)$ s with $k \in N_0$
12. b. $p = 20\cos\frac{2\pi}{3}t$ **c.** 10 cm from CD closest to B

Chapter 9

Exercise 9.1

1. a. 7°, 173° **b.** 70°, 290° **c.** 58°, 238° **d.** 229°, 311° **e.** 142°,
218° **f.** 163°, 343° **2. a.** 0.32, 2.82 **b.** 0.75, 5.53 **c.** 1.26, 4.41
d. 4.09, 5.33 **e.** 1.72, 4.56 **f.** 2.80, 5.94 **3. a.** 60°, 120° **b.** 45°,
315° **c.** 60°, 240° **d.** 120°, 240° **e.** 210°, 330° **f.** 135°, 315°

4. a. $\frac{\pi}{4}, \frac{3\pi}{4}$ **b.** $\frac{\pi}{6}, \frac{11\pi}{6}$ **c.** $\frac{\pi}{6}, \frac{7\pi}{6}$ **d.** $\frac{\pi}{2}$ **e.** $\frac{\pi}{2}, \frac{3\pi}{2}$ **f.** $\frac{3\pi}{4}, \frac{7\pi}{4}$
5. a. 194°, 346° **b.** 222°, 318° **c.** 109°, 251° **d.** 45°, 225°
6. a. no solution **b.** 1.05, 5.24 **c.** 1.37, 4.51 **d.** 0.52, 2.62 **e.** 3.71, 5.71 **f.** 1.05, 4.19, 2.09, 5.24 **g.** 1.57, 4.71, 1.23, 5.05 **h.** 1.23, 5.05 **7. a.** -0.52, 3.67 **b.** 2.36 **c.** 0.79, 5.50 **d.** 2.03, 5.18
8. $\sin \beta = \frac{18}{23}$, $\cos \alpha = \frac{19}{23}$ **9.** $\left(\left(\frac{2k+1}{4}\right)\pi, 1\right)$, $k \in I$
10. $\sin \beta = \frac{\sqrt{30}}{6}$, $\cos \beta = \frac{\sqrt{6}}{6}$ **11.** 0.44, 2.70, 6.72, 8.98
12. $\frac{\sqrt{5}+1}{4}$ **13.** $\frac{7\pi}{18}, \frac{19\pi}{18}, \frac{31\pi}{18}, \frac{11\pi}{18}, \frac{23\pi}{18}, \frac{35\pi}{18}$

Exercise 9.2

2. a. 1 **b.** $\sin \theta$ **c.** $\cos^2 \theta$ **d.** $\frac{1}{\sin \theta \cos \theta}$ **e.** $|\sin \theta|$ **f.** -1
g. $\frac{2}{\cos \theta}$ **h.** $1 + \cos^2 \theta$ or $2 - \sin^2 \theta$ **3. a.** $\sin^2 \theta$
b. $\tan^2 x + \sin^2 x$ **c.** $4(1 + 2\sin\theta\cos\theta)$ **d.** $\cos^2 x - \sin^2 x$ or $1 - 2\sin^2 x$ or $2\cos^2 x - 1$ **8.** $1 + \cot^2 \theta$ or $\csc^2 \theta$ **13.** 1

Exercise 9.3

1. 0, π, 2π **2.** 60°, 240°, 225°, 315° **3. a.** 90°, 270°, 30°, 150°
b. 60°, 300° **4. a.** $\frac{3\pi}{2}$ **b.** $\frac{\pi}{4}, \frac{3\pi}{4}$ **c.** 3.79, 5.64, $\frac{\pi}{6}, \frac{5\pi}{6}$ **d.** 0.46, 3.61, $\frac{3\pi}{4}, \frac{7\pi}{4}$ **5.** $\left(\frac{\pi}{3}, 0\right), \left(\frac{5\pi}{3}, 0\right)$ **6. a.** -0.73, 3.87, 5.55 **b.** $-\pi$, 0, π, 2π, 1.03, 4.17 **7. a.** $\left(\frac{\pi}{6}, 10.5\right), \left(\frac{5\pi}{6}, 10.5\right), \left(\frac{13\pi}{6}, 10.5\right),$
$\left(\frac{17\pi}{6}, 10.5\right)$ **8. a.** 1.44, 4.84 **b.** $\frac{2\pi}{3}, \frac{4\pi}{3}$, 0, 2π **c.** 1.38, 4.91 **d.** 0, 2π, 4.07 **9.** $\frac{\pi}{4}$ **10.** 12:00, 00:00 **11.** 06:39 **12.** (1.096, 1.914), (5.187, 1.914) **13.** 0, $\frac{\pi}{2}$, 2π **14.** 0, π, 2π, 0.83, 2.31, 4.27, 5.15
15. $\sin \theta \neq 0$, $\cos \theta \neq 0$

Exercise 9.4

4. a. $\frac{56}{65}$ **8. a.** $\frac{\sqrt{3}-1}{2\sqrt{2}}$ **b.** $\frac{\sqrt{3}-1}{2\sqrt{2}}$ **c.** $\frac{1-\sqrt{3}}{2\sqrt{2}}$ **d.** $-\left(\frac{\sqrt{3}+1}{2\sqrt{2}}\right)$
9. a. $\frac{\sqrt{3}\cos x + \sin x}{2}$ **b.** $-\frac{1}{\sqrt{2}}(\cos x + \sin x)$
c. $\frac{1}{\sqrt{2}}(\cos x + \sin x)$ **d.** $-\sin x$ **10. a.** $-\left(\frac{\sqrt{15}+2\sqrt{2}}{12}\right)$
b. $\frac{1-2\sqrt{30}}{12}$ **c.** $-\frac{7}{9}$ **d.** $-\frac{\sqrt{15}}{8}$ **11. a.** $\frac{3}{5}$ **b.** $\frac{4}{3}$; first

9.5 Review Exercise

1. a. 46°, 134° **b.** 59°, 301° **c.** 72°, 252° **d.** 188°, 352° **e.** 96°, 264° **f.** 157°, 337° **2. a.** 0.15, 2.99 **b.** 0.57, 5.71 **c.** 0.31, 3.45 **d.** 3.80, 5.62 **e.** 2.13, 4.15 **f.** 1.98, 5.12 **3. a.** 30°, 150° **b.** 180° **c.** 45°, 315°, 135°, 225° **d.** 42°, 138° **e.** 58°, 238° **f.** 0°, 180°, 360°, 199°, 341° **4. a.** 1.05, 5.24 **b.** 4.71 **c.** 0.52, 2.62, 3.67, 5.76 **d.** 1.23, 5.05, 1.05, 5.24 **e.** 1.37, 4.51 **f.** 1.57, 4.71, 1.05, 5.24 **5. a.** $\frac{\pi}{3}, \frac{5\pi}{3}$ **b.** 4.04, 5.39 **c.** $\frac{2\pi}{3}, \frac{4\pi}{3}$, 0, 2π **d.** 0.32, 3.46 **e.** $\frac{\pi}{6}, \frac{5\pi}{6}, \frac{\pi}{2}$ **f.** $\frac{\pi}{3}, \frac{5\pi}{3}$ **g.** 1.32, 4.97, 0, 2π **h.** $\frac{3\pi}{4}, \frac{7\pi}{4}$
7. $\left(-\frac{2\pi}{3}, \frac{3}{4}\right), \left(-\frac{\pi}{3}, \frac{3}{4}\right), \left(\frac{\pi}{3}, \frac{3}{4}\right), \left(\frac{2\pi}{3}, \frac{3}{4}\right), \left(\frac{4\pi}{3}, \frac{3}{4}\right), \left(\frac{5\pi}{3}, \frac{3}{4}\right)$
8. a. $\frac{8\sqrt{3}-1}{15}$ **b.** $\frac{-2\sqrt{6}+2\sqrt{2}}{15}$ **c.** $\frac{4\sqrt{6}}{25}$ **d.** $-\frac{7}{9}$ **11.** $\frac{3}{5}$, -1

Chapter 10

Exercise 10.1

2. a. $2x + 3y = 43$ **3. a.** $x^2 + y^2 - 12x + 32 = 0$
4. a. $x^2 - 12y + 36 = 0$ **5. a.** $y^2 - 8x + 16 = 0$

Exercise 10.2

4. a. 20 **b.** 16 **c.** 10 **d.** $OA_1{}^2 = OB_1{}^2 + OF_1{}^2$ **5. a.** 10 **b.** 6 **c.** 5
6. a. 10 **b.** $\sqrt{21}$ **d.** 5

Chapter 11

Exercise 11.1

1. a. (0, 0), 10 **b.** (0, 0), $5\sqrt{2}$ **c.** (0, 0), $\frac{1}{4}$ **d.** $(-3, 0)$, 9 **e.** (0, 4), $2\sqrt{6}$ **f.** (3, 1), 3 **g.** (0, 0), 4 **h.** (0, 0), $\frac{2}{3}$ **i.** (2, -2), $\sqrt{2}$
j. $(-3, 9)$, $2\sqrt{5}$ **2. a.** ± 7 **b.** ± 5 **c.** ± 1 **d.** $\pm 5\sqrt{2}$ **e.** $\pm\sqrt{34}$
f. $\pm\sqrt{41}$ **3. a.** inside **b.** on **c.** outside **d.** inside **e.** on **f.** inside
4. b. (2, -4), (-2, 4), (-2, -4) **6. a.** $x^2 + y^2 = 49$
b. $x^2 + y^2 = 3$ **c.** $x^2 + y^2 = \frac{9}{16}$ **d.** $(x-1)^2 + (y-2)^2 = 25$
e. $(x+4)^2 + y^2 = 16$ **f.** $(x-2)^2 + (y+4)^2 = 8$
7. domain: $-1 \le x \le 7$, $x \in R$; range: $0 \le y \le 8$, $y \in R$
8. a. (-3, 0), 6 **b.** (1, 1), $\sqrt{2}$ **c.** (1, -3), 4 **d.** (0, 4), 4
e. (-3, -4), $2\sqrt{2}$ **f.** (1, -2), $\frac{3}{2}$ **9. a.** $x^2 + y^2 = 13$
b. $x^2 + y^2 = 4$ **c.** $x^2 + y^2 = 64$ **d.** $(x-3)^2 + (y-4)^2 = 25$
e. $(x+1)^2 + (y-3)^2 = 20$ **f.** $(x+2)^2 + (y+2)^2 = 4$
10. a. $x^2 + y^2 = 34$ **b.** $(x-2)^2 + (y-5)^2 = 18$
12. a. x-intercepts 12, -2; y-intercepts 4, -6 **b.** x-intercepts 0, -8; y-intercepts 0, 6 **c.** x-intercepts 3; y-intercepts -1, -9
13. (0, 4), 5 **14. a.** $(x-2)^2 + (y+1)^2 = 25$ **b.** (2, 4), (2, -6)
16. b. (-1, 4) **17. b.** (-8, 6)

Exercise 11.2

1. a. 2 **b.** 1 **c.** 4 **d.** 3 **e.** $\frac{1}{4}$ **f.** $\frac{5}{4}$ **g.** $\frac{5}{4}$ **h.** $\frac{1}{2}$ **2. a.** (0, 2) **b.** (-1, 0)
c. (0, -4) **d.** (-3, 0) **e.** $\left(0, -\frac{1}{4}\right)$ **f.** $\left(\frac{5}{4}, 0\right)$ **g.** $\left(0, \frac{5}{4}\right)$ **h.** $\left(-\frac{1}{2}, 0\right)$
3. a. $y^2 = -12x$ **b.** $x^2 = -2y$ **c.** $y^2 = 18x$ **4. a.** x-intercepts ± 2, y-intercepts -1, vertex (0, -1) **b.** x-intercepts 4, y-intercepts ± 4, vertex (4, 0) **c.** x-intercept $-\frac{3}{2}$, y-intercepts 6 and -2, vertex (-2, 2) **d.** x-intercepts 7 and -3, y-intercept $\frac{21}{5}$, vertex (2, 5) **5. a.** vertex (3, 0), focus (3, 1)
b. vertex (0, -2), focus (1, -2) **c.** vertex (-1, 0), focus (-1, -2) **d.** vertex (0, -1), focus (0, -4)
e. vertex (3, -3), focus (3, -2) **f.** vertex (-6, 2), focus (-8, 2)
g. vertex (-3, -4), focus $\left(-2\frac{1}{2}, -4\right)$ **h.** vertex (3, -3), focus (3, -4) **6. a.** $x^2 = -8y$ **b.** $y^2 = 2x$ **c.** $y^2 = -16(x-4)$
d. $(x-1)^2 = y - 2$ **e.** $(x+2)^2 = -4y$ **f.** $x^2 = 24(y+3)$
7. a. $y^2 = -(x-4)$ **b.** $(x+2)^2 = -4(y-4)$ **c.** $x^2 = 12(y+1)$
d. $y^2 = 6\left(x - \frac{5}{2}\right)$ **8.** 15 m **9.** 70.315 cm **10.** 16.18 cm
11. $x^2 = 8b(y - b)$ **12.** (0, 4) **13. a.** 4 **b.** 12 **c.** 2

Exercise 11.3

1. a. 16, 8 **b.** 10, 8 **c.** $2\sqrt{3}$, $2\sqrt{2}$ **2. a.** (± 8, 0) **b.** (0, ± 5)
c. ($\pm\sqrt{3}$, 0) **3. a.** ($\pm 4\sqrt{3}$, 0) **b.** (0, ± 3) **c.** (± 1, 0) **4. a.** vertices (6, 0), (-6, 0); foci ($4\sqrt{2}$, 0), ($-4\sqrt{2}$, 0) **b.** vertices (0, 5), (0, -5); foci (0, 3), (0, -3) **c.** vertices ($\sqrt{6}$, 0), ($-\sqrt{6}$, 0); foci ($\sqrt{2}$, 0),($-\sqrt{2}$, 0) **5. a.** vertices (-3, 0), (7, 0); foci (2 + $2\sqrt{6}$, 0), (2 $- 2\sqrt{6}$, 0) **b.** vertices (0, 0), (0, 8); foci (0, 6), (0, 2) **c.** vertices (9, 3), (-11, 3); foci (7, 3), (-9, 3) **7. a.** $\frac{x^2}{36} + \frac{y^2}{20} = 1$ **b.** $\frac{x^2}{16} + \frac{y^2}{25} = 1$
c. $\frac{x^2}{16} + \frac{y^2}{80} = 1$ **d.** $\frac{x^2}{64} + \frac{y^2}{60} = 1$ **e.** $\frac{x^2}{25} + \frac{4y^2}{99} = 1$
8. a. $\frac{x^2}{400} + \frac{3y^2}{400} = 1$ **b.** $\frac{(x-2)^2}{16} + \frac{(y-4)^2}{4} = 1$
c. $\frac{(x-5)^2}{25} + \frac{y^2}{9} = 1$ **d.** $\frac{x^2}{45} + \frac{(y-2)^2}{36} = 1$
e. $\frac{(x-2)^2}{28} + \frac{(y+1)^2}{64} = 1$ **9.** 4.8 m
10. a. $\frac{x^2}{2685^2} + \frac{y^2}{652.54^2} = 1$ **b.** 80.5 million km
11. $\frac{x^2}{6811^2} + \frac{y^2}{6803^2} = 1$ **12. a.** $\frac{(x-2)^2}{9} + \frac{(y-3)^2}{4} = 1$

b. centre $(2, 3)$; vertices $(-1, 3)$, $(5, 3)$; foci $(2 - \sqrt{5}, 3)$, $(2 + \sqrt{5}, 3)$ **13.** 100 km, 500 km **14. a.** $\frac{18}{5}$

Exercise 11.4

1. a. 12 **b.** 8 **c.** $2\sqrt{2}$ **d.** 2 **2. a.** $(10, 0)$, $(-10, 0)$ **b.** $(0, 5)$, $(0, -5)$
c. $(2, 0)$, $(-2, 0)$ **d.** $(0, 3)$, $(0, -3)$ **3. a.** $4x + 3y = 0$,
$4x - 3y = 0$ **b.** $4x + 3y = 0$, $4x - 3y = 0$ **c.** $x + y = 0$,
$x - y = 0$ **d.** $x + 2\sqrt{2}y = 0$, $x - 2\sqrt{2}y = 0$ **4. a.** foci $(0, 3)$,
$(0, -3)$; vertices $(0, \sqrt{5})$, $(0, -\sqrt{5})$ **b.** foci $(\sqrt{6}, 0)$, $(-\sqrt{6}, 0)$;
vertices $(2, 0)$, $(-2, 0)$ **c.** foci $(6, 0)$, $(-6, 0)$; vertices $(4, 0)$,
$(-4, 0)$ **d.** foci $(0, 5\sqrt{2})$, $(0, -5\sqrt{2})$; vertices $(0, 7)$, $(0, -7)$
5. a. vertices $(0, 0)$, $(4, 0)$; foci $(-2, 0)$, $(6, 0)$
b. vertices $(0, -7)$, $(0, -1)$; foci $(0, -8)$, $(0, 0)$
c. vertices $(-7, 3)$, $(3, 3)$; foci $(-12, 3)$, $(8, 3)$
6. a. vertices $(\pm 3, 0)$; asymptotes $2x \pm y = 0$
b. vertices $(0, \pm 2)$; asymptotes $2x \pm 5y = 0$ **c.** vertices $(4, 0)$,
$(0, 0)$; asymptotes $3x - 2y = 6$, $3x + 2y = 6$
7. a. $\frac{x^2}{25} - \frac{y^2}{11} = 1$ **b.** $\frac{x^2}{39} - \frac{y^2}{25} = -1$ **c.** $\frac{x^2}{9} - \frac{y^2}{36} = 1$
d. $\frac{x^2}{4} - \frac{y^2}{21} = -1$ **9. a.** $\frac{(x-4)^2}{4} - \frac{(y-1)^2}{12} = 1$
b. $\frac{(x-3)^2}{5} - \frac{y^2}{4} = -1$ **11. a.** $\frac{x^2}{100} - \frac{y^2}{2400} = 1$
b. $\frac{x^2}{900} - \frac{y^2}{1600} = 1$ **c.** $\frac{x^2}{100} - \frac{y^2}{300} = 1$
12. $\frac{(x-1)^2}{90} - \frac{(y-3)^2}{45} = 1$; centre $(1, 3)$; vertices

$\left(1 \pm \frac{3\sqrt{70}}{7}, 3\right)$; asymptotes $\sqrt{7}(x-1) \pm \sqrt{2}(y-3) = 0$
13. b. $y = x$, $y = -x$ **c.** vertices $(2, 2)$, $(-2, -2)$;
foci $(2\sqrt{2}, 2\sqrt{2})$, $(-2\sqrt{2}, -2\sqrt{2})$

Exercise 11.5

1. a. $a = b$ **b.** $a \neq 0$, $b = 0$ or $a = 0$, $b \neq 0$ **c.** $ab > 0$ **d.** $ab < 0$
2. a. $(1, -2)$, $\sqrt{5}$ **b.** $(2, -4)$, $\frac{5}{2}$ **c.** $(0, -2)$, $\frac{\sqrt{5}}{3}$ **d.** $(-3, 4)$, $2\sqrt{3}$
e. $(-2, -2)$, $2\sqrt{2}$ **f.** $(4, 3)$, 2 **3. a.** hyperbola; $(-2, -2)$; $a = 4$,
$b = -1$, $c - 16$ **b.** ellipse; $(3, 3)$; $a = 1$, $b = 9$, $c = 81$
c. ellipse; $(1, -3)$; $a = 4$, $b = 9$, $c = 49$ **d.** hyperbola;
$(-3, -3)$; $a = 4$, $b = -1$, $c = 31$ **e.** parabola; $(5, -3)$; $a = 1$,
$b = 0$, $c = 1$ **f.** parabola; $(-2, -1)$; $a = 5$, $b = 0$, $c = 22$
g. parabola; $(1, -3)$; $a = 0$, $b = 1$, $c = 17$ **h.** hyberbola;
$(1, -7)$; $a = 4$, $b = -1$, $c = 16$ **4. a.** vertex $(2, -5)$;
focus $\left(2, -\frac{9}{2}\right)$ **b.** centre $(5, -3)$; vertices $(5, 2)$, $(5, -8)$
c. two lines $x - 5 = 0$ and $y + 3 = 0$ **d.** centre $\left(-\frac{1}{2}, \frac{3}{4}\right)$;
radius 1 **e.** vertex $(1, -3)$; focus $\left(\frac{4}{3}, -3\right)$ **f.** centre $(-5, 2)$;
vertices $(-5, 4)$, $(-5, 0)$ **5. a.** asymptotes $x + y = 0$,
$x - y = 0$; vertices $(-10, 0)$, $(10, 0)$ **b.** asymptotes $x \pm y = 0$;
vertices $(\pm 5, 0)$, $(\pm 3, 0)$, $(\pm 1, 0)$, $\left(\pm\frac{1}{2}, 0\right)$, $\left(\pm\frac{1}{5}, 0\right)$, $\left(\pm\frac{1}{10}, 0\right)$
c. two lines $x + y = 0$ and $x - y = 0$ **d.** asymptotes $x \pm y = 0$;
vertices $\left(0, \pm\frac{1}{10}\right)$, $\left(0, \pm\frac{1}{2}\right)$, $(0, \pm 1)$, $(0, \pm 5)$
6. a. two lines $x \pm 2y = 0$ intersecting at $(0, 0)$
b. two lines $x \pm y - 3 = 0$ intersecting at $(3, 0)$
c. two lines $2x + y - 5 = 0$ and $2x - y - 3 = 0$ intersecting
at $(2, 1)$ **d.** the point $(0, 0)$ **e.** the point $(1, -3)$
f. two lines $4x - 5y - 5 = 0$ and $4x + 5y + 5 = 0$
intersecting at $(0, -1)$

Exercise 11.6

1. a. $(4, 5)$, $(-1, 0)$ **b.** $(0, -5)$, $(1, 0)$ **c.** $(1, 2)$, $(9, 6)$ **d.** $(-4, 4)$
e. $(0, 5)$, $(5, 0)$ **f.** $(3, -4)$ **g.** $(0, 2)$, $(3, 0)$ **h.** $(3, 1)$
i. $\left(\frac{3\sqrt{10}}{2}, \sqrt{10}\right)$, $\left(\frac{-3\sqrt{10}}{2}, -10\right)$ **j.** $(-1, -4)$, $(4, 1)$
2. a. $\pm 5\sqrt{17}$ **b.** $-5\sqrt{17} < x < 5\sqrt{17}$
3. $4x + 3y - 25 = 0$ **4.** $x_1 x + y_1 y - r^2 = 0$

11.7 Review Exercise

1. a. $(0, 0)$, 12 **b.** $(0, 0)$, $\frac{3}{2}$ **c.** $(3, 2)$, $\sqrt{13}$ **d.** $(0, -3)$, 3
2. a. x-intercepts ± 12; y-intercepts ± 12 **b.** x-intercepts $\pm\frac{3}{2}$;
y-intercepts $\pm\frac{3}{2}$ **c.** x-intercepts 0, 6; y-intercepts 0, 4
d. x-intercepts 0; y-intercepts -6, 3 **3.** centre $(-4, 3)$; radius 5;
x-intercepts -8, 0; y-intercepts 0, 6 **5. a.** $x^2 + y^2 = \frac{4}{9}$
b. $x^2 + y^2 = 4$ **c.** $(x + 1)^2 + (y - 2)^2 = 9$ **d.** $(x - 2)^2 + y^2 = 8$
e. $(x - 1)^2 + (y - 1)^2 = 18$ **6. a.** $\left(0, \frac{1}{48}\right)$ **b.** $(3, 0)$
7. a. vertex $(0, 16)$, focus $\left(0, \frac{63}{4}\right)$ **b.** vertex $(4, 0)$, focus $(3, 0)$
c. vertex $\left(-\frac{1}{2}, 2\right)$, focus $\left(\frac{3}{2}, 2\right)$ **8. a.** $x^2 = 32y$ **b.** $y^2 = 8(x + 2)$
c. $x^2 = -(y - 4)$ **d.** $(y + 2)^2 = -4(x - 1)$ **9.** $x^2 = 4(y - 1)$,
$(y - 1)^2 = \frac{1}{2}x$ **10. a.** vertices $(\pm 10, 0)$; foci $(\pm 6, 0)$
b. vertices $(0, \pm 5)$; foci $(0, \pm 2\sqrt{6})$ **c.** vertices $\left(0, \pm\frac{5}{3}\right)$;
foci $(0, \pm 1)$ **d.** vertices $(\pm 3, 0)$; foci $(\pm 1, 0)$
e. vertices $(0, \pm 2\sqrt{2})$; foci $(0, \pm\sqrt{6})$ **f.** vertices $(9, -2)$,
$(-3, -2)$; foci $(7, -2)$, $(-1, -2)$ **11. a.** $\frac{x^2}{36} + \frac{y^2}{4} = 1$
b. $\frac{x^2}{81} + \frac{y^2}{225} = 1$ **c.** $\frac{x^2}{64} + \frac{(y-1)^2}{48} - 1$ **d.** $\frac{(x+6)^2}{36} + \frac{y^2}{32} - 1$
e. $\frac{(x-5)^2}{25} + \frac{(y-3)^2}{9} - 1$ **12. a.** vertices $(\pm 4, 0)$; foci $(\pm 5, 0)$;
asymptotes $3x \pm 4y = 0$ **b.** vertices $(0, \pm 2)$; foci $(0, \pm 3)$;
asymptotes $2x \pm \sqrt{5}y = 0$ **c.** vertices $\left(\pm\frac{5}{2}, 0\right)$;
foci $\left(\pm\frac{5\sqrt{2}}{2}, 0\right)$; asymptotes $x \pm y = 0$ **d.** vertices $(0, \pm\sqrt{21})$;
foci $(0, \pm 7)$; asymptotes $\sqrt{3}x \pm 2y = 0$ **e.** vertices $(-1, 2)$,
$(-7, 2)$; foci $(1, 2)$, $(-9, 2)$; asymptotes $4x + 3y + 10 = 0$,
$4x - 3y + 22 = 0$ **f.** vertices $(0, 2)$, $(2, 2)$; foci $(-4, 2)$, $(6, 2)$;
asymptotes $2\sqrt{6}(x - 1) + y - 2 = 0$,
$2\sqrt{6}(x - 1) - y + 4 = 0$ **13. a.** $\frac{x^2}{108} - \frac{y^2}{36} = -1$
b. $\frac{x^2}{16} - \frac{y^2}{100} = 1$ **c.** $\frac{x^2}{16} - \frac{y^2}{84} = 1$ **d.** $\frac{(x-4)^2}{1} - \frac{y^2}{15} = 1$
14. a. $40\sqrt{2}$ m **b.** $40\sqrt{3}$ m **15.** $(x - 120)^2 = -180(y - 80)$
16. a. $\frac{x^2}{7275^2} + \frac{y^2}{6551^2} = 1$ **b.** $\frac{x^2}{6913^2} + \frac{y^2}{6904^2} = 1$
17. $\frac{x^2}{400} - \frac{y^2}{3200} = 1$ **18. a.** circle; centre $(2, -3)$; radius 5
b. ellipse; centre $(2, -3)$; vertices $(7, -3)$, $(-3, -3)$
c. parabola; vertex $(2, 2)$; focus $(1, 2)$ **d.** ellipse; centre $(-3, 1)$;
vertices $(-3, 3)$, $(-3, -1)$ **e.** hyperbola; centre $(3, -5)$;
vertices $(8, -5)$, $(-2, -5)$
19. a. $\left(-1 + \sqrt{5}, \frac{-3+\sqrt{5}}{2}\right)$, $\left(-1 - \sqrt{5}, \frac{-3-\sqrt{5}}{2}\right)$
b. $(2, 4)$ **c.** $(0, 4)$, $(-2, 0)$ **20.** $(3, 1)$

Cumulative Review 4–11

1. a. $\theta = 62.2°$, $x = 13.3$ **b.** $\theta = 110°$, $x = 36.5$
c. $\theta = 23.8°$, $x = 32.2$ **2. a.** $142.5°$ or $217.5°$ **b.** $83.3°$ or
$263.3°$ **c.** $41.8°$, $138.2°$ or $210°$, $330°$ **3. a.** $\frac{3\pi}{2}$ **b.** $\frac{\pi}{6}$, $\frac{5\pi}{6}$, $\frac{7\pi}{6}$,
$\frac{11\pi}{6}$ **c.** $\frac{7\pi}{11}$, $\frac{11\pi}{6}$, $\frac{\pi}{2}$ **d.** $\frac{3\pi}{4}$, $\frac{7\pi}{4}$ **4. b.** $(\pi, -4)$, $(3\pi, -4)$

5. a. i) $y = 4\sin x$ **ii)** $y = 2 \sin 3\left(x - \frac{\pi}{6}\right)$ **iii)** $y = 10 \sin\left(\frac{x}{2}\right) + 10$

b. i) $y = 4 \cos\left(x - \frac{\pi}{2}\right)$ **ii)** $y = 2 \cos 3\left(\theta - \frac{\pi}{3}\right)$

iii) $y = 10 \cos \frac{1}{2}(x - \pi) + 10$

6.	Period	a	Sinusoidal Axis	Phase Shift
a.	π	3	$y = 0$	0
b.	4π	4	$y = 2$	0
c.	π	5	$y = 0$	$\frac{\pi}{6}$
d.	π	3	$y = -3$	$-\frac{\pi}{4}$
e.	$\frac{\pi}{3}$	2	$y = 5$	$\frac{\pi}{3}$
f.	0.4	0.5	$y = -0.25$	0

7. a. $\frac{1}{(\sin\theta \cos\theta)^2}$ **b.** 0 **c.** 0 **d.** $\frac{-1}{\sin\theta}$ **9. a.** 33 cm **b.** 396 cm^2
10. a. $AT = 15$ m, $TC = 41$ m **b.** 36.9° **c.** 7.2 m **d.** 20.5 m
11. 8.2 km **12. a.** Sunset Point 19.5 km, Rock Point 26.5 km
b. Rock Point boat is first, 0.23 h **13. a.** 20 cm **b.** 10 cm **c.** 4 s
d. $h = 10 \cos\left(\frac{\pi}{2}t\right) + 30$, where t is in seconds **e.** 20 cm
14. sunrise **a.** $y = 25$ **b.** $y = 40 \sin \frac{\pi}{12}(t - 6) + 25$

15. b. $P = 50 \sin \frac{2\pi}{29}\left(t - \frac{29}{4}\right) + 50$ **c.** 92.8% **d.** Rotation of
the earth causes the moon to be invisible for 12 out of 24 h
16. a. 1259 m **b.** 2201 m **17. a.** $(1, 0)$, $r = 7$ **b.** $(2, 2)$, $r = 4$
18. a. $x^2 + y^2 + 6x - 10y - 2 = 0$ **b.** $x^2 + y^2 - 8x - 4 = 0$

19.	Vertex	Focus	Directrix	Opens
a.	$V(0, 0)$	$(0, -2)$	$y = 2$	down
b.	$V\left(-\frac{5}{2}, -2\right)$	$\left(-\frac{3}{2}, -2\right)$	$x = -\frac{7}{2}$	right

20. $y^2 - 2y + 8x + 9 = 0$

21.	Vertices	Foci	Major	Minor
a.	$(0, \pm 5)$	$(0, \pm 3)$	10	8
b.	$(\pm\sqrt{3}, 0)$	$(\pm 1, 0)$	$2\sqrt{3}$	$2\sqrt{2}$

22. $9x^2 + 5y^2 = 45$

23.	Vertices	Foci	Transverse	Conjugate
a.	$(\pm 2, 0)$	$(\pm\sqrt{13}, 0)$	4	6
b.	$(0, \pm 5)$	$(0, \pm 13)$	10	24

24. $\frac{7(x - 1)^2}{81} - \frac{(y + 3)^2}{9} = -1$ **25. a.** parabola; vertex $(2, 1)$;
focus $(2, 0)$ **b.** parabola; vertex $(2, -4)$; focus $(1, -4)$
c. hyperbola; vertices $(-10, 2)$, $(-2, 2)$ **d.** ellipse;
vertices $(-6, -4)$, $(2, -4)$ **26.** $(0, 6)$, $(8, 2)$ **27. a.** $20x^2y^3$
b. 2^{5n+5} **28. a.** 27 **b.** $\frac{9}{4}$ **c.** 0, 2 **29. a.** $\frac{2}{3}$ **b.** $\frac{1}{8}$ **c.** -22
30. a. domain: $x \in R$; range: $y \geq 5$, $y \in R$ **b.** domain: $x \geq 3$,
$x \in R$; range: $y \geq 0$, $y \in R$

Index